BRYOPHYTE OF YACHANG

雅长苔藓

LIVERWORTS AND HORNWORTS
—— 苔类和角苔类

韦玉梅　唐启明　何文钏　罗成　**主编**
Yu-Mei Wei　Qi-Ming Tang　Boon-Chuan Ho　Cheng Luo　Editor-in-Chief

山东科学技术出版社
·济南·

图书在版编目（CIP）数据

雅长苔藓：苔类和角苔类 / 韦玉梅等主编. -- 济南：山东科学技术出版社, 2024.8
ISBN 978-7-5723-1976-1

Ⅰ.①雅… Ⅱ.①韦… Ⅲ.①苔藓植物 - 广西 Ⅳ.① Q949.350.8

中国国家版本馆 CIP 数据核字 (2024) 第 021862 号

雅长苔藓——苔类和角苔类
YACHANG TAIXIAN——TAILEI HE JIAOTAILEI

责任编辑：陈　昕　徐丽叶　庞　婕

主管单位：山东出版传媒股份有限公司
出 版 者：山东科学技术出版社
　　　　　　地址：济南市市中区舜耕路 517 号
　　　　　　邮编：250003　电话：（0531）82098088
　　　　　　网址：www.lkj.com.cn
　　　　　　电子邮件：sdkj@sdcbcm.com
发 行 者：山东科学技术出版社
　　　　　　地址：济南市市中区舜耕路 517 号
　　　　　　邮编：250003　电话：（0531）82098067
印 刷 者：济南新先锋彩印有限公司
　　　　　　地址：济南市工业北路 188-6 号
　　　　　　邮编：250100　电话：（0531）88615699

规格：16 开（210 mm × 285 mm）
印张：28.75　字数：400 千
版次：2024 年 8 月第 1 版　印次：2024 年 8 月第 1 次印刷
定价：298.00 元

编 委 会

主　编　韦玉梅　唐启明　何文钏　罗成

副主编　叶　文　黄伯高　杨秀星　李龙梅

编　委（按姓名拼音排序）

程夏芳　何文钏　黄　萍　黄伯高　李　满　李　微　李　雄　李龙梅
罗　成　彭　涛　秦伟志　舒　蕾　谭宏生　唐启明　王　健　王顺莉
韦玉梅　向友良　徐剑松　杨秀星　叶　文　张莉娜　赵祖壮　钟杰宽

参编人员单位

广西壮族自治区中国科学院广西植物研究所　韦玉梅　唐启明　黄　萍

新加坡植物园　何文钏

厦门大学　叶　文

广西雅长兰科植物国家级自然保护区管理中心　黄伯高　杨秀星　秦伟志
谭宏生　赵祖壮

广西壮族自治区林业勘测设计院　罗　成　李龙梅　钟杰宽　李　满　李　雄
徐剑松

贵州师范大学　彭　涛　向友良　王顺莉

华东师范大学　王　健　舒　蕾　程夏芳

海南大学　张莉娜

中国科学院沈阳应用生态研究所　李　微

基　金

国家自然科学基金（31960045，31400190）

广西自然科学基金（2020GXNSFBA297044，2015GXNSFBA139074）

广西喀斯特植物保育与恢复生态学重点实验室基金（22-035-26）

西南岩溶国家公园（广西）调查评估区植物和大型真菌考察项目（桂植转2023-22）

Editorial Committee

Editor-in-Chief　Yu-Mei Wei　Qi-Ming Tang　Boon-Chuan Ho　Cheng Luo

Assiociate Editors　Wen Ye　Bo-Gao Huang　Xiu-Xing Yang　Long-Mei Li

Editors (Sort alphabetically by Chinese names)

Xia-Fang Cheng	Boon-Chuan Ho	Ping Huang	Bo-Gao Huang
Man Li	Wei Li	Xiong Li	Long-Mei Li
Cheng Luo	Tao Peng	Wei-Zhi Qin	Lei Shu
Hong-Sheng Tan	Qi-Ming Tang	Jian Wang	Shun-Li Wang
Yu-Mei Wei	You-Liang Xiang	Jian-Song Xu	Xiu-Xing Yang
Wen Ye	Li-Na Zhang	Zu-Zhuang Zhao	Jie-Kuan Zhong

Affiliation of Editorial Board Members

Guangxi Institute of Botany, Guangxi Zhuang Autonomous Region and Chinese Academy of Sciences　Yu-Mei Wei　Qi-Ming Tang　Ping Huang

Singapore Botanic Gardens　Boon-Chuan Ho

Xiamen University　Wen Ye

Guangxi Yachang Orchid National Nature Reserve Management Center
Bo-Gao Huang　Xiu-Xing Yang　Wei-Zhi Qin　Hong-Sheng Tan　Zu-Zhuang Zhao

Guangxi Zhuang Autonomous Region Forest Inventory & Planning Institute
Cheng Luo　Long-Mei Li　Jie-Kuan Zhong　Man Li　Xiong Li　Jian-Song Xu

Guizhou Normal University　Tao Peng　You-Liang Xiang　Shun-Li Wang

East China Normal University　Jian Wang　Lei Shu　Xia-Fang Cheng

Hainan University　Li-Na Zhang

Institute of Applied Ecology, Chinese Academy of Sciences　Wei Li

Foundations

Natinal Natural Science Foundation of China (31960045, 31400190)

Guangxi Natural Science Foundation (2020GXNSFBA297044, 2015GXNSFBA139074)

The Fund of Guangxi Key Laboratory of Plant Conservation and Restoration Ecology in Karst Terrain (22-035-26)

Project on the plants and macrofungi biodiversity investigation in the survey and assessment area of the Southwest Karst National Park (Guangxi) (GZZ2023-22)

序一
FOREWORD

苔藓植物包含苔类植物、藓类植物和角苔类植物三大类群，有2万余种，其物种数量仅次于被子植物，是陆生植物的第二大类群，在全球生态系统中具有不可替代的作用。确切了解苔藓植物的物种多样性是生物多样性利用和保护的基础。由于迷你的外观以及有限的形态特征，苔藓植物分类和鉴定一直是个难题。在世界范围内，高质量的苔藓植物鉴定工具书，尤其是苔类和角苔类植物方面的工具书更显不足。

广西是我国植物多样性最丰富的省份之一。广西雅长兰科植物国家级自然保护区（以下简称"雅长保护区"）是我国唯一一个以兰科植物为保护对象并命名的国家级自然保护区，位于广西壮族自治区西北部乐业县境内，地处云贵高原东南缘，属于云贵高原向广西丘陵过渡的山原地带。优越的自然环境孕育了丰富且独特的植物资源。很多苔藓植物与兰科植物具有类似的生境喜好，苔藓植物为兰科植物的生长孕育了独特的生境，因此，系统调查雅长保护区的苔藓植物多样性具有特别重要的意义。《雅长苔藓——苔类和角苔类》是作者基于5年野外考察获得的约3500份标本的研究成果，该书的最大特点是所有照片和描述是基于新鲜材料，并且双语出版。这是我国第一本双语出版的地方苔藓植物图志，该书记载了雅长保护区产的苔类和角苔类植物，合计31科59属180种，其中苔类174种、角苔类6种。每个种下列有中文名、拉丁名、常见异名、形态特征描述、生境信息、雅长分布信息以及中国分布信息。书内的每个种均附有彩色图版。难能可贵的是，该书的绝大部分种都配有野外居群、显微结构（含油体）以及扫描电镜照片，这些由多层次照片构成的图版对读者极其有用。该书的照片都标注了凭证标本，这为今后开展全国苔藓植物志的修订提供了重要的资料。

《雅长苔藓——苔类和角苔类》是广西迄今为止最系统、最深入的一项苔藓植物研究成果，该书不仅是国内外苔藓植物分类学和多样性研究的基础资料，更是自然爱好者不可多得的一本工具书。

华东师范大学终身教授　朱瑞良

2023 年 12 月 14 日于上海

序 二
FOREWORD

Yachang Orchid National Nature Reserve, with an area of approximately 221 km^2, is situated in NW Guangxi province, China, and is home to a variety of plants, including bryophytes. It was designated as a national reserve and is the only nature reserve for Orchidaceae in the country, due to the orchid's remarkably high abundance that dominates the forest understory of the broad-leaved sub-tropical forest. Ranging from 400 to 1971 meters in elevation in this limestone mountain, there are about 174 species and sixty-four genera of orchid plants wreathing rocky outcrops and forming large communities that are quite special in the world, well-known as "The home of orchids in China".

Investigations and reports on vascular plants as well as macro-fungi, insects, and vertebrates within the karst and non-karst forests in the reserve have been conducted and published, while study of bryophytes in Yachang Nature Reserve was blank before the publication of this book. The publication of *Bryophytes of Yachang — Liverworts and Hornworts* fills a gap in bryological research in such an important and interesting area. Together with upcoming *Bryophytes of Yachang —Mosses*, they provide additional bryological data for our complete understanding of the plant diversity in Guangxi and more broadly in China. The book serves to present baseline botanical information for protecting and utilizing Guangxi's bryophytes resource.

This book documents 174 species of liverworts and six hornworts in fifty-nine genera and thirty-one families with excellent design and high bryological quality. It provides accurate descriptions in both Chinese and English for each species, along with relevant information on habitat and distribution. Exceptional color photographs of the habit and microscopic images suitable for practical interpretation accompany each species.

Research on the bryophytes of Yachang Nature Reserve began in 2018, and this publication is the result of five years of arduous amount of work and five extensive field trips, during which over 3500 specimens were collected from all around the reserve. What sets this book apart is that it is a specimen-based reference book containing outstanding illustrations. This means that it is an especially useful reference for researchers and botanists who need to accurately identify liverwort and hornwort species. The book is an illustrated guide to Guangxi bryophytes.

Overall, the *Bryophytes of Yachang — Liverworts and Hornworts* is a major achievement of bryological research in Guangxi province and will undoubtedly contribute to a greater understanding of the plant diversity in Yachang Orchid National Nature Reserve and the broader Guangxi region. Its comprehensive coverage and specimen-based approach make it an invaluable resource for researchers, botanists, and conservationists interested in the flora of the region.

Curator, Missouri Botanical Garden, St. Louis, USA

前言 PREFACE

雅长保护区成立于2005年,是中国唯一一个以兰科植物为保护对象并命名的国家级自然保护区。保护区内保存着大面积典型的南亚热带森林生态系统和大量的珍稀野生保护物种,尤其有大面积的野生兰科植物居群,其密度之高、分布之广、原生性之强、保存之完好世界罕见,是名副其实的"中国兰花之乡",也是"全球兰科植物热点"地区。

雅长保护区位于广西壮族自治区西北部乐业县境内,地处东经106°11′31″—106°27′04″、北纬24°44′16″—24°53′58″之间,处于云贵高原东南缘,为云贵高原向广西丘陵过渡的山原地带。东接大石围天坑群风景旅游区,西毗南盘江,北起狗论山,南至草王山,东西长26.2 km,南北宽18.0 km,总面积22 062 hm²。保护区内喀斯特地貌与非喀斯特地貌交会交融,最高点盘古王海拔1971 m,最低处位于一沟,海拔400 m,相对高差1571 m。保护区地处桂西中亚热带季风气候区,深受季风环流和焚风效应的影响,夏季盛行海洋湿润气团,冬季盛行大陆寒冷气团,年平均气温16.3 ℃,最高气温38.0 ℃,最低气温 −3.0 ℃,年平均降水量1051.7 mm。优越的自然环境孕育了丰富且独特的生物资源。

保护区现有已知野生兰科植物64属174种(含2变种),是中国野生兰科植物的重要分布区及基因库,是全球兰科植物分布热点区域。同时,保护区现有已知野生维管束植物199科900属2045种、大型真菌90科259属644种、陆生脊椎动物4纲28目91科320种、昆虫12目99科509种,是具有重要科学意义和保护价值的区域。

调查自 2018 年开始，共进行了 5 次实地考察，考察范围涉及保护区各主要线路，基本覆盖整个保护区，共采集标本约 3500 份，均存放于广西植物研究所标本馆（IBK）。考虑到篇幅的实际情况，将在雅长保护区调查所得的苔藓资源分为《雅长苔藓——苔类和角苔类》和《雅长苔藓——藓类》两卷出版。本书为《雅长苔藓——苔类和角苔类》卷，包括雅长保护区范围内的苔类和角苔类植物共计 31 科 59 属 180 种，其中苔类 174 种、角苔类 6 种。

The Yachang Orchid National Nature Reserve (hereafter as Yachang Orchid Reserve), established in 2005, has been the only national-level nature reserve named after orchids in China and also one of the few areas on the planet dedicated to the *in-situ* conservation of threatened orchids. The Yachang Orchid Reserve is often affectionately regarded as the Chinese hometown of the orchids and it is one of the world's biodiversity hotspots for the orchid family because the orchids there grow in surprisingly high densities, festooning rocky outcrops and forming large patches, that dominate the forest understory. Apart from the high orchid diversity, the reserve with its vast area of primary broad-leaved sub-tropical forest, is also home to numerous rare and threatened flora and fauna.

The Yachang Orchid Reserve, located in Leye County within Guangxi Zhuang Autonomous Region, is situated between 106°11′31″–106°27′04″E and 24°44′16″–24°53′58″N. It encompasses a mountainous area in the southeastern edge of the Yunnan-Guizhou Plateau transiting between the hills of Guangxi, bordered by the Dashiwei Tiankeng Group Scenic Spot to the east, the Nanpan River to the west, the "Goulunshan" to the north and Caowang Mountain to the south, stretching 26.2 km from east to west and 18.0 km north to south, covering an area of 85-square-mile (220-sq-km). The reserve sits on a mosaic of karst and non-karst landforms with elevation ranges from 400 to 1971 m, spanning 1571 m. It has a subtropical monsoon climate, greatly affected by monsoons and foehn winds, with prevalence of maritime tropical air masses during summer and continental polar air masses during winter. The mean annual temperature is 16.3 °C with maximum and minimum temperatures of 38.0 °C and −3.0 °C, respectively. has a well development karst landform.The karst area is permeated with several sinkholes and caves while the non-karst area is filled with high mountains, steep slopes, and deep and narrow valleys. The superior geologic and climatic conditions ensure that a biodiversity is supported. Its natural condition nurtures a rich and unique biodiversity resource.

With 174 species in 64 genera of orchids recorded in Yachang Orchid Reserve thus far, the area represents an important distribution range and genetic resource for Chinese orchids, while globally a hotspot for orchids. On the other hand, 2045 species in 900 genera and 199 families of vascular plants, 644 species in 259 genera and 90 families of macrofungi, 320 species in 91 families, 28 orders and 4 classes of land vertebrates, 509 species in 99 families and 12 orders of insects, have been recorded as native to the reserve, thus it represents an important region of scientific and conservation values.

Since 2018, five field surveys have been conducted along all major routes, essentially covering the entire boundary of the reserve. The 3500 vouchers collected during the surveys are deposited in the Herbarium of the Guangxi Institute of Botany (IBK). For practicality, results of our bryophyte diversity surveys of the Yachang Orchid Reserve will be published in two volumes: *Bryophyte of Yachang — Liverworts and Hornworts* and *Bryophyte of Yachang — Mosses*. This book, *Bryophyte of Yachang — Liverworts and Hornworts*, included 180 species in 59 genera and 31 families of liverwort and hornwort species found in the Yachang Orchid Reserve, of which 174 species belong to liverwort and 6 are hornworts.

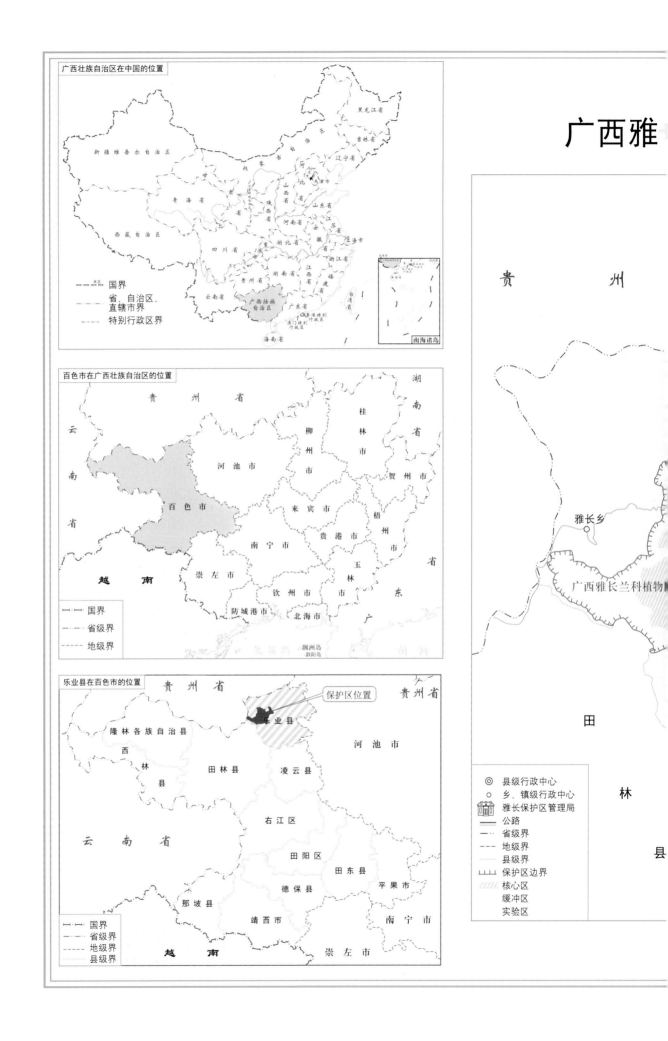

科植物国家级自然保护区位置图

审图号：桂S(2024)10-7号

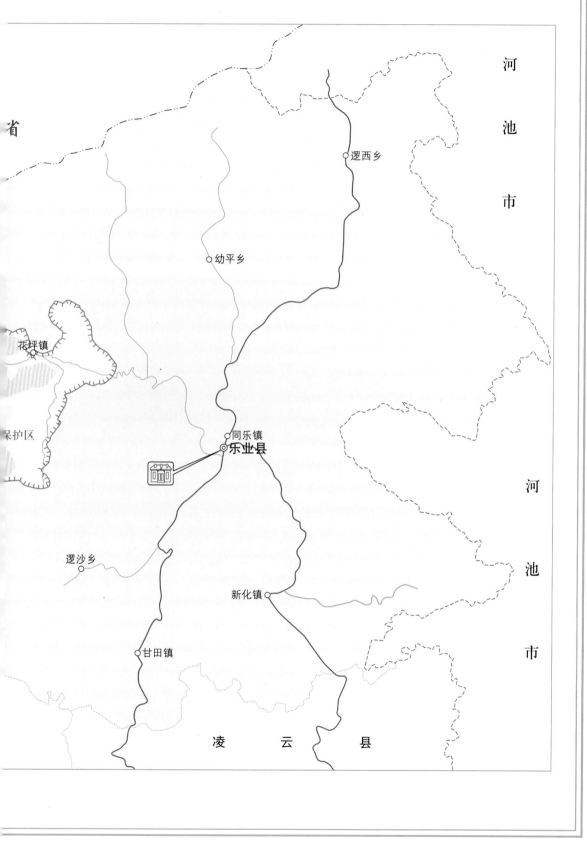

盘古王山顶（陈海玲 摄）
Summit of Panguwang mountains (Photo by Hailing Chen)

喀斯特地貌
Karst Landform

天坑生境（黄猄洞天坑）
Tiankeng Habitat (Huangjingdong Tiankeng)

洞穴生境
Cave Habitat

洞穴苔藓群落
Bryophyte community in Cave

石生群落
Petrophytia

叶附生苔类
Epiphyllitia

▶ 腐木群落
Putridae Epixylophytia

内容简介
SUMMARY

《雅长苔藓——苔类和角苔类》共记载了苔类和角苔类植物 31 科 59 属 180 种，其中苔类 174 种、角苔类 6 种。每个种下列有中文名、拉丁名、常见异名、形态特征描述、生境信息、雅长分布信息以及中国分布信息，部分种类还附有备注。本书所记载的所有苔类和角苔类植物均附有图版，绝大部分图版由野外种群照片以及显微结构照片组成。

本书采用中英双语编写，不仅可作为苔藓植物分类学和多样性研究的基础资料，供植物学、林学等科研、教学工作者参考，同时也可满足自然爱好者和相关园艺工作者在野外对苔藓植物进行识别的需求。

Bryophyte of Yachang — Liverworts and Hornworts describes 180 species of liverworts and hornworts recorded in Yachang, of which 174 species in 55 genera and 28 families are liverworts, and 6 species in 4 genera and 3 families are hornworts. The information of each species includes Chinese name, scientific name, common synonyms, morphological description, habitat, selected vouchers and geographic distribution in China. Additional notes are provided for selected species. All species are accompanied with microscopic images, and what is more, about 95% of the species included field images.

Written in both Chinese and English, this book would be a good reference for students and researchers in taxonomy and diversity studies of bryophytes. It also meets the needs of nature lovers and horticulturists to discern bryophytes in the field.

编写说明
AUTHORS' NOTES

[1] 本书参照 Frey（2009）出版的 *Syllabus of Plant Families (Part 3 Bryophytes and Seedless Vascular Plants)* 一书中科的顺序进行排列，属和种在科的排列基础上按其拉丁学名首字母顺序排列（细鳞苔科按腹叶全缘亚科和细鳞苔亚科划分后，再按其拉丁学名首字母顺序排列）。

[2] 种的中文名称主要参照贾渝和何思（2013）出版的《中国生物物种名录 第一卷 植物 苔藓植物》，对于一些新记录种需要拟中文名的文中均有标注，拉丁学名参照 Söderström 等（2016）发表的世界苔类和角苔类植物名录，少部分种类根据最新的研究成果进行变更。

[3] 文内物种的异名仅选择在国内志书中常见的异名列出。

[4] *唐启明等 202107xx-xx* 和 *黄萍等 2107xx-xx* 全部采集人为唐启明、黄萍、蔡叮叮、胡文昱、廖云标；*韦玉梅等 2011xx-xx* 和 *唐启明等 202011xx-xx* 全部采集人为韦玉梅、唐启明、廖云标；*韦玉梅等 19xx-xx* 全部采集人为韦玉梅、唐启明、廖云标；*韦玉梅等 2211xx-xx* 和 *唐启明等 202211xx-xx* 全部采集人为韦玉梅、唐启明、黄萍、王顺莉。

[5] 所有标本均存放在广西植物研究所标本馆（IBK）。引证标本中，同一地点采集的一般只选择其中一份列出。

[6] 每个种在中国的分布主要参考王利松等（2018）出版的《中国生物物种名录 第一卷 植物 总名录（上）》，在此基础上，查阅相关文献进行补充。分布范围覆盖全国 1/2 的省份则记载为"全国大部分省区均有分布"，分布范围覆盖全国 2/3 的省份则记载为"全国范围广布"。

[7] 野外种群照片由第一采集人拍摄，显微解剖照片由唐启明和韦玉梅拍摄，图版的制作由韦玉梅和唐启明完成。

[8] 若无特别标注，叶状体苔类和角苔类植物体显微结构图均为背面观，茎叶体苔类植物体显微结构图均为腹面观。

[1] The taxonomic arrangement of the families follows Frey et al. (2009) *Syllabus of Plant Families (Part 3 Bryophytes and Seedless Vascular Plants)*. Within each family, genera and species are arranged alphabetically except the genera of Lejeuneaceae where they are grouped first by the two subfamilies.

[2] Chinese names of the species follow Jia & He (2013) *Species Catalogue of China, Vol. 1. Plants: Bryophytes*. Those for new records proposed in this work are indicated accordingly. Latin names and nomenclature follow Söderström et al. (2016) *World checklist of hornworts and liverworts*, and to a small number of groups, adjusted according to most recent taxonomic works published thereafter.

[3] Synonyms included only those used in Chinese floras.

[4] Full list of collectors for *Tang et al. 202107xx-xx* and *Huang et al. 2107xx-xx* are Qi-Ming Tang, Ping Huang, Ding-Ding Cai, Wen-Yu Hu, and Yun-Biao Liao. Those of *Wei et al. 2011xx-xx*, *Tang et al. 202011xx-xx*, and *Wei et al. 19xx-xx* are Yu-Mei Wei, Qi-Ming Tang, and Yun-Biao Liao. Those of *Wei et al. 2211xx-xx* and *Tang et al. 202211xx-xx* are Yu-Mei Wei, Qi-Ming Tang, Ping Huang, and Shun-Li Wang.

[5] All the specimens are deposited in the herbarium of Guangxi Institute of Botany (IBK). Only one of the specimens from the same location is cited.

[6] The species distribution information in China are based on Wang et al. (2018) and recently published regional floras and inventory works. If a species occurs in more than half of the provinces in China (or 17 provinces or more), it will be noted as "Widely distributed in most provinces of China". If a species occurs in more than two-thirds of the provinces in China (or 22 provinces or more), it will be noted as "Widely distributed in China".

[7] Field images were photographed by the first collector. The morphological and anatomical characters were photographed by Qi-Ming Tang and Yu-Mei Wei. The illustrations were made by Yu-Mei Wei and Qi-Ming Tang.

[8] Microscopic images of Thalliod liverworts and hornworts were taken in dorsal view, whereas those leafy liverowrts were taken in ventral view, unless otherwise stated.

目录 CONTENTS

苔类 1

疣冠苔科 Aytoniaceae

花萼苔属 Asterella

十字花萼苔 Asterella cruciata 2

加萨花萼苔 Asterella khasyana 4

瓦氏花萼苔 Asterella wallichiana 6

紫背苔属 Plagiochasma

钝鳞紫背苔 Plagiochasma appendiculatum 8

无纹紫背苔 Plagiochasma intermedium ... 10

石地钱属 Reboulia

石地钱 Reboulia hemisphaerica 12

蛇苔科 Conocephalaceae

蛇苔属 Conocephalum

小蛇苔 Conocephalum japonicum 14

暗色蛇苔 Conocephalum salebrosum 16

地钱科 Marchantiaceae

地钱属 Marchantia

楔瓣地钱东亚亚种 Marchantia emarginata subsp. tosata 18

粗裂地钱凤兜亚种 Marchantia paleacea subsp. diptera 20

地钱土生亚种 Marchantia polymorpha subsp. ruderalis 22

毛地钱科 Dumortieraceae

毛地钱属 Dumortiera

毛地钱 Dumortiera hirsuta 24

单月苔科 Monosoleniaceae

单月苔属 Monosolenium

单月苔 Monosolenium tenerum 26

光苔科 Cyathodiaceae

光苔属 Cyathodium

光苔 Cyathodium cavernarum ············ 28

芽胞光苔 Cyathodium tuberosum ········ 30

钱苔科 Ricciaceae

钱苔属 Riccia

无翼钱苔 Riccia billardieri ············· 32

稀枝钱苔 Riccia hueberiana ············ 34

印尼钱苔 Riccia junghuhniana ·········· 36

厚壁钱苔 Riccia oryzicola ·············· 38

花边钱苔 Riccia rhenana ··············· 40

小叶苔科 Fossombroniaceae

小叶苔属 Fossombronia

日本小叶苔 Fossombronia japonica ······ 42

小叶苔 Fossombronia pusilla ··········· 44

带叶苔科 Pallaviciniaceae

带叶苔属 Pallavicinia

带叶苔 Pallavicinia lyellii ·············· 46

溪苔科 Pelliaceae

异溪苔属 Apopellia

异溪苔 Apopellia endiviifolia ··········· 48

叶苔科 Jungermanniaceae

狭叶苔属 Liochlaena

短萼狭叶苔 Liochlaena subulata ········ 50

无褶苔属 Mesoptychia

秩父无褶苔 Mesoptychia chichibuensis ··· 52

中华无褶苔 Mesoptychia chinensis ······· 54

玉山无褶苔 Mesoptychia morrisoncola ··· 56

被蒴苔属 Nardia

南亚被蒴苔 Nardia assamica ············ 58

假苞苔属 Notoscyphus

假苞苔 Notoscyphus lutescens ··········· 60

管口苔属 Solenostoma

偏叶管口苔 Solenostoma comatum ······· 62

截叶管口苔 Solenostoma truncatum ······ 64

护蒴苔科 Calypogeiaceae

护蒴苔属 Calypogeia

刺叶护蒴苔 Calypogeia arguta ·········· 66

全缘护蒴苔 Calypogeia japonica ········ 68

双齿护蒴苔 Calypogeia tosana ·········· 70

隐蒴苔科 Adelanthaceae

对耳苔属 Syzygiella

东亚对耳苔 Syzygiella nipponica（雌株）··· 72

东亚对耳苔 Syzygiella nipponica（雄株）··· 74

大萼苔科 Cephaloziaceae

大萼苔属 Cephalozia

弯叶大萼苔 Cephalozia hamatiloba ······· 76

拳叶苔属 Nowellia

拳叶苔 Nowellia curvifolia ·············· 78

裂齿苔属 Odontoschisma
合叶裂齿苔 Odontoschisma denudatum ··· 80

拟大萼苔科 Cephaloziellaceae
拟大萼苔属 Cephaloziella
粗齿拟大萼苔 Cephaloziella dentata ······ 82

小叶拟大萼苔 Cephaloziella microphylla
················· 84

筒萼苔属 Cylindrocolea
鳞叶筒萼苔 Cylindrocolea kiaeri ········ 86

甲克苔科 Jackiellaceae
甲克苔属 Jackiella
甲克苔 Jackiella javanica ················ 88

折叶苔科 Scapaniaceae
合叶苔属 Scapania
柯氏合叶苔 Scapania koponenii ············ 90

睫毛苔科 Blepharostomataceae
睫毛苔属 Blepharostoma
小睫毛苔 Blepharostoma minus ············ 92

指叶苔科 Lepidoziaceae
鞭苔属 Bazzania
卵叶鞭苔 Bazzania angustistipula ·········· 94

三裂鞭苔 Bazzania tridens ··············· 96

指叶苔属 Lepidozia
指叶苔 Lepidozia reptans ················· 98

剪叶苔科 Herbertaceae
剪叶苔属 Herbertus
长角剪叶苔 Herbertus dicranus ········· 100

羽苔科 Plagiochilaceae
树羽苔属 Chiastocaulon
树羽苔 Chiastocaulon dendroides ······ 102

羽苔属 Plagiochila
埃氏羽苔 Plagiochila akiyamae ········ 104

阿萨羽苔 Plagiochila assamica ········ 106

中华羽苔 Plagiochila chinensis ········ 108

树生羽苔 Plagiochila corticola ········ 110

德氏羽苔 Plagiochila delavayi ········· 112

裂叶羽苔 Plagiochila furcifolia ········ 114

裸茎羽苔 Plagiochila gymnoclada ······ 116

容氏羽苔 Plagiochila junghuhniana ······ 118

加萨羽苔 Plagiochila khasiana ········· 120

昆明羽苔 Plagiochila kunmingensis ··· 122

尼泊尔羽苔 Plagiochila nepalensis ······ 124

卵叶羽苔 Plagiochila ovalifolia ········ 126

圆头羽苔 Plagiochila parvifolia ········ 128

刺叶羽苔 Plagiochila sciophila ········ 130

大耳羽苔 Plagiochila subtropica ········ 132

短齿羽苔 Plagiochila vexans ··········· 134

韦氏羽苔 Plagiochila wightii ··········· 136

齿萼苔科 Lophocoleaceae
裂萼苔属 Chiloscyphus
裂萼苔 Chiloscyphus polyanthos ········ 138

异萼苔属 Heteroscyphus

四齿异萼苔 Heteroscyphus argutus 140

双齿异萼苔 Heteroscyphus coalitus ... 142

平叶异萼苔 Heteroscyphus planus 144

南亚异萼苔 Heteroscyphus zollingeri ... 146

齿萼苔属 Lophocolea

尖叶齿萼苔 Lophocolea bidentata 148

拟异叶齿萼苔 Lophocolea concreta ... 150

疏叶齿萼苔 Lophocolea itoana 152

芽胞齿萼苔 Lophocolea minor 154

光萼苔科 Porellaceae

光萼苔属 Porella

尖瓣光萼苔原亚种 Porella acutifolia ... 156

尖瓣光萼苔东亚亚种 Porella acutifolia subsp. tosana 158

丛生光萼苔原变种 Porella caespitans 160

丛生光萼苔心叶变种 Porella caespitans var. cordifolia 162

密叶光萼苔原亚种 Porella densifolia ... 164

密叶光萼苔长叶亚种 Porella densifolia subsp. appendiculata 166

大叶光萼苔 Porella grandifolia 168

尾尖光萼苔 Porella handelii 170

日本光萼苔 Porella japonica 172

基齿光萼苔 Porella madagascariensis ... 174

亮叶光萼苔 Porella nitens 176

钝叶光萼苔鳞叶变种 Porella obtusata var. macroloba 178

毛边光萼苔原变种 Porella perrottetiana 180

毛边光萼苔狭叶变种 Porella perrottetiana var. angustifolia 182

毛边光萼苔齿叶变种 Porella perrottetiana var. ciliatodentata 184

小瓣光萼苔 Porella plumosa 186

齿边光萼苔 Porella stephaniana 188

多瓣光萼苔 Porella ulophylla 190

扁萼苔科 Radulaceae

扁萼苔属 Radula

尖舌扁萼苔 Radula acuminata 192

大瓣扁萼苔 Radula cavifolia 194

扁萼苔 Radula complanata 196

爪哇扁萼苔 Radula javanica 198

尖叶扁萼苔 Radula kojana 200

刺边扁萼苔 Radula lacerata 202

芽胞扁萼苔 Radula lindenbergiana 204

星苞扁萼苔 Radula stellatogemmipara 206

耳叶苔科 Frullaniaceae

耳叶苔属 Frullania

黑耳叶苔 Frullania amplicrania 208

细茎耳叶苔 Frullania bolanderi 210

达乌里耳叶苔 Frullania davurica 212

皱叶耳叶苔 Frullania ericoides 214

细瓣耳叶苔 Frullania hypoleuca ……… 216

石生耳叶苔 Frullania inflata ………… 218

列胞耳叶苔 Frullania moniliata ……… 220

羊角耳叶苔喙尖变种 Frullania monocera var. acutiloba ………… 222

盔瓣耳叶苔 Frullania muscicola ……… 224

尼泊尔耳叶苔 Frullania nepalensis …… 226

大隅耳叶苔 Frullania osumiensis …… 228

钟瓣耳叶苔 Frullania parvistipula …… 230

喙瓣耳叶苔 Frullania pedicellata……… 232

大萌耳叶苔 Frullania physantha ……… 234

微齿耳叶苔 Frullania rhytidantha …… 236

陕西耳叶苔 Frullania schensiana……… 238

欧耳叶苔长叶变种 Frullania tamarisci var. elongatistipula ………… 240

云南耳叶苔密叶变种 Frullania yunnanensis var. siamensis ………… 242

汤泽耳叶苔 Frullania yuzawana ……… 244

细鳞苔科 Lejeuneaceae

顶鳞苔属 Acrolejeunea

南亚顶鳞苔 Acrolejeunea sandvicensis 246

中华顶鳞苔 Acrolejeunea sinensis …… 248

冠鳞苔属 Lopholejeunea

大叶冠鳞苔 Lopholejeunea eulopha … 250

黑冠鳞苔 Lopholejeunea nigricans …… 252

皱萼苔属 Ptychanthus

皱萼苔 Ptychanthus striatus …………… 254

多褶苔属 Spruceanthus

东亚多褶苔 Spruceanthus kiushianus … 256

疣叶多褶苔 Spruceanthus mamillilobulus 258

多褶苔 Spruceanthus semirepandus …… 260

毛鳞苔属 Thysananthus

南亚毛鳞苔 Thysananthus repletus …… 262

异鳞苔属 Tuzibeanthus

异鳞苔 Tuzibeanthus chinensis ……… 264

唇鳞苔属 Cheilolejeunea

圆叶唇鳞苔 Cheilolejeunea intertexta … 266

粗茎唇鳞苔 Cheilolejeunea trapezia … 268

卷边唇鳞苔 Cheilolejeunea xanthocarpa ………… 270

疣鳞苔属 Cololejeunea

单胞疣鳞苔 Cololejeunea kodamae…… 272

狭瓣疣鳞苔 Cololejeunea lanciloba…… 274

阔瓣疣鳞苔 Cololejeunea latilobula … 276

阔体疣鳞苔 Cololejeunea latistyla …… 278

鳞叶疣鳞苔 Cololejeunea longifolia … 280

大瓣疣鳞苔 Cololejeunea magnilobula 282

粗柱疣鳞苔 Cololejeunea ornata …… 284

粗齿疣鳞苔 Cololejeunea planissima … 286

尖叶疣鳞苔 Cololejeunea pseudocristallina ………… 288

拟疣鳞苔 Cololejeunea raduliloba …… 290

全缘疣鳞苔 Cololejeunea schwabei … 292

卵叶疣鳞苔 Cololejeunea shibiensis … 294

刺疣鳞苔 Cololejeunea spinosa ……… 296
疣瓣疣鳞苔 Cololejeunea subkodamae
　　……………………………………… 298

管叶苔属 Colura
管叶苔 Colura calyptrifolia ………… 300
细角管叶苔 Colura tenuicornis ……… 302

角鳞苔属 Drepanolejeunea
狭叶角鳞苔 Drepanolejeunea angustifolia
　　……………………………………… 304
日本角鳞苔 Drepanolejeunea erecta … 306
单齿角鳞苔 Drepanolejeunea ternatensis
　　……………………………………… 308
短叶角鳞苔 Drepanolejeunea vesiculosa
　　……………………………………… 310

细鳞苔属 Lejeunea
狭瓣细鳞苔 Lejeunea anisophylla …… 312
拟细鳞苔 Lejeunea apiculata ………… 314
双齿细鳞苔 Lejeunea bidentula ……… 316
瓣叶细鳞苔 Lejeunea cocoes ………… 318
弯叶细鳞苔 Lejeunea curviloba ……… 320
神山细鳞苔 Lejeunea eifrigii ………… 322
黄色细鳞苔 Lejeunea flava …………… 324
巨齿细鳞苔 Lejeunea kodamae ……… 326
科诺细鳞苔 Lejeunea konosensis …… 328
麦氏细鳞苔 Lejeunea micholitzii …… 330
暗绿细鳞苔 Lejeunea obscura ……… 332
角齿细鳞苔 Lejeunea otiana ………… 334
疣萼细鳞苔 Lejeunea tuberculosa …… 336

薄鳞苔属 Leptolejeunea
巴氏薄鳞苔 Leptolejeunea balansae … 338

纤鳞苔属 Microlejeunea
斑叶纤鳞苔 Microlejeunea punctiformis
　　……………………………………… 340
疏叶纤鳞苔 Microlejeunea ulicina …… 342

拟多果苔属 Myriocoleopsis
圆叶拟多果苔 Myriocoleopsis minutissima
　　……………………………………… 344

绿片苔科 Aneuraceae
绿片苔属 Aneura
大绿片苔 Aneura maxima …………… 346

片叶苔属 Riccardia
波叶片叶苔 Riccardia chamaedryfolia
　　……………………………………… 348
黄片叶苔 Riccardia flavovirens ……… 350
长崎片叶苔 Riccardia nagasakiensis … 352
掌状片叶苔 Riccardia palmata ……… 354
纤细片叶苔 Riccardia pusilla ………… 356

叉苔科 Metzgeriaceae
叉苔属 Metzgeria
狭尖叉苔 Metzgeria consanguinea …… 358
林氏叉苔 Metzgeria lindbergii………… 360

角苔类 363

角苔科 Anthocerotaceae
角苔属 Anthoceros
塔拉加角苔 Anthoceros telaganus 364

短角苔科 Notothyladaceae
大角苔属 Megaceros
东亚短角苔 Notothylas japonica 366

短角苔 Notothylas orbicularis 368

黄角苔属 Phaeoceros
高领黄角苔 Phaeoceros carolinianus ... 370

小黄角苔 Phaeoceros exiguus 372

树角苔科 Dendrocerotaceae
大角苔属 Megaceros
东亚大角苔 Megaceros flagellaris 374

致　谢 376

主要参考文献 377

凭证标本 381

中文名索引 414

拉丁名索引 417

苔类
LIVERWORTS

▶ 圆头羽苔

Plagiochila parvifolia Lindenb.

十字花萼苔（柔叶花萼苔）

Asterella cruciata (Steph.) Horik., Hikobia 1(2): 79. 1951.

Asterella mitsuminensis Shimizu & S. Hatt., J. Hattori Bot. Lab. 8: 48. 1952.

叶状体扁平柔质，背面黄绿色至绿色，揉碎时散发浓郁的鱼腥味。叶状体横切面具2—3层大气室，气室中无营养丝。气孔简单，高出表皮细胞。腹鳞片白色至淡紫色，位于叶状体腹面中肋两侧，每侧各1行，每个腹鳞片具单个披针形附器。假根二型，具光滑假根和瘤状假根。雌雄同株。雄器位于雌托基部靠后的位置，线形或Y形垫状。雌托位于叶状体先端，雌托柄短，0.8—1.2 mm，具1条假根槽，托盘扁平，表面无或稍具疣状突起，具3—5裂瓣。苞膜瓣状，假蒴萼白色。孢子球形，远极面具二级网状纹饰，4—6个大网孔横穿整个远极面；近极面具二级网状纹饰，并具明显的三射线。弹丝2螺旋加厚。

生境： 潮湿石生，石生，土生。海拔491—1429 m。

雅长分布： 二沟，拉雅沟，盘古王，一沟。

中国分布： 重庆，广西，贵州，四川，云南。

Thalli dorsally yellowish green to green when fresh, exuding strong smell of rotten fish when crushed. Air chambers 2–3 layers, without photosynthetic filament. Epidermal pores simple, raised above epidermis. Ventral scales colourless to light purple, in 1 row on either side of midrib; appendage 1 per scale, lanceolate. Rhizoids of 2 types, either smooth or tuberculate. Monoicous. Male receptacles on main thallus behind female receptacles, cushion-like, linear or Y-shaped. Female receptacles at apical region of thallus, stalk short, 0.8–1.2 mm, with single rhizoidal furrow; disc flat without or with slight verrucose above, deeply 3–5-lobed. Involucres flap-like. Pseudoperianths white. Spores globose, distal surface with 4–6 large areolae across the diameter, areolae containing smaller alveoli and pits, proximal surface reticulate, triradiate mark strongly distinct. Elaters with 2 spiral thickenings.

Habitat: On (wet) rocks and soil. Elev. 491–1429 m.

Distribution in Yachang: Ergou, Layagou, Panguwang, Yigou.

Distribution in China: Chongqing, Guangxi, Guizhou, Sichuan, Yunnan.

▶ A. 种群；B. 叶状体；C. 叶状体横切面；D. 叶状体横切面中部部分；E. 叶状体横切面边缘部分；F. 弹丝；G. 腹鳞片和附器；H. 雌托盘；I. 孢子近极面观；J. 孢子远极面观。（凭证标本：唐启明等 20201109-255）

A. Population; B. Thallus; C. Transverse section of thallus; D. Transverse section of thallus at median part; E. Transverse section of thallus at marginal part; F. Elaters; G. Ventral scale and appendage; H. Female disc; I. Proximal view of spore; J. Distal view of spore. (All from *Tang et al.* 20201109-255)

图 1 十字花萼苔

Fig. 1 *Asterella cruciata* (Steph.) Horik.

加萨花萼苔

Asterella khasyana (Griff.) Grolle, Khumbu Himal 1(4): 267. 1966.

叶状体扁平柔质，背面黄绿色或淡绿色，揉碎时散发浓郁的鱼腥味。叶状体横切面具 1—2 层大气室，气室中无营养丝。气孔简单，高出表皮细胞。腹鳞片紫色，位于叶状体腹面中肋两侧，每侧各 1 行，每个腹鳞片具单个披针形附器。假根二型，具光滑假根和瘤状假根。雌雄同株。雄器位于雌托基部靠后的位置，椭圆形或线形垫状。雌托位于叶状体先端，雌托柄长，12—18 mm，具 1 条假根槽；托盘扁平或稍凸起，表面具疣状突起，不裂。苞膜瓣状，假蒴萼白色。孢子球形，远极面具二级网状纹饰，3—6 个大网孔横穿整个远极面；近极面具二级网状纹饰，并具明显的三射线。弹丝 2 螺旋加厚。

生境： 石生，土生，岩面薄土生。海拔 1020—1717 m。

雅长分布： 草王山。

中国分布： 湖南，四川，云南。首次记录于广西。

Thalli dorsally yellowish green to light green when fresh, exuding strong smell of rotten fish when crushed. Air chambers 1–2 layers, without photosynthetic filament. Epidermal pores simple, raised above epidermis. Ventral scales purple, in 1 row on either side of midrib; appendage 1 per scale, lanceolate. Rhizoids of 2 types, either smooth or tuberculate. Monoicous. Male receptacles on main thallus behind female receptacles, cushion-like, elliptical or linear. Female receptacles at apical region of thallus, stalk 12–18 mm long, with single rhizoidal furrow; disc flat to slightly convex with strong papillose above, unlobed. Involucres flap-like. Pseudoperianths white. Spores globose, distal surface with 3–6 large areolae across the diameter, areolae containing smaller alveoli and pits, proximal surface reticulate, triradiate mark strongly distinct. Elaters with 2 spiral thickenings.

Habitat: On rocks, soil and on rocks with a thin layer of soil. Elev. 1020–1717 m.

Distribution in Yachang: Caowangshan.

Distribution in China: Hunan, Sichuan, Yunnan. New to Guangxi.

▶ A. 种群；B. 叶状体；C. 叶状体横切面；D. 叶状体横切面中部部分；E. 叶状体横切面边缘部分；F. 弹丝；G. 腹鳞片和附器；H. 雌托盘；I. 孢子近极面观；J. 孢子远极面观。（凭证标本：*唐启明 & 韦玉梅 20191013-96*）

A. Population; B. Thallus; C. Transverse section of thallus; D. Transverse section of thallus at median part; E. Transverse section of thallus at marginal part; F. Elaters; G. Ventral scales and appendages; H. Female disc; I. Proximal view of spore; J. Distal view of spore. (All from *Tang & Wei 20191013-96*)

图 2 加萨花萼苔
Fig. 2 *Asterella khasyana* (Griff.) Grolle

瓦氏花萼苔（狭叶花萼苔、单纹花萼苔、卷边花萼苔）

Asterella wallichiana (Lehm. et Lindenb.) Grolle, Khumbu Himal 1(4): 262. 1966.

Asterella angusta (Steph.) Pandé, K. P. Sirvast et Sultan Khan, J. Hattori Bot. Lab. 11: 8. 1954.
Asterella monospiris (Horik.) Horik., Hikobia 1: 79. 1951.
Asterella reflexa (Herzog) P. C. Chen ex Piippo, J. Hattori Bot. Lab. 68: 9. 1990.

叶状体扁平革质，背面深绿色，揉碎时散发鱼腥味。叶状体横切面具 3—4 层小气室，气室中有营养丝。气孔简单，高出表皮细胞。腹鳞片紫色，位于叶状体腹面中肋两侧，每侧各 1 行，每个腹鳞片具 1—2 个披针形附器。假根二型，具光滑假根和瘤状假根。雌雄异株。雄器圆形或线形垫状。雌托位于叶状体先端，雌托柄 5—9 mm 长，具 1 条假根槽，托盘扁平，表面具疣状突起，2—6 瓣浅裂。苞膜瓣状，假蒴萼紫红色。孢子球形，远极面具带状纹饰，近极面具带状纹饰，并具三射线。弹丝单螺旋加厚。

生境：土生环境常见，潮湿石生和石生环境也有分布。海拔 1020—1774 m。

雅长分布：草王山，李家坨屯，里郎天坑，逻家田屯，盘古王，深洞，下岩洞屯，中井天坑，中井屯。

中国分布：北京，甘肃，贵州，广西，四川，台湾，西藏，云南。

Thalli dorsally dark green when fresh, exuding smell of rotten fish when crushed. Air chambers 3−4 layers, with photosynthetic filaments. Epidermal pores simple, raised above epidermis. Ventral scales purple, in 1 row on either side of midrib; appendage 1−2 per scale, lanceolate. Rhizoids of 2 types, either smooth or tuberculate. Dioicous. Male receptacles cushion-like, circular or linear. Female receptacles at apical region of thallus, stalk 5−9 mm long, with single rhizoidal furrow; disc flat with strong papillose above, shallowly 2−6-lobed. Involucres flap-like. Pseudoperianths purplish red. Spores globose, distal surface with conspicuous narrow, wavy ridges, proximal surface with similar ornamentation, triradiate mark distinct. Elaters with 1 spiral thickening.

Habitat: Often on soil, sometimes on (wet) rocks. Elev. 1020−1774 m.

Distribution in Yachang: Caowangshan, Lijiatuo Tun, Lilang Tiankeng, Luojiatian Tun, Panguwang, Shendong, Xiayandong Tun, Zhongjing Tiankeng, Zhongjing Tun.

Distribution in China: Beijing, Gansu, Guizhou, Guangxi, Sichuan, Taiwan, Tibet, Yunnan.

▶ A. 种群；B. 叶状体一段；C. 叶状体横切面；D. 叶状体横切面中部部分；E. 叶状体横切面边缘部分；F. 弹丝；G. 腹鳞片和附器；H. 雌托盘；I. 孢子近极面观；J. 孢子远极面观。（凭证标本：*唐启明等 20201110-264*）

A. Population; B. Portion of thallus; C. Transverse section of thallus; D. Transverse section of thallus at median part; E. Transverse section of thallus at marginal part; F. Elaters; G. Ventral scales and appendages; H. Female disc; I. Proximal view of spore; J. Distal view of spore. (All from *Tang et al. 20201110-264*)

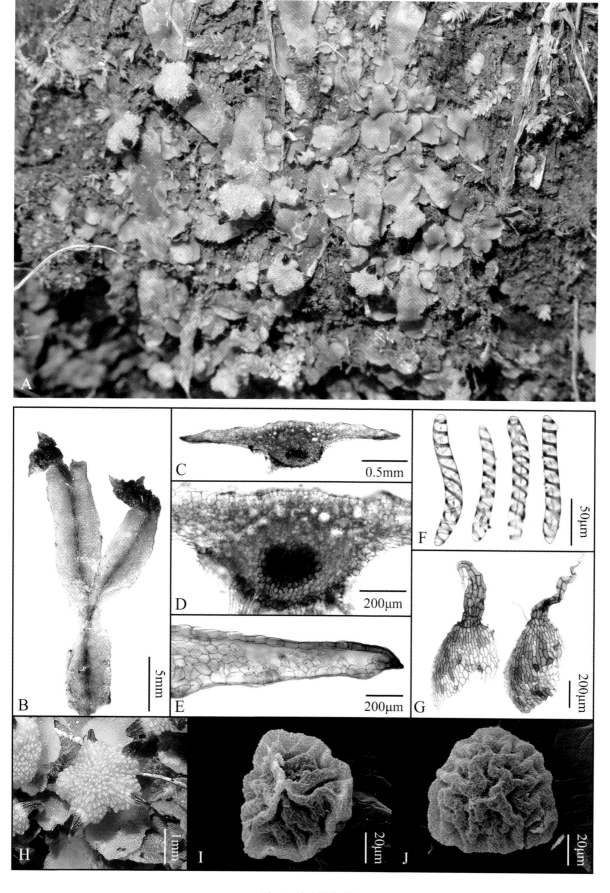

图 3 瓦氏花萼苔

Fig. 3 *Asterella wallichiana* (Lehm.) Grolle

钝鳞紫背苔

Plagiochasma appendiculatum Lehm. et Lindenb., Nov. Stirp. Pug. 4: 14. 1832.

叶状体扁平革质，背面深绿色，边缘紫色且凹凸不平呈波纹状。叶状体横切面具 4—6 层气室，气室中无营养丝。气孔简单，高出表皮细胞。腹鳞片紫色，位于叶状体腹面中肋两侧，每侧各 1 行，每个腹鳞片具 1 个无色透明卵圆形附器。假根二型，具光滑假根和瘤状假根。生殖器官未见。

生境：潮湿石生，石生，土生，钙华基质，岩面薄土生。海拔 793—1726 m。

雅长分布：白岩坨屯，草王山，大宴坪竖井，黄猄洞天坑，拉雅沟，蓝家湾天坑，盘古王，下岩洞屯，中井天坑。

中国分布：安徽，福建，甘肃，广西，贵州，湖南，四川，台湾，云南。

Thalli dorsally dark green when fresh, margins purple, slightly undulate. Air chambers 4–6 layers, without photosynthetic filament. Epidermal pores simple, raised above epidermis. Ventral scales purple, in 1 row on either side of midrib; appendage 1 per scale, hyaline, ovate. Rhizoids of 2 types, either smooth or tuberculate. Male and Female receptacles not seen.

Habitat: On (wet) rocks, soil, calcareous substrates and on rocks with a thin layer of soil. Elev. 793–1726 m.

Distribution in Yachang: Baiyantuo Tun, Caowangshan, Dayanping Tun, Huangjingdong Tiankeng, Layagou, Lanjiawan Tiankeng, Panguwang, Xiayandong Tun, Zhongjing Tiankeng.

Distribution in China: Anhui, Fujian, Gansu, Guangxi, Guizhou, Hunan, Sichuan, Taiwan, Yunnan.

▶ A. 种群；B. 叶状体；C. 叶状体横切面；D. 叶状体横切面中部部分；E. 叶状体横切面边缘部分；F. 叶状体表面气孔；G. 腹鳞片和附器。（凭证标本：唐启明等 20201110-295）

A. Population; B. Thallus; C. Transverse section of thallus; D. Transverse section of thallus at median part; E. Transverse section of thallus at median part; F. Epidermal pores; G. Ventral scale and appendage. (All from *Tang et al.* 20201110-295)

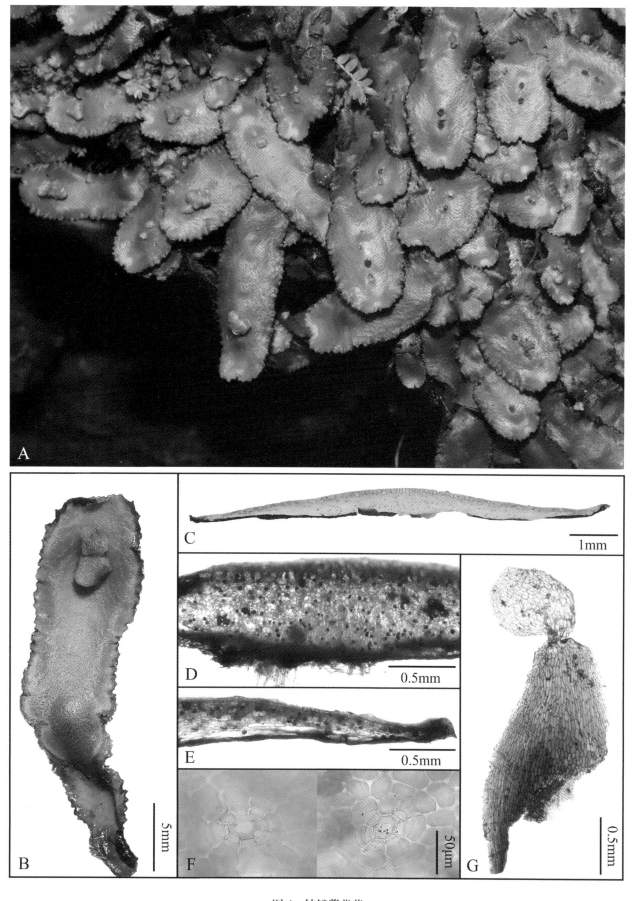

图 4 钝鳞紫背苔

Fig. 4 *Plagiochasma appendiculatum* Lehm. & Lindenb.

无纹紫背苔

Plagiochasma intermedium Lindenb. et Gottsche, Syn. Hepat. 4: 513. 1846.

叶状体扁平革质，背面绿色至黄绿色，边缘紫色且凹凸不平呈波纹状。叶状体横切面具 2—4 层气室，气室中无营养丝。气孔简单，高出表皮细胞。腹鳞片紫色，位于叶状体腹面中肋两侧，每侧各 1 行，每个腹鳞片具 1—3 个紫色披针形附器。假根二型，具光滑假根和瘤状假根。雌雄同株。雄器 V 形垫状。雌托位于叶状体中部，雌托柄短，不具假根槽，雌托盘扁平，1—4 瓣深裂。苞膜两唇形，不具假蒴萼。孢子球形，远极面具规则网状纹饰，4—5 个网孔横穿整个远极面，近极面具规则网状纹饰，并具三射线。弹丝不具螺旋加厚。

生境： 土生环境常见，石生、砂土、岩面薄土生环境也有分布。海拔 1115—1715 m。

雅长分布： 草王山，大宴坪天坑漏斗，吊井天坑，黄猄洞天坑，蓝家湾天坑，老屋基天坑，里郎天坑，盘古王，悬崖天坑，中井屯。

中国分布： 全国大部分省区均有分布。

Thalli dorsally green to yellowish green when fresh, margins purple, slightly undulate. Air chambers 2–4 layers, without photosynthetic filament. Epidermal pores simple, raised above epidermis. Ventral scales purple, in 1 row on either side of midrib; appendage 1–3 per scale, purple, lanceolate. Rhizoids of 2 types, either smooth or tuberculate. Monoicous. Male receptacles cushion-like, V-shaped. Female receptacles at median region of thallus, stalk short, without rhizoidal furrow; disc flat, deeply 1–4-lobed. Involucres bilabiate. Pseudoperianths absent. Spores globose, distal surface regularly reticulate forming 4–5 areolae across the diameter, proximal surface reticulate, triradiate mark distinct. Elaters without spiral thickening.

Habitat: Often on soil, sometimes on rocks, sandy soil and on rocks with a thin layer of soil. Elev. 1115–1715 m.

Distribution in Yachang: Caowangshan, Dayanping Tiankeng, Diaojing Tiankeng, Huangjingdong Tiankeng, Lanjiawan Tiankeng, Laowuji Tiankeng, Lilang Tiankeng, Panguwang, Xuanya Tiankeng, Zhongjing Tun.

Distribution in China: Widely distributed in most provinces of China.

▶ A. 种群；B. 叶状体；C. 叶状体横切面；D. 叶状体横切面中部部分；E. 叶状体横切面边缘部分；F. 腹鳞片和附器；G. 叶状体表面气孔；H. 弹丝；I. 孢子近极面观；J. 孢子远极面观。（凭证标本：*韦玉梅等 191018-297*）

A. Population; B. Thallus; C. Transverse section of thallus; D. Transverse section of thallus at median part; E. Transverse section of thallus at marginal part; F. Ventral scale and appendages; G. Epidermal pore; H. Elaters; I. Proximal view of spore; J. Distal view of spore. (All from *Wei et al. 191018-297*)

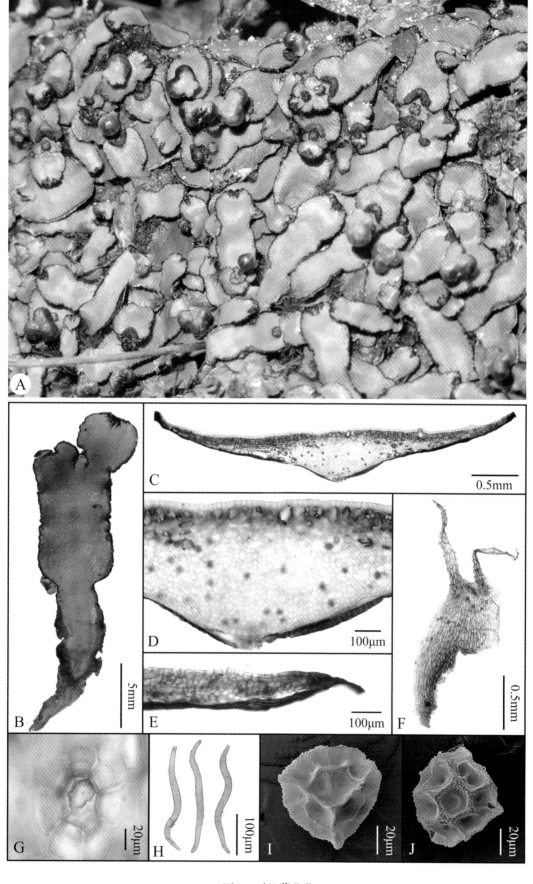

图5 无纹紫背苔

Fig. 5 *Plagiochasma intermedium* Lindenb. et Gottsche

石地钱

Reboulia hemisphaerica (L.) Raddi, Opusc. Sci. 2(6): 357. 1818.

叶状体扁平革质，背面深绿色，边缘凹凸不平呈波纹状。叶状体横切面具 1—3 层气室，气室中无营养丝。气孔简单，高出表皮细胞。腹鳞片紫色，位于叶状体腹面中肋两侧，每侧各 1 行，每个腹鳞片具 1—3 个紫色近线形附器。假根二型，具光滑假根和瘤状假根。雌雄同株或异株。雄器马蹄形垫状。雌托长在主叶状体上，雌托柄具 1 条假根槽，雌托盘半球形，3—6 瓣深裂。苞膜两唇形，不具假蒴萼。孢子球形，远极面具网状纹饰，3—4 个网孔横穿整个远极面；近极面具网状纹饰，并具明显的三射线。弹丝 2—3 螺旋加厚。

生境： 潮湿土生，土生。海拔 1038—1220 m。
雅长分布： 吊井天坑，黄猄洞天坑。
中国分布： 全国范围广布。

Thalli dorsally dark green when fresh, margins slightly undulate. Air chambers 1–3 layers, without photosynthetic filament. Epidermal pores simple, raised above epidermis. Ventral scales purple, in 1 row on either side of midrib; appendage 1–3 per scale, purple, linear. Rhizoids of 2 types, either smooth or tuberculate. Monoicous or dioicous. Male receptacles cushion-like, U-shaped. Female receptacles on main part of thallus, stalk with single rhizoidal furrow; disc hemispherical, deeply 3–6-lobed. Involucres bilabiate. Pseudoperianths absent. Spores globose, distal surface reticulate forming 3–4 areolae across the diameter, proximal surface reticulate, triradiate mark strongly distinct. Elaters with 2–3 spiral thickenings.

Habitat: On (wet) soil. Elev. 1038–1220 m.
Distribution in Yachang: Diaojing Tiankeng, Huangjingdong Tiankeng.
Distribution in China: Widely distributed in China.

▶ A. 种群；B. 叶状体；C. 腹鳞片和附器；D. 叶状体横切面；E. 叶状体横切面中部部分；F. 叶状体横切面边缘部分；G. 孢子近极面观；H. 孢子远极面观；I. 雌托柄横切面；J. 弹丝。（凭证标本：*唐启明等 20201104-11*）

A. Population; B. Thallus; C. Ventral scale and appendages; D. Transverse section of thallus; E. Transverse section of thallus at median part; F. Transverse section of thallus at marginal part; G. Proximal view of spore; H. Distal view of spore; I. Transverse section of female stalk; J. Elaters. (All from *Tang et al. 20201104-11*)

图 6 石地钱

Fig. 6 *Reboulia hemisphaerica* (L.) Raddi

小蛇苔

Conocephalum japonicum (Thunb.) Grolle, J. Hattori Bot. Lab. 55: 501. 1984.

叶状体扁平革质，背面绿色至黄绿色，先端常长有大量芽胞。叶状体横切面具1层气室，气室中有营养丝。气孔简单，高出表皮细胞。腹鳞片无色至紫色，位于叶状体腹面中肋两侧，每侧各1行，每个腹鳞片具1个淡紫色近圆形附器。假根二型，具光滑假根和瘤状假根。雌雄器官未见。

生境：土生。海拔1138—1465 m。

雅长分布：吊井天坑，李家坨屯，逻家田屯，旁墙屯，全达村，山干屯，中井屯。

中国分布：全国范围广布。

备注：Akiyama & Odrzykoski（2020）将小蛇苔从蛇苔属中移出重新置于 *Sandea* Lindb. 属下，作为 *Sandea japonica* Steph. ex Yoshin。但鉴于当前划分还未获得广泛接受，为便于读者理解，本书仍保留其作为蛇苔属的一员。

Thalli dorsally green to yellowish green when fresh, the tips usually produce subrotund gemmae. Air chambers 1 layer, with photosynthetic filaments. Epidermal pores simple, raised above epidermis. Ventral scales light purple to colourless, in 1 row on either side of midrib; appendage 1 per scale, light purple, subrounded. Rhizoids of 2 types, either smooth or tuberculate. Male and Female receptacles.

Habitat: On soil. Elev. 1138−1465 m.

Distribution in Yachang: Diaojing Tiankeng, Lijiatuo Tun, Luojiatian Tun, Pangqiang Tun, Quanda Village, Shangan Tun, Zhongjing Tun.

Distribution in China: Widely distributed in China.

Note: Although all three species of *Conocephalum* are together shown to be monophyletic, Akiyama & Odrzykoski (2020) have chosen to reinstated subg. *Sandea* (Lindb.) Inoue to the genus level with *Sandea japonica* Steph. ex Yoshin. as the only species. This classification has yet to receive wide acceptance and therefore the species is still retained as *Conocephalum japonicum*.

▶ A. 种群；B. 叶状体；C. 叶状体横切面；D. 叶状体横切面中部部分；E. 叶状体横切面边缘部分；F. 气孔横切；G. 叶状体表面气孔；H. 腹鳞片和附器。（凭证标本：*唐启明 & 韦玉梅 20191018-308*）

A. Population; B. Thallus; C. Transverse section of thallus; D. Transverse section of thallus at median part; E. Transverse section of thallus at marginal part; F. Transverse section of epidermal pore; G. Epidermal pore; H. Ventral scale and appendage. (All from *Tang & Wei 20191018-308*)

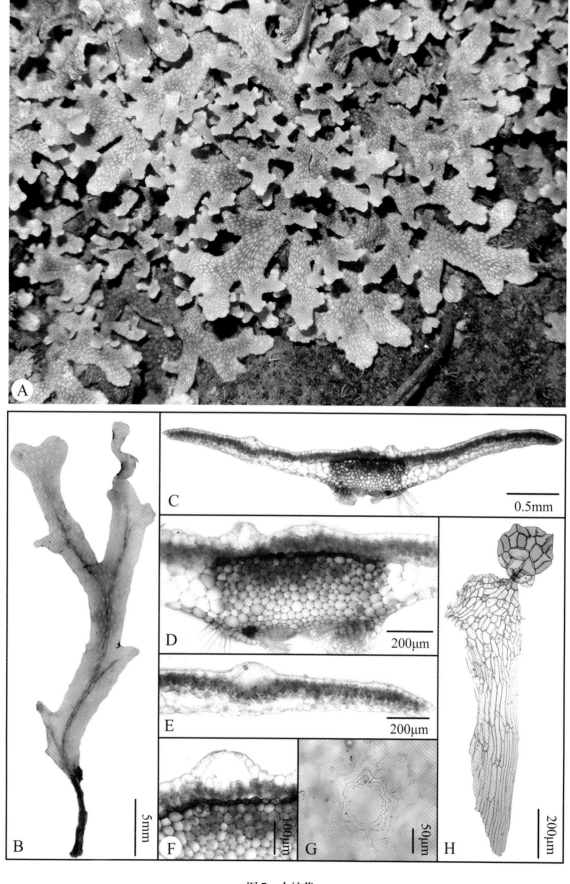

图 7 小蛇苔

Fig. 7 *Conocephalum japonicum* (Thunb.) Grolle

暗色蛇苔

Conocephalum salebrosum Szweyk., Buczk. et Odrzyk., Pl. Syst. Evol. 253: 146. 2005.

叶状体扁平革质，背面暗绿色，气室边界明显。叶状体横切面具1层气室，气室中有营养丝。气孔简单，高出表皮细胞。表皮细胞1层，椭圆形，横切面排列起伏不平。腹鳞片淡紫色，位于叶状体腹面中肋两侧，每侧各1行，每个腹鳞片具1个淡紫色肾形或近圆形附器。假根二型，具光滑假根和瘤状假根。雌雄异株。雄器椭圆形垫状。雌托未见。

生境：潮湿石生，石生，潮湿土生，土生。海拔1241—1755 m。

雅长分布：草王山，里郎天坑，盘古王，中井屯，霄罗湾洞穴。

中国分布：安徽，贵州，河南，湖北，青海，陕西，四川，云南。首次记录于广西。

备注：该种形态上与蛇苔 *Conocephalum conicum* (L.) Dumort. 极为相似，它们的区别主要是：1）前者叶状体一般为暗绿色，后者一般为亮绿色；2）前者叶状体中部横切面背部表皮细胞常为椭圆形，排列起伏不平，后者叶状体中部横切面背部表皮细胞近长方形，排列整齐；3）前者雌托盘横切中托盘表皮细胞单层，后者双层。

Thalli dorsally dark green when fresh, Epidermis at border between particular air chambers distinctly furrowed, dorsal surface in cross-section clearly uneven. Air chambers 1 layer, with photosynthetic filaments. Epidermal pores simple, raised above epidermis. Epidermal cells elliptical. Ventral scales light purple, in 1 row on either side of midrib; appendage 1 per scale, light purple, reniform or subrounded. Rhizoids of 2 types, either smooth or tuberculate. Dioicous. Male receptacles cushion-like, elliptical. Female receptacles not seen.

Habitat: On (wet) rocks and (wet) soil. Elev. 1241−1755 m.

Distribution in Yachang: Caowangshan, Lilang Tiankeng, Panguwang, Zhongjing Tun, Xiaoluowan Cave.

Distribution in China: Anhui, Guizhou, Henan, Hubei, Qinghai, Shaanxi, Sichuan, Yunnan. New to Guangxi.

Note: Superficially *Conocephalum salebrosum* is very similar to *C. conicum*. The later, however, differs in its 1) light green color of Thalli (dark green in *C. salebrosum*); 2) outer epidermal cell subrectangular, dorsal thallus surface usually even (outer epidermal cell oblong, dorsal thallus surface uneven in *C. salebrosum*); 3) epidermis of archegoniophore air chambers usually bistratose (unistratose in *C. salebrosum*).

A. 种群；B. 叶状体一段；C. 叶状体表面气孔；D. 叶状体横切面；E. 叶状体横切面中部部分；F. 叶状体横切面近中部部分；G. 叶状体横切面边缘部分；H. 腹鳞片和附器；I. 气孔横切面。（凭证标本：*唐启明等 20201110-259*）

A. Population; B. Portion of thallus; C. Epidermal pore; D. Transverse section of thallus; E. Transverse section of thallus at median part; F. Transverse section of thallus near median part; G. Transverse section of thallus at marginal part; H. Ventral scale and appendage; I. Transverse section of epidermal pores. (All from *Tang et al. 20201110-259*)

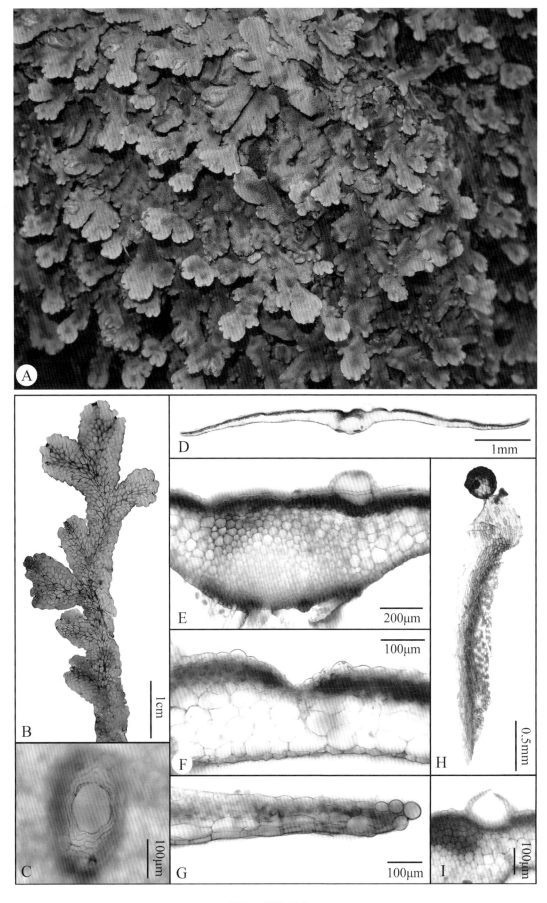

图 8 暗色蛇苔

Fig. 8 *Conocephalum salebrosum* Szweyk., Buczkowska & Odrzykoski

楔瓣地钱东亚亚种（东亚地钱）

Marchantia emarginata subsp. **tosana** (Steph.) Bischl., Cryptog. Bryol. Lichénol. 10(1): 77. 1989.

Marchantia tosana Steph., Bull. Herb. Boissier 5(2): 99. 1897.

叶状体扁平革质，背面深绿色。叶状体横切面具1层气室，气室中有营养丝。气孔复式，气孔内面观呈四边形或五边形。腹鳞片紫色，位于叶状体腹面中肋两侧，每侧各2行，每个腹鳞片具1个紫色心形或肾形附器，附器边缘具大量小齿。芽胞杯杯状，边缘具1—2个细胞宽、1—4个细胞长的齿。假根二型，具光滑假根和瘤状假根。雌雄异株。雄托位于叶状体先端，雄托柄具2或4条假根槽，托盘圆盘状，4—10瓣深裂。雌托位于叶状体先端，雌托柄具2条假根槽，托盘圆盘状，5—11瓣深裂，裂瓣楔形。苞膜双瓣状，假蒴萼存在。孢子体未见。

生境： 土生环境常见，石生和潮湿土生环境也有分布。海拔565—1730 m。

雅长分布： 白岩坨屯，草王山，吊井天坑，二沟，拉雅沟，逻家田屯。

中国分布： 广泛分布于我国南方各省区。

Thalli dorsally dark green when fresh. Air chambers 1 layer, with photosynthetic filaments. Epidermal pores compound, inner opening tetragonal or pentagonal. Ventral scales purple, in 2 rows on either side of midrib; appendage 1 per scale, purple, cordiform or reniform, margin with abundant unicellular teeth. Cupules ciliate, cilia 1–4 cells long, 1–2 cells wide basally. Rhizoids of 2 types, either smooth or tuberculate. Dioicous. Male receptacles at apical region of thallus, stalk with 2 or 4 rhizoidal furrows; disc discoid, deeply 4–10-lobed. Female receptacles at apical region of thallus, stalk with 2 rhizoidal furrows; disc flat, deeply 5–11-lobed, lobes cuneate. Involucres bivalved. Pseudoperianths present. Sporophytes not seen.

Habitat: Often on soil, sometimes on rocks and wet soil. Elev. 565–1730 m.

Distribution in Yachang: Baiyantuo Tun, Caowangshan, Diaojing Tiankeng, Ergou, Layagou, Luojiatian Tun.

Distribution in China: Widely distributed in southern provinces of China.

▶ A. 种群；B. 叶状体；C. 叶状体横切面；D. 叶状体横切面中部部分；E. 叶状体横切面边缘部分；F. 腹鳞片和附器；G. 附器；H. 雌托柄横切；I. 芽胞；J. 雌托盘；K. 芽胞杯边缘齿；L. 叶状体表面气孔，内面观。（凭证标本：*唐启明等 20190521-420*）

A. Population; B. Thallus; C. Transverse section of thallus; D. Transverse section of thallus at median part; E. Transverse section of thallus at marginal part; F. Ventral scale and appendage; G. Appendage; H. Transverse section of female stalk; I. Gemmae; J. Female discs; K. Cupule with cilia; L. Epidermal pore, inner opening. (All from *Tang et al. 20190521-420*)

图 9 楔瓣地钱东亚亚种

Fig. 9 *Marchantia emarginata* subsp. *tosata* (Steph.) Bischl.

粗裂地钱风兜亚种（风兜地钱）

Marchantia paleacea subsp. **diptera** (Nees et Mont.) Inoue, J. Jap. Bot. 64(7): 194. 1989.

Marchantia diptera Nees et Mont., Ann. Sci. Nat. Bot. (sér. 2) 19: 243. 1843.

叶状体扁平革质，背面深绿色。叶状体横切面具1层气室，气室中有营养丝。气孔复式，气孔内面观呈十字。腹鳞片紫色，位于叶状体腹面中肋两侧，每侧各2行，每个腹鳞片具1个紫色心形至圆形附器，附器顶端圆钝。芽胞杯杯状，边缘具6—9个细胞宽、12—19个细胞长的裂片状齿。假根二型，具光滑假根和瘤状假根。雌雄异株。雄托位于叶状体先端，雄托柄具2或4条假根槽，托盘圆盘状，5—8瓣浅裂。雌托位于叶状体先端，雌托柄具2条假根槽，托盘风兜状。苞膜双瓣状，假蒴萼存在。

生境： 潮湿土生，土生。海拔778—1343 m。

雅长分布： 吊井天坑，拉雅沟，逻家田屯，中井屯。

中国分布： 广泛分布于除东北以外的我国大部分省区。

Thalli dorsally dark green when fresh. Air chambers 1 layer, with photosynthetic filaments. Epidermal pores compound, inner opening cruciate. Ventral scales purple, in 2 rows on either side of midrib; appendage 1 per scale, purple, cordiform to circular, apex obtuse. Cupules laciniate, lacina 12−19 cells long, 6−9 cells wide basally. Rhizoids of 2 types, either smooth or tuberculate. Dioicous. Male receptacles at apical region of thallus, stalk with 2 or 4 rhizoidal furrows; disc discoid, shallowly 5−8-lobed. Female receptacles at apical region of thallus, stalk with 2 rhizoidal furrows; disc dipterous. Involucres bivalved. Pseudoperianths present.

Habitat: On (wet) soil. Elev. 778−1343 m.

Distribution in Yachang: Diaojing Tiankeng, Layagou, Luojiatian Tun, Zhongjing Tun.

Distribution in China: Widely distributed in most provinces of China except Northeast Region.

▶ A. 种群；B. 叶状体一段；C. 叶状体横切面；D. 叶状体横切面中部部分；E. 叶状体横切面边缘部分；F. 芽胞杯边缘裂片；G. 芽胞；H. 雌托盘；I. 气孔横切面；J. 叶状体表面气孔，内面观；K. 腹鳞片；L. 腹鳞片和附器。（凭证标本：*唐启明 & 韦玉梅 20191016-216*）

A. Population; B. Portion of thallus; C. Transverse section of thallus; D. Transverse section of thallus at median part; E. Transverse section of thallus at marginal part; F. Cupule with lobes; G. Gemmae; H. Female discs; I. Transverse section of epidermal pore; J. Epidermal pore, inner opening; K. Appendage; L. Ventral scale and appendage. (All from *Tang & Wei 20191016-216*)

图 10 粗裂地钱风兜亚种

Fig. 10 *Marchantia paleacea* subsp. *diptera* (Nees & Mont.) Inoue

地钱土生亚种

Marchantia polymorpha subsp. **ruderalis** Bischl. et Boissel.-Dub., J. Bryol. 16(3): 364. 1991.

叶状体扁平柔质，背面青绿色。叶状体横切面具1层气室，气室中有营养丝。气孔复式，气孔内面观十字状。腹鳞片紫色，位于叶状体腹面中肋两侧，每侧各3行，每个腹鳞片具1个无色到淡紫色心形至卵圆形附器，边缘具小齿。芽胞杯杯状，边缘具6—12个细胞宽、11—15个细胞长的裂片状齿。假根二型，具光滑假根和瘤状假根。雌雄异株。雄托位于叶状体先端，雄托柄具2条假根槽，托盘圆盘状6—10瓣浅裂。雌托位于叶状体先端，雌托柄具2条假根槽，托盘深裂具5—11个指状裂瓣。苞膜双瓣状，假蒴萼存在。

生境：土生。海拔1465—1762 m。

雅长分布：全达村。

中国分布：安徽，青海，云南。首次记录于广西。

备注：通过叶状体的外观形态即可很好地识别地钱的3个亚种：叶状体中间有明显深黑色条带的是地钱原亚种 Marchantia polymorpha subsp. polymorpha L.，叶状体中间深黑色条带不怎么明显的是地钱土生亚种，叶状体中间完全没有深黑色条带的是地钱高山亚种 Marchantia polymorpha subsp. montivagans Bischl. et Boissel.-Dub.。

Thalli dorsally green when fresh. Air chambers 1 layer, with photosynthetic filaments. Epidermal pores compound, inner opening cruciate. Ventral scales purple, in 3 rows on either side of midrib; appendage 1 per scale, colorless to light purple, cordiform to oval, margin with unicellular teeth. Cupules laciniate, lacina 11−15 cells long, 6−12 cells wide basally. Rhizoids of 2 types, either smooth or tuberculate. Dioicous. Male receptacles at apical region of thallus, stalk with 2 rhizoidal furrows; disc palmate, shallowly 6−10-lobed. Female receptacles at apical region of thallus, stalk with 2 rhizoidal furrows; disc conical, deeply 5−11-lobed, lobes digitate. Involucres bivalved. Pseudoperianths present.

Habitat: On soil. Elev. 1465−1762 m.

Distribution in Yachang: Quanda Village.

Distribution in China: Anhui, Qinghai, Yunan. New to Guangxi.

Note: Superficially, the three subspecies of Marchantia polymorpha are easily separated from each other. A median line on the dorsal surface of thallus is present in both subsp. polymorpha and subsp. ruderalis, but completely absent in subsp. montivagans. Moreover, the median lines in subsp. polymorpha are usually broad, uninterrupted and conspicuous whereas those in subsp. ruderalis are irregularly developed and sometimes interrupted.

▶ A. 种群；B. 叶状体一段；C. 雌托盘；D. 叶状体横切面；E. 叶状体横切面中部部分；F. 叶状体横切面边缘部分；G. 芽胞；H. 芽胞杯边缘裂片；I. 叶状体表面气孔，内面观；J. 腹鳞片和附器。（凭证标本：*唐启明等 20201108-229*）

A. Population; B. Portion of thallus; C. Female disc; D. Transverse section of thallus; E. Transverse section of thallus at median part; F. Transverse section of thallus at marginal part; G. Gemmae; H. Cupule with lobes; I. Epidermal pore, inner opening; J. Ventral scale and appendage. (All from *Tang et al. 20201108-229*)

图 11　地钱土生亚种

Fig. 11　*Marchantia polymorpha* subsp. *ruderalis* Bischl. et Boissel.-Dub.

毛地钱

Dumortiera hirsuta (Sw.) Nees, Nova Acta Phys.-Med. Acad. Caes. Leop.-Carol. Nat. Cur. 12(1): 410. 1824.

叶状体扁平，背面亮绿色到深绿色，具退化的气室和营养丝残留，边缘具无色透明鳞毛。无气孔。腹鳞片退化。假根二型，具光滑假根和瘤状假根。雌雄异株。雄器位于叶状体先端，圆形垫状具白色鳞毛。雌托位于叶状体先端，雌托柄具2条假根槽，托盘半球形，6—8瓣浅裂，具大量无色鳞毛。苞膜管状，不具假蒴萼。孢子椭球形，远极面具短条纹、短棒状或蠕虫状纹饰，近极面与远极面纹饰相似，并具不明显的三射线。弹丝2（—3）螺旋加厚。

生境：潮湿石生，石生，潮湿土生，土生。海拔816—1424 m。

雅长分布：吊井天坑，拉雅沟，李家坨屯，里郎天坑，逻家田屯，二沟，中井天坑。

中国分布：全国范围广布。

Thalli dorsally light green to dark green when fresh, margin hairy. Air chamber and photosynthetic filament vestigial on dorsal surface. Epidermal pore absent. Ventral scale reduced. Rhizoids of 2 types, either smooth or tuberculate. Dioicous. Male receptacles at apical region of thallus, cushion-like, circular, with white hairs above. Female receptacles at apical region of thallus, stalk with 2 rhizoidal furrows; disc hemispherical, shallowly 6−8-lobed, with numerous colorless hairs above. Involucres tubular. Pseudoperianths absent. Spores ellipsoidal, distal surface with ornamented irregularly striate, short baculate, and vermiculate, proximal surface with similar ornamentation, triradiate mark indistinct. Elaters with 2(−3) spiral thickenings.

Habitat: On (wet) rocks and (wet) soil. Elev. 816−1424 m.

Distribution in Yachang: Diaojing Tiankeng, Layagou, Lijiatuo Tun, Lilang Tiankeng, Luojiatian Tun, Ergou, Zhongjing Tiankeng.

Distribution in China: Widely distributed in China.

▶ A. 种群；B. 雌托；C. 叶状体；D. 残留气室中的营养丝细胞；E. 叶状体横切面；F. 叶状体横切面中部部分；G. 叶状体近中部横切面；H. 弹丝；I. 孢子近极面观；J. 孢子远极面观。（凭证标本：*唐启明 & 韦玉梅 20191020-461*）

A. Population; B. Female receptacles; C. Thallus; D. Assimilatory filaments in reduced air chamber; E. Transverse section of thallus; F. Transverse section of thallus at median part; G. Transverse section of thallus near median part; H. Elaters; I. Proximal view of spore; J. Distal view of spore. (All from *Tang & Wei 20191020-461*)

图 12　毛地钱

Fig. 12　*Dumortiera hirsuta* (Sw.) Nees

单月苔

Monosolenium tenerum Griff., Not. Pl. Asiat. 2: 341. 1849.

叶状体扁平，背面绿色到鲜绿色，表皮细胞中具大量油胞。叶状体无气孔和气室的分化。腹鳞片无色透明，位于叶状体腹面中肋两侧，每侧各1行，每个腹鳞片具1个线形或披针形附器。假根二型，具光滑假根和瘤状假根。雌雄同株。雄器位于雌托后面，圆形或椭圆形垫状。雌托位于叶状体先端，雌托柄短，具2条假根槽，托盘方形，4—7瓣浅裂。苞膜管状，不具假蒴萼。孢子球形，远极面具网状纹饰，8—10个网孔横穿整个远极面；近极面具网状纹饰，并具三射线。弹丝单螺旋加厚。

生境： 石生，潮湿土生，土生。海拔468—1425 m。

雅长分布： 二沟，拉雅沟，一沟，中井天坑。

中国分布： 安徽，澳门，重庆，广东，贵州，上海，四川，台湾，云南。首次记录于广西。

备注： 广西新记录的科和属。

Thalli dorsally green to light green when fresh, with numerous scattered oil-cells in epidermal cells. Air chamber and epidermal pore absent. Ventral scales hyaline, in 1 row on either side of midrib; appendage 1 per scale, linear to lanceolate. Rhizoids of 2 types, either smooth or tuberculate. Monoicous. Male receptacles behind female receptacles, cushion-like, circular or elliptical. Female receptacles at apical region of thallus, stalk short, with 2 rhizoidal furrows; disc quadrate, shallowly 4−7-lobed. Involucres tubular. Pseudoperianths absent. Spores globose, distal surface reticulate forming 8−10 areolae across the diameter, proximal surface reticulate, triradiate mark distinct. Elaters with 1 spiral thickening.

Habitat: On rocks and (wet) soil. Elev. 468−1425 m.

Distribution in Yachang: Ergou, Layagou, Yigou, Zhongjing Tiankeng.

Distribution in China: Anhui, Macao, Chongqing, Guangdong, Guizhou, Shanghai, Sichuan, Taiwan, Yunnan. New to Guangxi.

Note: New family and genus records for Guangxi.

▶ A. 种群；B. 叶状体；C. 雄器和雌托；D. 雌托柄横切面；E. 叶状体横切面中部部分；F. 叶状体横切面边缘部分；G. 孢子近极面观；H. 孢子远极面观；I. 腹鳞片和附器；J. 弹丝；K. 叶状体一部分，示分散的油胞；L. 表皮细胞，示油胞。（凭证标本：*韦玉梅等 191014-145*）

A. Population; B. Thallus; C. Male and Female receptacles; D. Transverse section of female stalk; E. Transverse section of thallus at median part; F. Transverse section of thallus at marginal part; G. Proximal view of spore; H. Distal view of spore; I. Ventral scale and appendage; J. Elaters; K. Part of thallus showing scattered oil-cells; L. Epidermal cells showing oil-cells. (All from *Wei et al. 191014-145*)

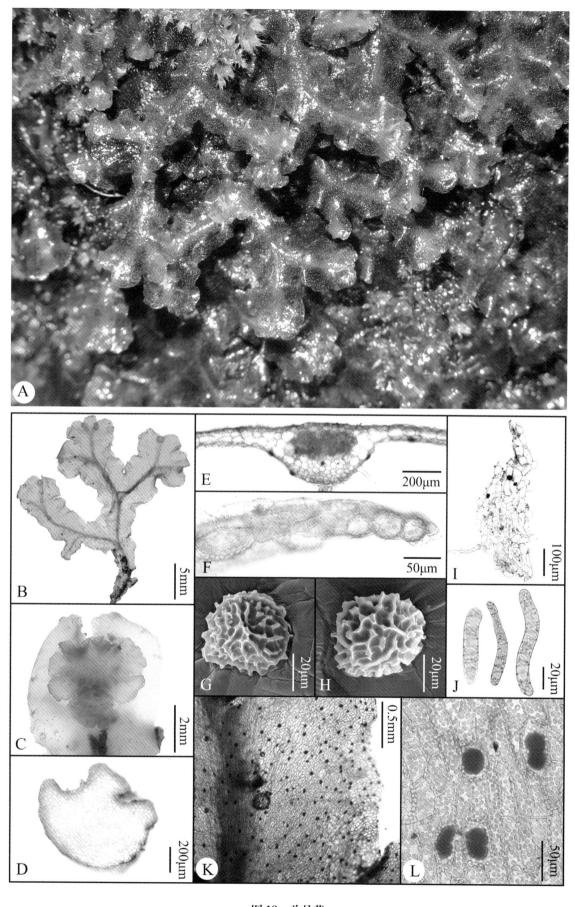

图 13 单月苔

Fig. 13 *Monosolenium tenerum* Griff.

光苔

Cyathodium cavernarum Kunze ex Lehm., Nov. Stirp. Pug. 6: 18. 1834.

叶状体扁平柔薄，背面淡绿色到黄绿色，多长在阴暗潮湿的环境，看上去呈现出绿色的荧光，顶部不产生块状芽。叶状体中部横切一般三层，包括上表皮和下表皮，以及中间的巨大气室，气室中无营养丝。气孔简单，高出表皮细胞或与表皮细胞平行。腹鳞片无色透明，片状，位于叶状体腹面，无附器。假根单型，仅具光滑假根。雌雄生殖器官未见。

生境： 潮湿土生。海拔 1241 m。

雅长分布： 霄罗湾洞穴。

中国分布： 广西，贵州，四川，云南。

Thalli dorsally light green to yellowish green, usually luminous in dark places, tips without tuber. Air chambers 1 layer, without photosynthetic filament. Epidermal pores simple, raised above or plane with epidermis. Ventral scales hyaline, plate-like, on ventral thallus, without appendage. Rhizoids of 1 type, smooth. Androecia and gynoecia not seen.

Habitat: On wet soil. Elev. 1241 m.

Distribution in Yachang: Xiaoluowan Cave.

Distribution in China: Guangxi, Guizhou, Sichuan, Yunnan.

▶ A. 种群；B. 叶状体；C–D. 叶状体一段；E. 腹鳞片；F. 叶状体表面气孔；G. 表皮细胞。（凭证标本：*唐启明 & 张仕燕 20181001-190*）

A. Population; B. Thallus; C–D. Portions of the thalli; E. Ventral scale; F. Epidermal pore; G. Epidermal cells. (All from *Tang & Zhang 20181001-190*)

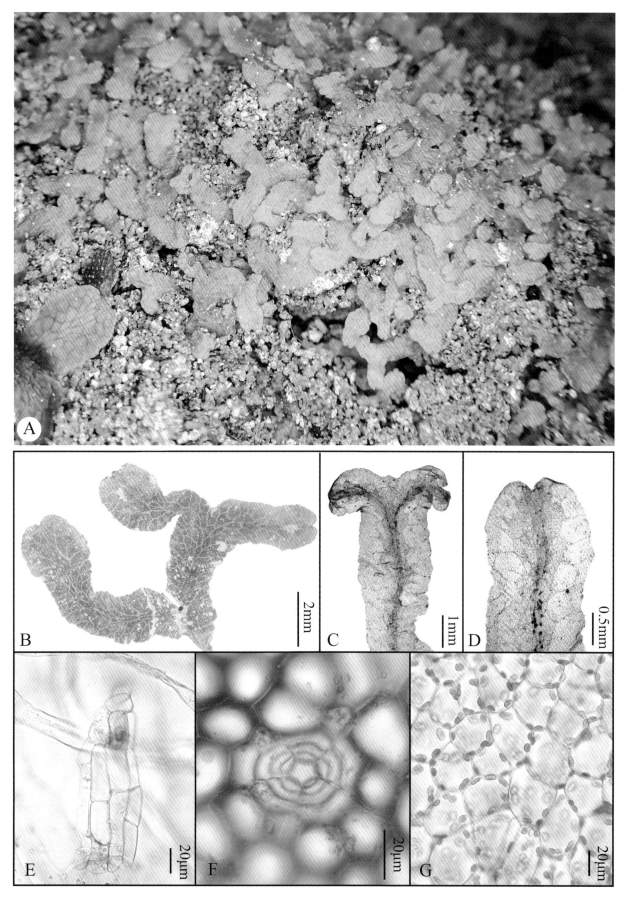

图 14 光苔

Fig. 14 *Cyathodium cavernarum* Kunze

芽胞光苔

Cyathodium tuberosum Kashyap, New Phytol. 13(6/7): 210. 1914.

叶状体扁平柔薄，背面淡绿色到黄绿色，多长在阴暗潮湿的环境，看上去呈现出绿色的荧光，顶端常长有深绿色被无色透明假根的块状芽。叶状体中部横切一般三层，包括上表皮和下表皮，以及中间的巨大气室，气室中无营养丝。气孔简单，高出表皮细胞或与表皮细胞平行。腹鳞片无色透明，线形，位于叶状体腹面，无附器。假根单型，仅具光滑假根。雌雄生殖器官未见。

生境：潮湿石生，石生，潮湿土生，土生。海拔 751—1860 m。

雅长分布：草王山，黄猄洞天坑，拉雅沟，李家坨屯，盘古王，下岩洞屯，悬崖天坑，中井屯竖井，中井屯，霄罗湾洞穴。

中国分布：贵州，陕西，云南。首次记录于广西。

Thalli dorsally light green to yellowish green, usually luminous in dark places, tips often produce dark green tubers covered with dense rhizoids. Air chambers 1 layer, without photosynthetic filament. Epidermal pores simple, raised above or plane with epidermis. Ventral scales hyaline, linear, on ventral thallus, without appendage. Rhizoids of 1 type, smooth. Androecia and gynoecia not seen.

Habitat: On (wet) rocks and (wet) soil. Elev. 751−1860 m.

Distribution in Yachang: Caowangshan, Huangjingdong Tiankeng, Layagou, Lijiatuo Tun, Panguwang, Xiayandong Tun, Xuanya Tiankeng, Zhongjing Tun, Xiaoluowan Cave.

Distribution in China: Guizhou, Shaanxi, Yunnan. New to Guangxi.

▶ A. 种群；B. 叶状体；C. 叶状体一段；D. 块茎；E. 叶状体表面气孔；F. 叶状体一段带块茎；G. 腹鳞片；H. 表皮细胞。（凭证标本：*唐启明等 20190515-34*）

A. Population; B. Thallus; C. Portion of thallus; D. Tuber; E. Epidermal pore; F. Portion of thallus with tuber; G. Ventral scale; H. Epidermal cells. (All from *Tang et al. 20190515-34*)

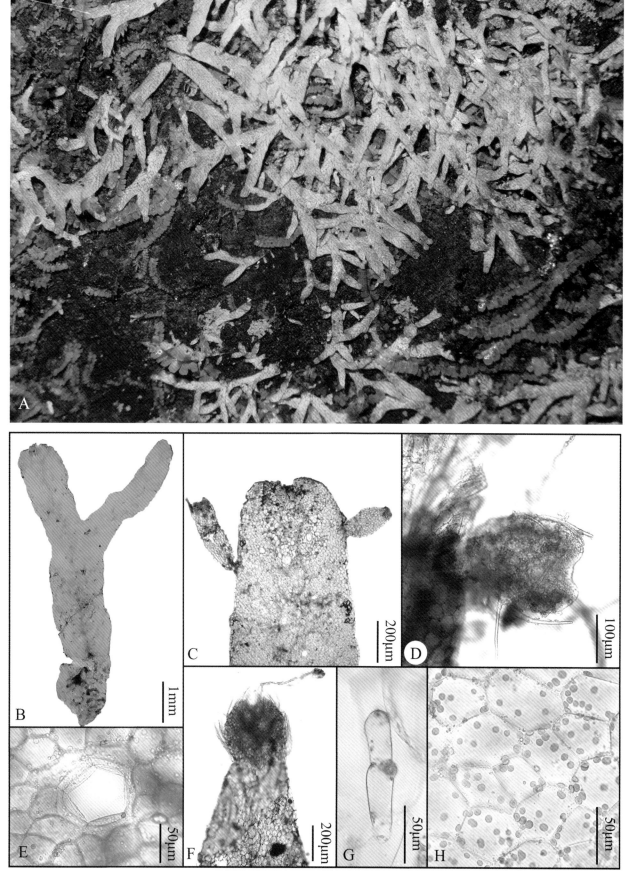

图 15 芽胞光苔

Fig. 15 *Cyathodium tuberosum* Kashyap

无翼钱苔

Riccia billardierei Mont. et Nees, Syn. Hepat. 4: 602. 1846.

叶状体扁平，背面绿色，丛生。整个叶状体背面可见线形凹槽。叶状体横切面宽为厚的3—5倍。叶状体无气孔和气室分化，仅具营养丝。腹鳞片明显，紫色，位于叶状体腹面侧翼，无附器。假根二型，具光滑假根和瘤状假根。雌雄同株。雄器和雌器均着生于叶状体内。不具苞膜和假蒴萼。孢子体大量，排列不规则。孢子球形，直径70—95 μm，远极面具网状纹饰，5—8个网孔横穿整个远极面；近极面具网状纹饰，并具不明显的三射线。孢子无翼边。弹丝缺。

生境：土生。海拔819 m。

雅长分布：隆合朝屯。

中国分布：海南，云南。首次记录于广西。

Thalli dorsally green, in mats. Dorsal furrows distinct, along the whole thallus. Thallus 3–5 times as broad as thick in transverse section. Air chambers and epidermal pores absent, photosynthetic filament present. Ventral scales distinct, purple, on ventral flanks of thallus, without appendage. Rhizoids of 2 types, either smooth or tuberculate. Monoicous. Androecia and gynoecia embedded within the thallus. Involucres and pseudoperianths absent. Sporangia numerous, irregularly arranged. Spores globose, 70–95 μm in diameter, distal surface reticulate forming 5–8 areolae across the diameter, proximal surface reticulate, triradiate mark indistinct. Spore wings absent. Elaters absent.

Habitat: On soil. Elev. 819 m.

Distribution in Yachang: Longhechao Tun.

Distribution in China: Hainan, Yunnan. New to Guangxi.

▶ A. 种群；B. 叶状体一段；C. 叶状体一段（腹面观）；D. 叶状体横切面；E. 叶状体横切面中部部分上半部；F. 叶状体横切面中部部分，示颈卵器；G. 叶状体横切面边缘部分；H. 孢子近极面观；I. 孢子远极面观。（凭证标本：黄萍等 210724-5A）

A. Population; B. Portion of thallus; C. Portion of thallus (ventral view); D. Transverse section of thallus; E. Transverse section of thallus at median part showing the upper portion; F. Transverse section of thallus at median part showing archegonium; G. Transverse section of thallus at marginal part; H. Proximal view of spore; I. Distal view of spore. (All from *Huang et al. 210724-5A*)

图 16 无翼钱苔

Fig. 16 *Riccia billardieri* Mont. & Nees

稀枝钱苔

Riccia huebeneriana Lindenb., Nova Acta Phys.-Med. Acad. Cacs. Leop.-Carol. Nat. Cur. 18(1): 504d. 1836.

叶状体扁平，背面淡绿色，通常形成环状或半环状。叶状体背面先端和中部可见线形凹槽。叶状体横切面宽为厚的 2.6—3.5 倍。气室大，尤其在叶状体边缘处，气室中无营养丝。气孔简单，极小。腹鳞片退化，无附器。假根二型，具光滑假根和瘤状假根。雌雄同株。雄器和雌器均着生于叶状体内。不具苞膜和假蒴萼。孢子体少，在叶状体腹面隆起。孢子球形，直径 55—70 μm，远极面具网状纹饰，5—6 个网孔横穿整个远极面；近极面具网状纹饰，并具明显的三射线。孢子具翼边，4—7 μm 宽，边缘有钝齿。弹丝缺。

生境：土生。海拔 1194 m。

雅长分布：逻家田屯。

中国分布：澳门，广西，贵州，湖北，吉林，辽宁，内蒙古，山东，云南。

Thalli dorsally light green, in rosettes or hemi-rosettes. Dorsal furrows distinct, along apical and median part of thallus. Thallus 2.6–3.5 times as broad as thick in transverse section. Air chambers large, especially in margin, without photosynthetic filament. Epidermal pores simple, tiny. Ventral scales reduced, without appendage. Rhizoids of 2 types, either smooth or tuberculate. Monoicous. Androecia and gynoecia embedded within the thallus. Involucres and pseudoperianths absent. Sporangia few, prominent on the dorsal surface. Spores globose, 55–70 μm in diameter, distal surface reticulate forming 5–6 areolae across the diameter, proximal surface reticulate, triradiate mark distinct. Spore wings 4–7 μm, slightly crenulate. Elaters absent.

Habitat: On soil. Elev. 1194 m.

Distribution in Yachang: Luojiatian Tun.

Distribution in China: Macao, Guangxi, Guizhou, Hubei, Jilin, Liaoning, Inner Mongolia, Shandong, Yunnan.

▶ A. 种群；B. 叶状体；C–D. 叶状体一段；E. 叶状体横切面；F. 叶状体横切面边缘部分；G. 孢子，近极面观（左），远极面观（右）。（凭证标本：唐启明 & 张仕艳 20181003-347）

A. Population; B. Thallus; C–D. Portions of the thalli; E. Transverse section of thallus; F. Transverse section of thallus at marginal part; G. Spores, Proximal view (left), Distal view (right). (All from *Tang & Zhang 20181003-347*)

图 17　稀枝钱苔

Fig. 17　*Riccia huebeneriana* Lindenb.

印尼钱苔

Riccia junghuhniana Nees et Lindenb., Syn. Hepat. 4: 609. 1846.

叶状体扁平，背面淡绿色，丛生。叶状体仅背面先端见线形凹槽。叶状体横切面宽为厚的2.6—3.5倍。气室狭小，无营养丝。气孔简单，极小。腹鳞片退化，无附器。假根二型，具光滑假根和瘤状假根。雌雄同株。雄器和雌器着生于叶状体内。不具苞膜和假蒴萼。孢子体大量，排列不规则。孢子球形，直径60—85 μm，远极面具网状纹饰，7—10个网孔横穿整个远极面；近极面具网状纹饰，并具明显的三射线。孢子具翼边，2—3 μm宽，边缘有钝齿。弹丝缺。

生境：土生。海拔819 m。

雅长分布：隆合朝屯。

中国分布：浙江。首次记录于广西。

Thalli dorsally light green, in mats. Dorsal furrows distinct, along the apical part of thallus. Thallus 2.6–3.5 times as broad as thick in transverse section. Air chambers narrow, without photosynthetic filament. Epidermal pores simple, tiny. Ventral scales reduced, without appendage. Rhizoids of 2 types, either smooth or tuberculate. Monoicous. Androecia and gynoecia embedded within the thallus. Involucres and pseudoperianths absent. Sporangia numerous, irregularly arranged. Spores globose, 60−85 μm in diameter, distal surface reticulate forming 7–10 areolae across the diameter, proximal surface reticulate, triradiate mark distinct. Spore wings 2−3 μm, slightly crenulate. Elaters absent.

Habitat: On soil. Elev. 819 m.

Distribution in Yachang: Longhechao Tun.

Distribution in China: Zhejiang. New to Guangxi.

▶ A. 种群；B. 叶状体；C. 叶状体横切面；D. 叶状体横切面中部部分上半部；E. 叶状体横切面中部部分；F. 叶状体横切面边缘部分；G. 孢子近极面观；H. 孢子远极面观。（凭证标本：黄萍等 *210724-7B*）

A. Population; B. Thallus; C. Transverse section of thallus; D. Transverse section of thallus at median part showing the upper portion; E. Transverse section of thallus at median part; F. Transverse section of thallus at marginal part; G. Proximal view of spore; H. Distal view of spore. (All from *Huang et al. 210724-7B*)

图 18 印尼钱苔

Fig. 18 *Riccia junghuhniana* Nees & Lindenb.

厚壁钱苔

Riccia oryzicola T. Tominaga et Furuki, Hikobia 17(3): 181–186. 2017.

叶状体扁平，背面鲜绿色，丛生。叶状体仅背面先端见线形凹槽。叶状体横切面宽为厚的2—3倍。无气孔和气室分化，仅具营养丝。腹鳞片无色透明，位于叶状体腹面侧翼，无附器。假根二型，具光滑假根和瘤状假根。雌雄同株。雄器和雌器着生于叶状体内。不具苞膜和假蒴萼。孢子体大量，排列不规则。孢子球形，直径60—80 μm，远极面具网状纹饰，8—10个网孔横穿整个远极面，网孔节点处有瘤状凸起；近极面具网状纹饰，并具明显的三射线。孢子具翼边，3—8 μm 宽，边缘有钝齿。弹丝缺。

生境：土生。海拔819—1039 m。

雅长分布：隆合朝屯，悬崖屯。

中国分布：湖南，陕西，上海，四川，浙江。首次记录于广西。

Thalli dorsally light green, in mats. Dorsal furrows distinct, along the apical part of thallus. Thallus 2–3 times as broad as thick in transverse section. Air chambers and epidermal pores absent, photosynthetic filament present. Ventral scales hyaline, on ventral flanks of thallus, without appendage. Rhizoids of 2 types, either smooth or tuberculate. Monoicous. Androecia and gynoecia embedded within the thallus. Involucres and pseudoperianths absent. Sporangia numerous, irregularly arranged. Spores globose, 60–80 μm in diameter, distal surface reticulate forming 8–10 areolae across the diameter, lamellae with tubercules at the angles, proximal surface reticulate, triradiate mark distinct. Spore wings 3–8 μm, slightly crenulate. Elaters absent.

Habitat: On soil. Elev. 819–1039 m.

Distribution in Yachang: Longhechao Tun, Xuanya Tun.

Distribution in China: Hunan, Shaanxi, Shanghai, Sichuan, Zhejiang. New to Guangxi.

▶ A. 种群；B. 叶状体；C. 叶状体一段；D. 叶状体横切面；E. 叶状体横切面中部部分；F. 叶状体横切面近中部部分；G. 叶状体横切面边缘部分；H. 孢子近极面观；I. 孢子远极面观。（凭证标本：*黄萍等 210724-5B*）

A. Population; B. Thalli; C. Portion of thallus; D. Transverse section of thallus; E. Transverse section of thallus at median part; F. Transverse section of thallus near median part; G. Transverse section of thallus at marginal part; H. Proximal view of spore; I. Distal view of spore. (All from *Huang et al. 210724-5B*)

图 19 厚壁钱苔

Fig. 19 *Riccia oryzicola* T. Tominaga & Furuki

花边钱苔

Riccia rhenana Lorb. ex Müll. Frib., Hedwigia 80(1/2): 94. 1941.

叶状体扁平，背面绿色至黄绿色，丛生。叶状体背面凹槽不明显。叶状体横切面宽为厚的4—8倍。气室大，无营养丝。气孔简单，小。腹鳞片退化，无附器。假根二型，具光滑假根和瘤状假根。雌雄同株。雄器和雌器着生于叶状体内。不具苞膜和假蒴萼。孢子体极少。孢子球形，直径55—75 μm，远极面具网状纹饰，3—5个网孔横穿整个远极面，网孔中间具刺状凸起；近极面具网状纹饰，并具明显的三射线。孢子具翼边，3—7 μm宽，边缘有钝齿。弹丝缺。

生境：潮湿土生。海拔1421 m。

雅长分布：大宴坪屯。

中国分布：云南。首次记录于广西。

Thalli dorsally green to yellowish green, in mats. Dorsal furrows indistinct. Thallus 4–8 times as broad as thick in transverse section. Air chambers large, without photosynthetic filament. Epidermal pores simple, tiny. Ventral scales reduced, without appendage. Rhizoids of 2 types, either smooth or tuberculate. Monoicous. Androecia and gynoecia embedded within the thallus. Involucres and pseudoperianths absent. Sporangia few. Spores globose, 55–75 μm in diameter, distal surface reticulate forming 3–5 areolae across the diameter, in the middle of areolae with spinate projections, proximal surface reticulate, triradiate mark distinct. Spore wings 3–7 μm, slightly crenulate. Elaters absent.

Habitat: On wet soil. Elev. 1421 m.

Distribution in Yachang: Dayanping Tun.

Distribution in China: Yunnan. New to Guangxi.

▶ A. 种群；B. 叶状体；C. 叶状体横切面；D. 叶状体表面气孔；E. 叶状体横切面中部部分；F. 叶状体横切面中部部分下半部；G. 孢子近极面观；H. 孢子远极面观。（凭证标本：*唐启明等 20201105-127*）

A. Population; B. Thallus; C. Transverse section of thallus; D. Epidermal pore; E. Transverse section of thallus at median part; F. Transverse section of thallus at median part showing the lower portion; G. Proximal view of spore; H. Distal view of spore. (All from *Tang et al. 20201105-127*)

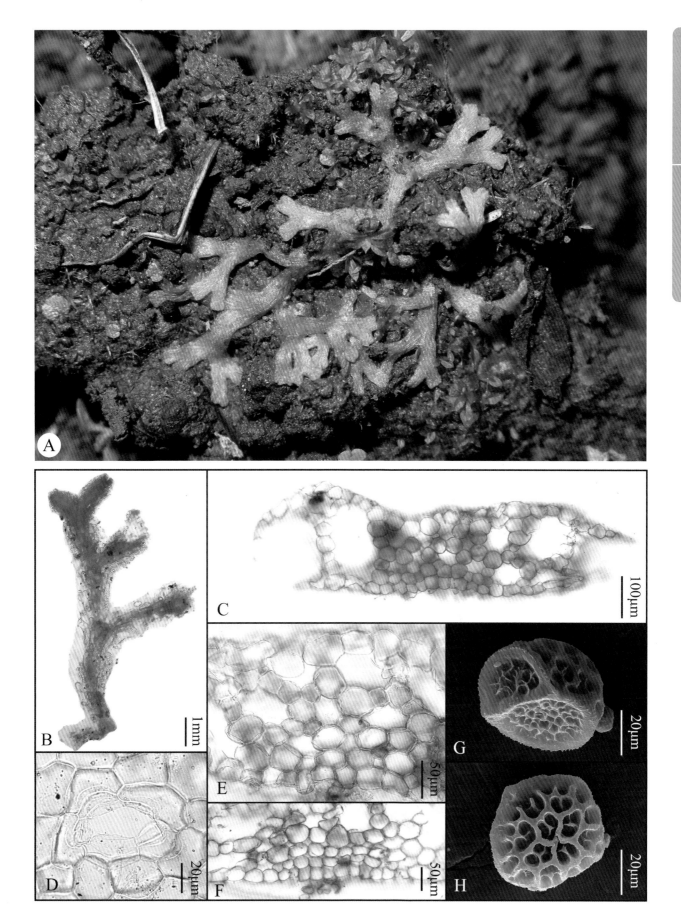

图 20 花边钱苔

Fig. 20 *Riccia rhenana* Lorb. ex Müll. Frib.

日本小叶苔

Fossombronia japonica Schiffn., Österr. Bot. Z. 49(11): 389. 1899.

植物体绿色至黄绿色，匍匐着生。侧叶生于茎两侧，蔽后式排列，覆瓦状至毗邻，方形至长方形，边缘波状，具不规则齿突。叶细胞薄壁，无三角体，角质层平滑。油体聚合型。不具腹叶。假根多，紫红色，生于茎腹面。雌雄同株。颈卵器和精子器混生，裸露生于茎背面。假蒴萼钟形，边缘波曲。孢蒴球形，成熟时不规则裂开。孢子近球形，远极面具网状纹饰，5—8个网孔横穿整个远极面；近极面具脊状纹饰。弹丝1（—2）螺旋加厚。

生境： 潮湿土生，土生。海拔1153—1774 m。

雅长分布： 草王山，李家坨屯，逻家田屯，盘古王，旁墙屯。

中国分布： 澳门，安徽，福建，湖北，广东，台湾，香港。首次记录于广西。

Plants green to yellowish green, prostrate. Leaves obliquely inserted on the stem in two lateral rows, succubous, imbricate to contiguous, oblate to quadrate, margin undulate and irregularly dentate. Leaf cells thin-walled, trigones absent, cuticle smooth. Oil bodies segmented. Underleaves absent. Rhizoids purplish red, scattered along the ventral surface of the stem. Monoicous. Archegonia and antheridia mixed, naked, scattered on dorsal side of the stem. Caulocalices (Pseudoperianths) campanulate, margin undulate. Capsules spherical, dehiscence irregular at maturity. Spores subglobose, distal surface reticulate forming 5−8 areolae across the diameter, proximal surface cristate. Elaters with 1(−2) spiral thickenings.

Habitat: On (wet) soil. Elev. 1153−1774 m.

Distribution in Yachang: Caowangshan, Lijiatuo Tun, Luojiatian Tun, Panguwang, Pangqiang Tun.

Distribution in China: Macao, Anhui, Fujian, Hubei, Guangdong, Taiwan, Hongkong. New to Guangxi.

▶ A. 种群；B. 侧叶；C. 弹丝；D. 孢蒴内壁；E. 叶中部细胞，示油体；F. 孢子近极面观；G. 孢子远极面观。（凭证标本：*专玉梅等 191017-273*）

A. Population; B. Leaf; C. Elaters; D. the inner layer of capsule wall; E. Median cells of leaf showing oil bodies; F. Proximal view of spore; G. Distal view of spore. (All from *Wei et al. 191017-273*)

图 21 日本小叶苔

Fig. 21 *Fossombronia japonica* Schiffn.

小叶苔

Fossombronia pusilla (L.) Nees, Naturgesch. Eur. Leberm. 3: 319. 1838.

植物体绿色至黄绿色，匍匐着生。侧叶生于茎两侧，蔽后式排列，覆瓦状至毗邻，方形至长方形，边缘平滑至波状。叶细胞薄壁，无三角体，角质层平滑。油体聚合型。不具腹叶。假根多，紫红色，生于茎腹面。雌雄同株。颈卵器和精子器混生，裸露生于茎背面。假蒴萼钟形或倒钟形，边缘明显波曲。孢蒴球形，成熟时不规则裂开。孢子近球形，远极面具薄片状纹饰，偶尔会形成1—3个网孔，近极面具脊状纹饰。弹丝2—4螺旋加厚。

生境： 潮湿石生，土生。海拔1822—1862 m。

雅长分布： 草王山。

中国分布： 甘肃，广西，贵州，河北，黑龙江，湖南，吉林，辽宁，山东，四川，台湾，西藏，云南，浙江。

Plants green to yellowish green, prostrate. Leaves obliquely inserted on the stem in two lateral rows, succubous, imbricate to contiguous, oblate to quadrate, margin plane to undulate. Leaf cells thin-walled, trigones absent, cuticle smooth. Oil bodies segmented. Underleaves absent. Rhizoids purplish red, scattered along the ventral surface of the stem. Monoicous. Archegonia and antheridia mixed, naked, scattered on dorsal side of the stem. Caulocalices (Pseudoperianthes) campanulate or inverted bell-shaped, margin highly undulate. Capsules spherical, dehiscence irregular at maturity. Spores subglobose, distal surface lamellate, with the lamellae sometimes forming 1–3 areolae, proximal surface cristate. Elaters with 2–4 spiral thickenings.

Habitat: On wet rocks and soil. Elev. 1822−1862 m.

Distribution in Yachang: Caowangshan.

Distribution in China: Gansu, Guangxi, Guizhou, Hebei, Heilongjiang, Hunan, Jilin, Liaoning, Shandong, Sichuan, Taiwan, Tibet, Yunnan, Zhejiang.

▶ A. 种群；B. 侧叶；C. 弹丝；D. 孢蒴内壁；E. 叶中部细胞，示油体；F. 孢子近极面观；G. 孢子远极面观。（凭证标本：*韦玉梅等 191013-114*）

A. Population; B. Leaf; C. Elaters; D. the inner layer of capsule wall; E. Median cells of leaf showing oil bodies; F. Proximal view of spore; G. Distal view of spore. (All from *Wei et al. 191013-114*)

图 22 小叶苔

Fig. 22 *Fossombronia pusilla* (L.) Nees

带叶苔

Pallavicinia lyellii (Hook.) Gray, Nat. Arr. Brit. Pl. 1: 775. 1821.

叶状体背面绿色至深绿色，宽带状，匍匐着生。中肋明显，占叶状体宽约 1/8，中央有由厚壁细胞组成的中轴。翼部单层细胞，边缘具纤毛状齿，有时齿不明显或缺。叶翼细胞五边形或六边形，薄壁。油体聚合型。假根淡褐色，生于叶状体腹面先端或中肋腹面。雌雄异株。精子器两列，生于叶状体背面中肋两侧，被苞叶覆盖。颈卵器聚生于中肋背面，周围有杯状的苞膜包围。假蒴萼圆筒形，口部具纤毛状齿。孢蒴长椭圆形，成熟时 4 瓣纵裂。孢子球形，具网格状纹饰。弹丝 2 螺旋加厚。

生境： 石生，土生。海拔 527—620 m。

雅长分布： 二沟，拉雅沟，旁墙屯。

中国分布： 全国大部分省区均有分布。

Thalli dorsally green to dark green when fresh, widely ribbon-like, prostrate. Midrib sharply defined, ca. 1/8 of thallus width, with one central strand of thick-walled cells. Thallus wings unistratose, margins with or without long ciliate teeth. Wing cells 5–6 angled, thin-walled. Oil bodies segmented. Rhizoids pale brown, distributed near ventral thallus apex, or throughout the ventral midrib. Dioicous. Antheridia covered by bracts, arranged in 1 row along each side of dorsal midrib. Archegonia clustered on dorsal surface of midrib, surrounded by cup-like involucres. Pseudoperianths cylindrical, with ciliate mouth. Capsules long-elliptical, longitudinally dehiscing by 4 regular valves at maturity. Spores globose, surface finely reticulate. Elaters with 2 spiral thickenings.

Habitat: On rocks and soil. Elev. 527–620 m.

Distribution in Yachang: Ergou, Layagou, Pangqiang Tun.

Distribution in China: Widely distributed in most provinces of China.

▶ A. 种群；B. 叶状体横切面；C. 叶状体横切面边缘部分；D. 叶状体横切面中部部分；E. 弹丝；F. 翼部细胞，示油体；G. 孢子。（凭证标本：韦玉梅等 201109-17）

A. Population; B. Transverse section of thallus; C. Transverse section of thallus at marginal part; D. Transverse section of thallus at median part; E. Elaters; F. Wing cells showing oil bodies; G. Spore. (All from *Wei et al. 201109-17*)

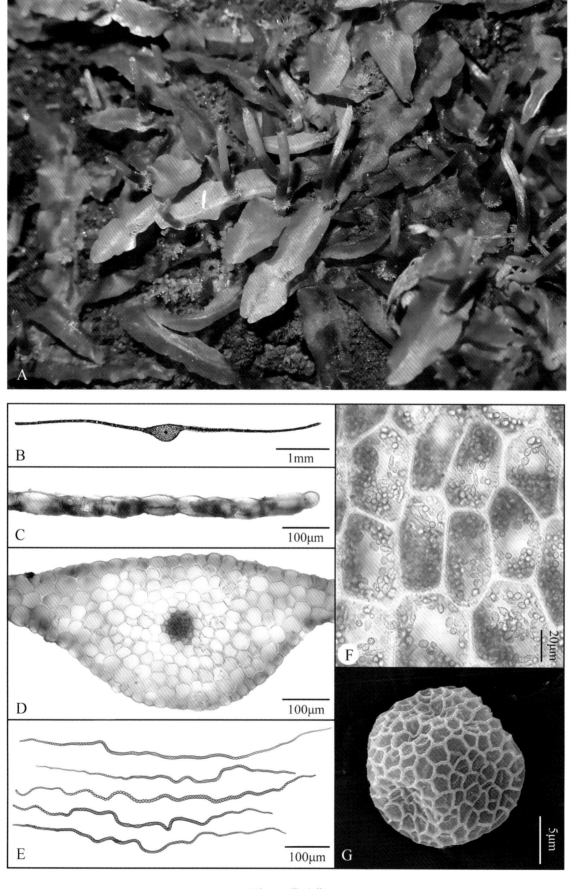

图 23　带叶苔

Fig. 23　*Pallavicinia lyellii* (Hook.) Gray

异溪苔（鹿角苔，花叶溪苔）

Apopellia endiviifolia (Dicks.) Nebel et D. Quandt, Taxon 65(2): 230. 2016.

Pellia endiviifolia (Dicks.) Dumort., Recueil Observ. Jungerm.: 27. 1835.

叶状体背面淡绿色至褐绿色，顶端心脏形，秋季或初冬末端常大量产生易落的花状分瓣。中肋界限不明显。叶状体横切面中央可达11个细胞厚，向边缘逐渐变薄为单层细胞，边缘平展或呈波状。表皮细胞五边形或六边形，薄壁。油体聚合型。假根褐色，生于叶状体腹面。雌雄异株。精子器散生于中肋背面的精子器腔中。颈卵器聚生于中肋背面，周围有杯状的苞膜包围。孢子体未见。

生境： 潮湿土生和土生环境常见，潮湿石生、石生、钙华基质环境也有分布。海拔1138—1755 m。

雅长分布： 草王山，吊井天坑，里郎天坑，逻家田屯，盘古王，旁墙屯，全达村，下岩洞屯，中井天坑。

中国分布： 安徽，福建，甘肃，广西，贵州，河南，黑龙江，湖北，吉林，江西，陕西，山东，台湾，新疆，浙江。

Thalli dorsally light green to brownish green when fresh, apex emarginate, usually proliferating caducous furcate branches at apex in autumn and early winter. Midribs not clearly defined. Transverse section of thallus up to 11 cells thick in middle, gradually becomes single-layered towards margin. Margins flat or undulate. Epidermal cells 5–6 angled, thin-walled. Oil bodies segmented. Rhizoids brown, distributed throughout the ventral midrib. Dioicous. Antheridia scattered in small cavities in the dorsal midribs. Archegonia clustered on dorsal surface of midrib, surrounded by cup-like involucres. Sporophytes not seen.

Habitat: Often on (wet) soil, sometimes on (wet) rocks and calcareous substrates. Elev. 1138–1755 m.

Distribution in Yachang: Caowangshan, Diaojing Tiankeng, Lilang Tiankeng, Luojiatian Tun, Panguwang, Pangqiang Tun, Quanda Village, Xiayandong Tun, Zhongjing Tiankeng.

Distribution in China: Anhui, Fujian, Gansu, Guangxi, Guizhou, Henan, Heilongjiang, Hubei, Jilin, Jiangxi, Shaanxi, Shandong, Taiwan, Xinjiang, Zhejiang.

▶ A. 种群；B. 叶状体；C. 花状分瓣；D. 叶状体横切面；E. 叶状体横切面中部部分；F. 叶状体横切面近中部部分；G. 叶状体横切面边缘部分；H. 叶状体雌株一段；I. 表皮细胞，示油体。（凭证标本：*唐启明等 20201108-228*）

A. Population; B. Thallus; C. Furcate branches; D. Transverse section of thallus; E. Transverse section of thallus at median part; F. Transverse section of thallus near median part; G. Transverse section of thallus at marginal part; H. Portion of the female thallus; I. Epidermal cells showing oil bodies. (All from *Tang et al. 20201108-228*)

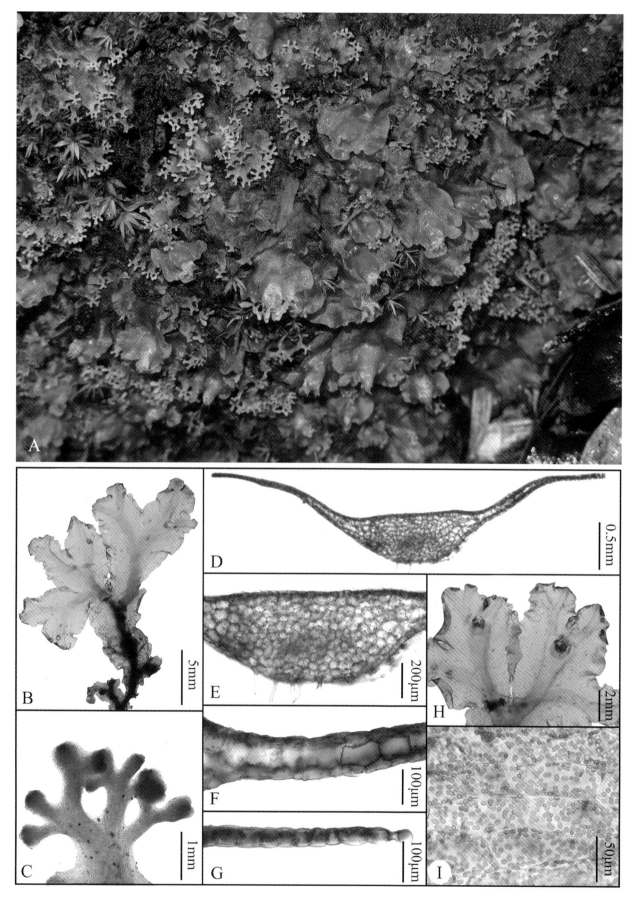

图 24 异溪苔

Fig. 24 *Apopellia endiviifolia* (Dicks.) Nebel & D. Quandt

短萼狭叶苔（狭叶叶苔）

Liochlaena subulata (A. Evans) Schljakov, Pečen. Mchi Sev. SSSR 4: 71. 1981.
Jungermannia subulata A. Evans, Trans. Connecticut Acad. Arts 8(15): 258. 1891.

植物体绿色，带叶宽 3.25—4.15 mm，分枝间生型，不育枝先端常具鞭状枝。茎横切面 11—12 个细胞厚，内外细胞不明显分化，但背腹面的表皮细胞略有不同，腹面表皮细胞通常比背面的小，薄壁。假根散生于茎腹面。侧叶覆瓦状排列，椭圆状卵形，顶端圆，边缘全缘，背侧基部下延。侧叶细胞薄壁，角质层具条纹状疣，三角体大，简单三角形或肿胀形，中部球状加厚缺。油体聚合型，每个细胞 5—10 个。腹叶缺。雄苞和雌苞未见。芽胞单细胞，簇生于鞭状枝顶端。

生境：土生和腐木环境常见，石生环境也有分布。海拔 1048—1589 m。

雅长分布：草王山，黄猄洞天坑，蓝家湾天坑，盘古王，旁墙屯，下岩洞屯，中井屯。

中国分布：福建，广西，贵州，河南，黑龙江，湖南，吉林，江西，辽宁，台湾，西藏，云南，浙江。

Plants green when fresh, 3.25−4.15 mm wide, branches intercalary, flagelliform branches emerging from the sterile shoot apex. Stem in transverse section 11−12 cells thick, without differentiated cortex, but epidermal cells on ventral side usually smaller than cells on dorsal side, thin-walled. Rhizoids scattered on ventral side of stem. Leaves imbricate, oblong-ovate, apex rounded, margin entire, dorsal base decurrent. Leaf cells thin-walled, trigones large, simple-triangulate or nodulate, intermediate thickenings absent. Cuticle striate-verrucose. Oil bodies segmented, 5−10 per median cell. Underleaves absent. Androecia and gynoecia not seen. Gemmae 1-celled, abundantly occurring on the tip of flagelliform branches.

Habitat: Often on soil and rotten logs, sometimes on rocks. Elev. 1048−1589 m.

Distribution in Yachang: Caowangshan, Huangjingdong Tiankeng, Lanjiawan Tiankeng, Panguwang, Pangqiang Tun, Xiayandong Tun, Zhongjing Tun.

Distribution in China: Fujian, Guangxi, Guizhou, Henan, Heilongjiang, Hunan, Jilin, Jiangxi, Liaoning, Taiwan, Tibet, Yunnan, Zhejiang.

▶ A. 种群；B. 植物体（背面观）；C. 侧叶；D. 芽胞；E. 茎横切面；F. 叶中部细胞，示角质层疣；G. 叶中部细胞，示油体。（凭证标本：*韦玉梅等 201110-37A*）

A. Population; B. Plant (dorsal view); C. Leaves; D. Gemmae; E. Transverse section of stem; F. Median cells of leaf showing verrucose cuticle; G. Median cells of leaf showing oil bodies. (All from *Wei et al. 201110-37A*)

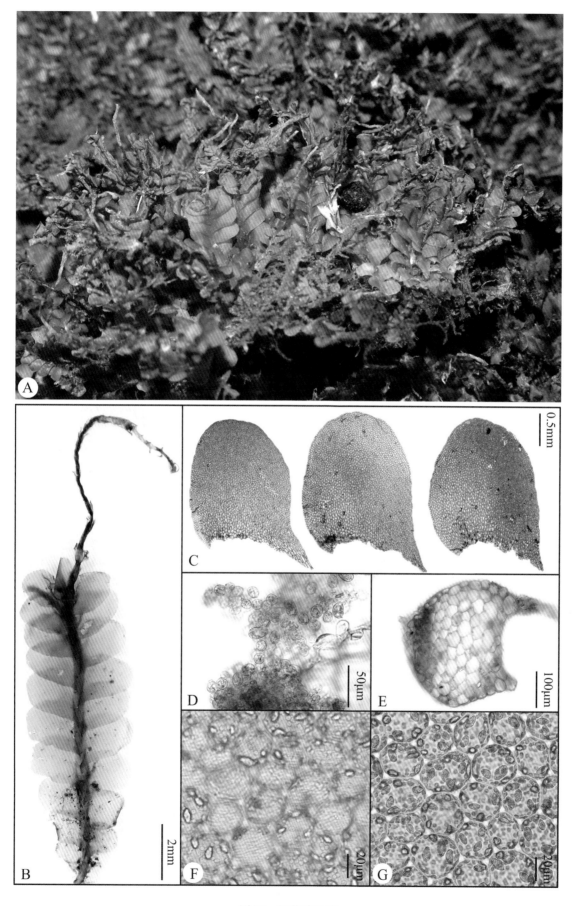

图 25 短萼狭叶苔

Fig. 25 *Liochlaena subulata* (A. Evans) Schljakov

秩父无褶苔

Mesoptychia chichibuensis (Inoue) L. Söderstr. et Váňa, Phytotaxa 65: 53. 2012.

植物体浅绿色，带叶宽 0.56—1.12 mm，分枝少，间生型。茎横切面 6—7 个细胞厚，内外细胞不明显分化，但背腹面的表皮细胞略有不同，腹面表皮细胞通常比背面的小，薄壁。假根散生于茎腹面。侧叶远生，偶尔毗邻，卵圆形，顶端 2 裂至叶长的 1/4—1/3，侧缘全缘，基部不下延。侧叶细胞薄壁，角质层具乳头状或条纹状疣，三角体小，简单三角形，中部球状加厚缺。油体聚合型，每个细胞(2—)3—5（—7）个。腹叶常可见，披针形，偶尔线形，3—6 个细胞长，基部 1—3 个细胞宽；近雌苞的腹叶较大，可达 14 个细胞长，基部 5 个细胞宽。雌雄同株。雄苞生于雌苞下，雄苞叶 2—3 对。雌苞顶生，具有 1—2 个新生枝。蒴萼圆柱形，近口部具少数弱的褶皱，口部略收缩，边缘具加长细胞组成的齿突。

生境：石生。海拔 931—1240 m。

雅长分布：蓝家湾天坑，深洞。

中国分布：广西，新疆。

Plants light green when fresh, 0.56–1.12 mm wide, sparingly branched, branches intercalary. Stem in transverse section 6–7 cells thick, without differentiated cortex, but epidermal cells on ventral side usually smaller than cells on dorsal side, thin-walled. Rhizoids scattered on ventral side of stem. Leaves distant, sometimes contiguous, rounded-ovate, apex bilobed to 1/4–1/3 leaf length, margin entire, base not decurrent. Leaf cells thin-walled, trigones small, simple-triangulate, intermediate thickenings absent. Cuticle papillose or striate-verrucose. Oil bodies segmented, (2–) 3–5 (–7) per median cell. Underleaves frequently present, lanceolate, sometimes linear, 3–6 cells long, 1–3 cells wide at base, usually larger near gynoecia, up to 14 cells long, 5 cells wide at base. Monoicous.. Androecia below the gynoecium, bracts in 2–3 pairs. Gynoecia terminal, with 1–2 innovations. Perianths cylindrical, weakly pluriplicate near the mouth, mouth slightly constrated, marginal cells elongate, margin crenulate.

Habitat: On rocks. Elev. 931–1240 m.

Distribution in Yachang: Lanjiawan Tiankeng, Shendong.

Distribution in China: Guangxi, Xinjiang.

▶ A. 种群；B. 植物体一段（背面观）；C. 植物体一段带雄苞和蒴萼；D. 蒴萼口部；E. 腹叶茎横切面；F. 侧叶，示角质层的疣；G–H. 腹叶；I. 侧叶中部细胞，示油体；J. 雌苞叶；K. 雌苞腹叶；L. 近雌苞的腹叶。（凭证标本：A 拍摄自韦玉梅等 *191017-263*，其余来自韦玉梅等 *191019-391*）

A. Population; B. Portions of plants (dorsal view); C. Portion of plant with an androecium and a perianth; D. Perianth mouth; E. Transverse section of stem; F. Leaf showing papillose or striate-verrucose cuticle; G–H. Underleaves; I. Median cells of leaf showing oil bodies; J. Bracts; K. Bracteole; L. Underleaf near the gynoecium. (A from *Wei et al. 191017-263*, the others from *Wei et al. 191019-391*)

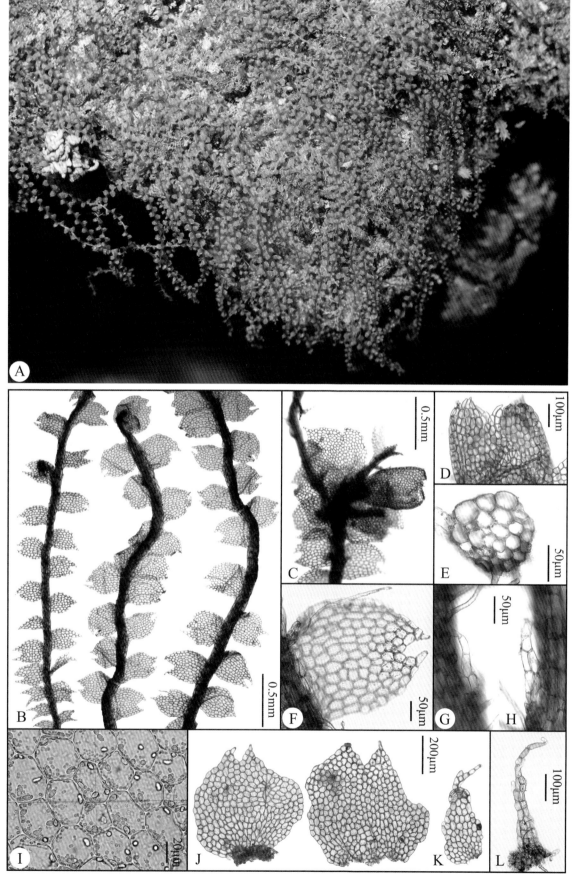

图 26 秩父无褶苔

Fig. 26 *Mesoptychia chichibuensis* (Inoue) L. Söderstr. et Váňa

中华无褶苔（中华斜裂苔）

Mesoptychia chinensis Bakalin, Vilnet et Y. X. Xiong, J. Bryol. 37(3): 196. 2015.

植物体浅绿色，带叶宽 0.40—1.04 mm，分枝少，间生型。茎横切面 5—6 个细胞厚，内外细胞不明显分化，但背腹面的表皮细胞略有不同，腹面表皮细胞通常比背面的小，薄壁。假根散生于茎腹面。侧叶远生，卵圆形或椭圆形，顶端圆或近截形，全缘或微凹，有时浅 2 裂，侧缘全缘，背侧基部略下延。侧叶细胞薄壁，角质层具乳头状或条纹状疣，有时平滑，三角体小，不明显，中部球状加厚缺。油体聚合型，每个细胞 2—4 个。腹叶远生，常退化，丝状，1—2 个细胞长，有时缺失。雌雄异株。雄苞间生，雄苞叶与邻近的侧叶相似，2—3 对。雌苞未见。

生境： 潮湿土生，钙华基质，岩面薄土生。海拔 931—1048 m。
雅长分布： 黄猄洞天坑，盘古王，深洞。
中国分布： 广西，贵州。

Plants light green when fresh, 0.40–1.04 mm wide, sparingly branched, branches intercalary. Stem in transverse section 5–6 cells thick, without differentiated cortex, but epidermal cells on ventral side usually smaller than cells on dorsal side, thin-walled. Rhizoids scattered on ventral side of stem. Leaves distant, rounded-ovate or oblong, apex rounded to subtruncate, entire or retuse, sometimes shallowly bilobed, lateral margin entire, dorsal base slightly decurrent. Leaf cells thin-walled, trigones small, indistinct, intermediate thickenings absent. Cuticle papillose or striate-verrucose to smooth. Oil bodies segmented, 2–4 per median cell. Underleaves vestigial, filiform, 1–2 cells long, sometimes absent. Dioicous. Androecia intercalary, bracts simillar to adjacent leaves, in 2–3 pairs. Gynoecia not seen.

Habitat: On wet soil, calcareous substrates and on rocks with a thin layer of soil. Elev. 931–1048 m.
Distribution in Yachang: Huangjingdong Tiankeng, Panguwang, Shendong.
Distribution in China: Guangxi, Guizhou.

A. 种群；B. 植物体一段（背面观）；C. 侧叶，示角质层疣；D. 侧叶；E. 侧叶中部细胞，示油体；F. 腹叶；G. 茎横切面。（凭证标本：*韦玉梅等 201104-50*）

A. Population; B. Portions of plants (dorsal view); C. Leaf showing papillose or striate-verrucose cuticle; D. Leaf; E. Median cells of leaf showing oil bodies; F. Underleaf; G. Transverse section of stem. (All from *Wei et al. 201104-50*)

图 27 中华无褶苔

Fig. 27 *Mesoptychia chinensis* Bakalin, Vilnet et Y. X. Xiong

玉山无褶苔（异瓣裂叶苔、玉山裂叶苔）

Mesoptychia morrisoncola (Horik.) L. Söderstr. et Váňa, Phytotaxa 65: 54. 2012.

Lophozia diversiloba S. Hatt., J. Jap. Bot. 20: 265. f. 48. 1944.
Lophozia morrisoncola Horik., J. Sci. Hiroshima Univ., Ser. B, Div. 2, Bot. 2: 150. 1934.

植物体黄绿色至浅绿色，带叶宽 1.20—1.60 mm，分枝少，间生型。茎横切面 11—13 个细胞厚，内外细胞不明显分化，但背腹面的表皮细胞略有不同，腹面表皮细胞通常比背面的小，薄壁。假根散生于茎腹面。侧叶覆瓦状排列，圆方形，顶端 2 裂至叶长的 1/4—2/5，偶尔 3 裂，裂瓣通常不对称，长渐尖，尖部通常由 2—6 个单列细胞组成，边缘全缘，或偶尔具少量的短刺齿，背缘直，腹缘拱起呈弧形，背侧基部略下延。侧叶细胞薄壁，角质层具乳头状或条纹状疣，三角体大，肿胀形，中部球状加厚缺。油体聚合型，每个细胞 4—7 个。腹叶缺失。雄苞和雌苞未见。

生境： 石生环境常见，土生、钙华基质、腐木环境也有分布。海拔 931—1424 m。

雅长分布： 大宴坪竖井，拉洞天坑，蓝家湾天坑，李家坨屯，深洞，中井天坑，中井屯竖井。

中国分布： 重庆，甘肃，广西，贵州，湖北，陕西，四川，台湾，云南。

Plants yellowish green to light green when fresh, 1.20–1.60 mm wide, sparingly branched, branches intercalary. Stem in transverse section 11–13 cells thick, without differentiated cortex, but epidermal cells on ventral side usually smaller than cells on dorsal side. Rhizoids scattered on ventral side of stem. Leaves imbricate, rounded-quadrate, apex bilobed to 1/4–2/5 leaf length, sometimes 3-lobed, lobes asymmetrical, apices 2–6 cells in a uniseriate row, margin entire, or occasionally with few spinose teeth, dorsal margin straight, ventral margin arched, dorsal base slightly decurrent. Leaf cells thin-walled, trigones large, nodulate, intermediate thickenings absent. Cuticle papillose or striate-verrucose. Oil bodies segmented, 4–7 per median cell. Underleaves absent. Androecia and gynoecia not seen.

Habitat: often on rocks, sometimes on soil, calcareous substrates and rotten logs. Elev. 931–1424 m.

Distribution in Yachang: Dayanping Shaft, Ladong Tiankeng, Lanjiawan Tiankeng, Lijiatuo Tun, Shendong, Zhongjing Tiankeng, Zhongjing Tun.

Distribution in China: Chongqing, Gansu, Guangxi, Guizhou, Hubei, Shaanxi, Sichuan, Taiwan, Yunnan.

▶ A. 种群；B. 植物体一段（背面观）；C. 侧叶；D. 植物体一段（腹面观）；E. 侧叶裂瓣尖部；F. 叶中部细胞，示油体；G. 叶中部细胞，示角质层疣；H. 茎横切面。（凭证标本：*韦玉梅等 191016-232*）

A. Population; B. Portion of plant (dorsal view); C. Leaves; D. Portion of plant (ventral view); E. Apex of leaf lobe; F. Median cells of leaf showing oil bodies; G. Median cells of leaf showing papillose or striate-verrucose cuticle; H. Transverse section of stem. (All from *Wei et al. 191016-232*)

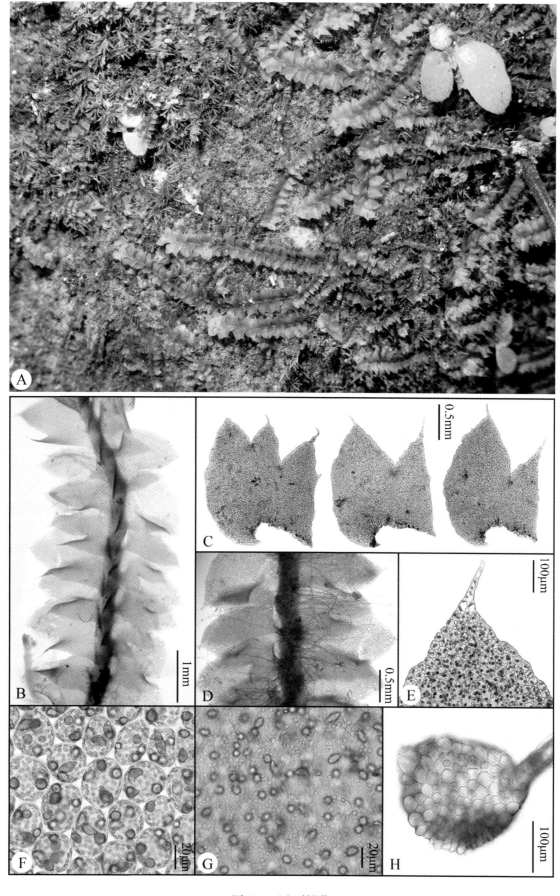

图 28 玉山无褶苔

Fig. 28 *Mesoptychia morrisoncola* (Horik.) L. Söderstr. et Váňa

南亚被蒴苔

Nardia assamica (Mitt.) Amakawa, J. Hattori Bot. Lab. 26: 23. 1963.

植物体浅绿色，带叶宽0.30—0.45 mm，分枝少，顶生型。茎横切面7—8个细胞厚，内外细胞不分化，薄壁。假根生于腹叶基部。侧叶远生，雌株上通常毗邻，肾形至卵圆形，顶端钝至圆钝，边缘全缘，基部几乎不下延。侧叶细胞薄壁，角质层具浅条纹状疣，三角体小，简单三角形，中部球状加厚缺。油体聚合型，每个细胞1（—2）个。腹叶远生，阔三角形，顶端钝，偶尔钝尖。雌雄异株。雄苞未见。雌苞顶生。蒴萼纺锤形，略高出雌苞叶，口部收缩，具5—6条纵褶，下半部分与苞叶合生，形成一个肉质的蒴囊。孢蒴椭球形，成熟时4瓣开裂。孢子球形，表面具密集的蠕虫状饰。弹丝2螺旋加厚。

生境： 土生。海拔1491—1862 m。

雅长分布： 草王山，九十九堡。

中国分布： 安徽，重庆，福建，贵州，湖北，湖南，江苏，江西，辽宁，四川，台湾，香港，云南，浙江。首次记录于广西。

Plants light green when fresh, 0.30–0.45 mm wide, sparingly branched, branches terminal. Stem in transverse section 7–8 cells thick, thin-walled, without differentiated cortex. Rhizoids at base of underleaves. Leaves distant, contigous in female shoots, reniform to rounded-ovate, apex obtuse to rounded-obtuse, margin entire, base hardly decurrent. Leaf cells thin-walled, trigones small, simple-triangulate, intermediate thickenings absent. Cuticle shallow striate-verrucose. Oil bodies segmented, 1(–2) per median cell. Underleaves distant, broad triangular, apex obtuse, occasionally obtuse-acute. Diocious. Androecia not seen. Gynoecia terminal. Perianths fusiform, emergent from bracts, 5–6-plicate near the mouth, mouth slightly constrated, the lower half united with the bracts forming a fleshy perigynium. Capsules ellipsoidal, longitudinally dehiscing by 4 regular valves at maturity. Spores globose, surface densely vermiculate. Elaters with 2 spiral thickenings.

Habitat: On soil. Elev. 1491–1862 m.

Distribution in Yachang: Caowangshan, Jiushijiubao.

Distribution in China: Anhui, Chongqing, Fujian, Guizhou, Hubei, Hunan, Jiangsu, Jiangxi, Liaoning, Sichuan, Taiwan, Hongkong, Yunnan, Zhejiang. New to Guangxi.

▶ A. 种群；B. 植物体一段（背面观）；C. 植物体一段，示腹叶；D. 茎横切面；E. 侧叶；F. 叶中部细胞，示油体；G. 雌苞带成熟孢子体；H. 雌苞纵切面；I. 弹丝和孢子。（凭证标本：韦玉梅等 201108-30）

A. Population; B. Portions of plants (dorsal view); C. Portion of plant showing underleaves; D. Transverse section of stem; E. Leaves; F. Median cells of leaf showing oil bodies; G. Gynoecium with a mature sporophyte; H. Longitudinal section of a gynoecium; I. Elater and spores. (All from *Wei et al. 201108-30*)

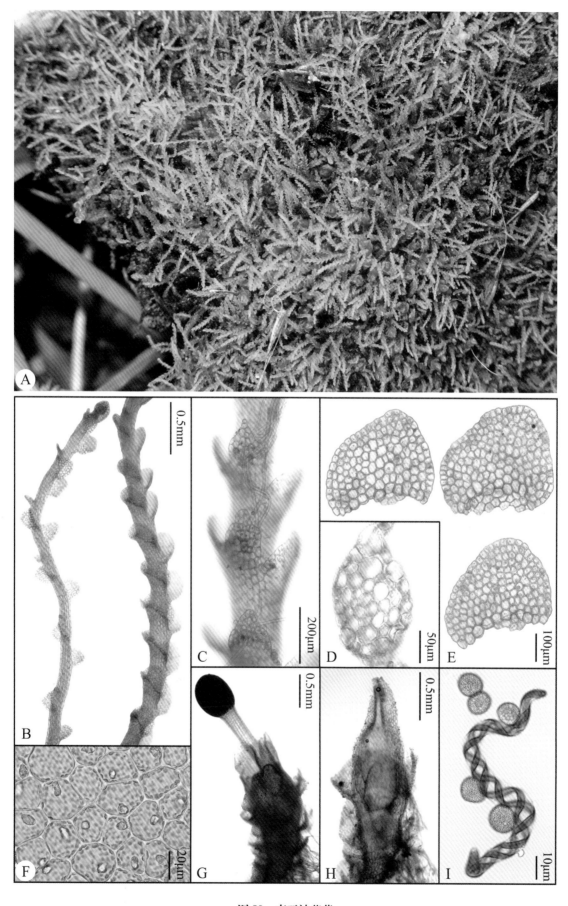

图 29 南亚被蒴苔

Fig. 29 *Nardia assamica* (Mitt.) Amakawa

假苞苔（黄色杯囊苔、厚角杯囊苔、小杯囊苔）

Notoscyphus lutescens (Lehm. et Lindenb.) Mitt., Fl. Vit.: 407. 1871 [1873].

Notoscyphus collenchymatosus C. Gao, X. Y. Jia & T. Cao, Bull. Bot. Res., Harbin 19: 366. 1999.
Notoscyphus parvus C. Gao, X. Y. Jia & T. Cao, Bull. Bot. Res., Harbin 19: 362. 1999.

植物体浅绿色，带叶宽 1.05—1.50 mm，分枝少，间生型。茎横切面约 7 个细胞厚，内外细胞不分化，稍厚壁。假根生于腹叶基部。侧叶覆瓦状排列至远生，圆方形至椭圆形，顶端平截或圆，边缘全缘，基部不下延。侧叶细胞薄或稍厚壁，角质层具乳头状或条纹状疣，三角体小至大，简单三角形或肿胀形，中部球状加厚稀少。油体聚合型，每个细胞 2—5 个。腹叶远生，顶端 2 裂约达 2/3，裂瓣 3—5 个细胞长，基部 1—2 个细胞宽，两侧边缘全缘。雌雄异株。雄苞未见。雌苞顶生，雌苞叶 2—3 对，最内层雌苞叶大，上部分裂成不规则裂片，雌苞腹叶 2—3 对。蒴囊半圆形，下垂，肉质，具有大量假根附着。孢蒴椭球形，成熟时 4 瓣开裂。孢子球形，表面具颗粒状或短条纹状纹饰。弹丝 2 螺旋加厚。

生境：土生。海拔 837—1124 m。

雅长分布：二沟，黄猄洞天坑，旁墙屯。

中国分布：澳门，福建，广东，广西，贵州，海南，湖北，湖南，江西，四川，台湾，香港，云南，浙江。

Plants light green when fresh, 1.05–1.50 mm wide, branches few, terminal. Stem in transverse section ca. 7 cells thick, slightly thick-walled, without differentiated cortex. Rhizoids at base of underleaves. Leaves imbricate to distant, rounded-quadrate to oblong, apex truncate to rounded, margin entire, base not decurrent. Leaf cells thin-walled or slightly thick-walled, trigones small to large, simple-triangulate or nodulate, intermediate thickenings scarce. Cuticle papillose or striate-verrucose. Oil bodies segmented, 2–5 per median cell. Underleaves distant, deeply bilobed to ca. 2/3 underleaf length, lobes 3–5 cells long, 1–2 cells wide at base. Diocious. Androecia not seen. Gynoecia terminal, with 2–3 pairs of bracts and 2–3 bracteoles, innermost bracts large, irregularly lobed. Perigynium bulbous, pendent, freshy, and with rhizoids. Capsules ellipsoidal, longitudinally dehiscing by 4 regular valves at maturity. Spores globose, surface densely granulate and short striate. Elaters with 2 spiral thickenings.

Habitat: On soil. Elev. 837–1124 m.

Distribution in Yachang: Ergou, Huangjingdong Tiankeng, Pangqiang Tun.

Distribution in China: Macao, Fujian, Guangdong, Guangxi, Guizhou, Hainan, Hubei, Hunan, Jiangxi, Sichuan, Taiwan, Hongkong, Yunnan, Zhejiang.

▶ A. 种群；B. 植物体（背面观）；C. 植物体一段带成熟的孢子体（背面观）；D. 侧叶；E. 最内层雌苞叶和雌苞腹叶；F. 腹叶；G. 孢子；H. 植物体一段，示腹叶；I. 茎横切面；J. 弹丝；K. 叶中部细胞，示角质层疣；L. 叶中部细胞，示油体。（凭证标本：*韦玉梅等 201104-27*）

A. Population; B. Plant (dorsal view); C. Portion of plant with a mature sporophyte (dorsal view); D. Leaves; E. Innermost bracts and bracteole; F. Underleaves; G. Spores; H. Portion of plant showing underleaves; I. Transverse section of stem; J. Elaters; K. Median cells of leaf showing papillose and verrucose cuticle; L. Median cells of leaf showing oil bodies. (All from *Wei et al. 201104-27*)

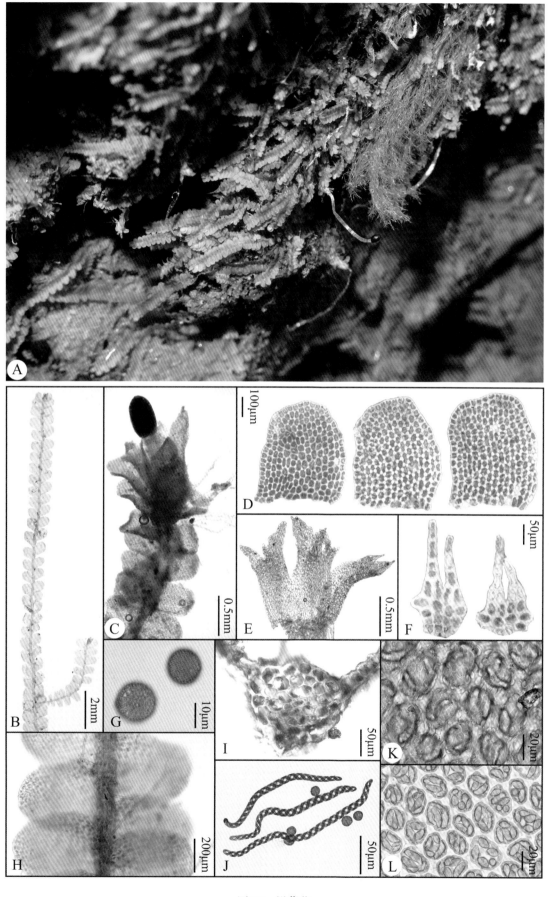

图 30　假苞苔

Fig. 30　*Notoscyphus lutescens* (Lehm. & Lindenb.) Mitt.

偏叶管口苔（偏叶叶苔）

Solenostoma comatum (Nees) C. Gao, Fl. Hepat. Chin. Boreali-Orient.: 73. 1981.

Jungermannia comata Nees, Enum. Pl. Crypt. Javae: 78. 1830.

植物体白绿色至浅绿色，带叶宽 1.86—2.76 mm，分枝少，间生型。茎横切面 14—15 个细胞厚，皮部细胞略分化，稍厚壁，内部细胞薄壁。假根散生于茎腹面或侧叶基部。侧叶覆瓦状排列，长舌形或长方舌形，顶端圆，边缘全缘，背侧边缘内弯，背侧基部稍下延。侧叶细胞薄壁，角质层具大的粗圆疣，三角体小至大，简单三角形，偶尔肿胀形，中部球状加厚缺。油体聚合型，每个细胞具 3—6 个小型的白色油体，除此外，还有大型的暗棕色油体散生在叶细胞中，每个细胞具 2—4 个。腹叶缺。雄苞和雌苞未见。

生境： 土生。海拔 500—620 m。

雅长分布： 二沟，拉雅沟。

中国分布： 安徽，重庆，福建，广东，广西，贵州，海南，湖南，吉林，江西，辽宁，四川，台湾，西藏，云南，浙江。

Plants whitish green to light green when fresh, 1.86–2.76 mm wide, branches few, lateral-intercalary. Stem in transverse section 14–15 cells thick, cortex weakly differentiated, cortical cells slightly thick-walled, medullary cells thin-walled. Rhizoids scattered on ventral side of stem and basal part of leaves. Leaves imbricate, ligulate to ligulate-rectangular, apex rounded, margin entire, dorsal margin incurved, dorsal base slightly decurrent. Leaf cells thin-walled, trigones small to large, simple-triangulate, occasionally nodulate, intermediate thickenings absent. Cuticle coarsely verrucose. Oil bodies segmented, large and dark brown oil bodies scattered in leaf cells, 2–4 per cell, small and white ones in all leaf cells, 3–6 per cell. Underleaves absent. Androecia and gynoecia not seen.

Habitat: On soil. Elev. 500–620 m.

Distribution in Yachang: Ergou, Layagou.

Distribution in China: Anhui, Chongqing, Fujian, Guangdong, Guangxi, Guizhou, Hainan, Hunan, Jilin, Jiangxi, Liaoning, Sichuan, Taiwan, Tibet, Yunnan, Zhejiang.

▶ A. 种群；B. 植物体（背面观）；C. 植物体一段（背面观）；D. 侧叶；E. 茎横切面；F. 叶中部细胞，示角质层疣；G. 叶中部细胞，示油体。（凭证标本：*韦玉梅等 201109-20*）

A. Population; B. Plant (dorsal view); C. Portion of plant (dorsal view); D. Leaves; E. Transverse sections of stems; F. Median cells of leaf showing verrucose cuticle; G. Median cells of leaf showing oil bodies. (All from *Wei et al. 201109-20*)

图 31 偏叶管口苔

Fig. 31 *Solenostoma comatum* (Nees) C. Gao

截叶管口苔（截叶叶苔）

Solenostoma truncatum (Nees) R. M. Schust. ex Váňa et D. G. Long, Nova Hedwigia 89(3/4): 509. 2009.

Jungermannia truncata Nees, Enum. Pl. Crypt. Javae: 29. 1830.

植物体绿色，带叶宽 2.0—2.8 mm，分枝少，间生型。茎横切面 11—12 个细胞厚，皮部细胞几乎不分化，内外细胞均薄壁。假根散生于茎腹面。侧叶覆瓦状排列，卵舌形，顶端圆钝或平截，边缘全缘，背侧基部下延。侧叶细胞薄壁，角质层具条纹状疣，三角体小至大，简单三角形或肿胀形，中部球状加厚缺。油体聚合型，每个细胞 2—5 个。腹叶缺。雌雄异株。雄苞顶生或间生，雄苞叶 4—10 对。雌苞顶生。蒴萼短，略伸出苞叶，纺锤形，上部具不规则纵褶，口部收缩，下半部与苞叶合生，形成一个肉质的蒴囊。孢蒴椭球形，成熟时 4 瓣开裂。孢子球形，表面具细疣。弹丝 2 螺旋加厚。

生境： 石生和土生环境常见，潮湿土生、树基、岩面薄土生环境也有分布。海拔 590—1553 m。

雅长分布： 草王山，吊井天坑，黄猄洞天坑，九十九堡，拉雅沟，瞭望台，逻家田屯，盘古王，山干屯，塘英村，下棚屯，下岩洞屯，一沟。

中国分布： 全国大部分省区均有分布。

Plants green when fresh, 2.0−2.8 mm wide, branches few, lateral-intercalary. Stem in transverse section 11−12 cells thick, cortex hardly differentiated, thin-walled. Rhizoids scattered on ventral side of stem. Leaves imbricate, ligulate-ovate, apex rounded-obtuse or truncate, margin entire, dorsal base decurrent. Leaf cells thin-walled, trigones small to large, simple-triangulate or nodulate, intermediate thickenings absent. Cuticle striate-verrucose. Oil bodies segmented, 2−5 per median cell. Underleaves absent. Diocious. Androecia terminal or intercalary, bracts in 4−10 pairs. Gynoecia terminal. Perianths fusiform, slightly emergent from bracts, pluriplicate near the mouth, mouth constrated, the lower half united with the bracts forming a fleshy perigynium. Capsules ellipsoidal, longitudinally dehiscing by 4 regular valves at maturity. Spores globose, surface verrucose. Elaters with 2 spiral thickenings.

Habitat: Often on rocks and soil, sometimes on wet soil, tree bases and rocks with a thin layer of soil. Elev. 590−1553 m.

Distribution in Yachang: Caowangshan, Diaojing Tiankeng, Huangjingdong Tiankeng, Jiushijiubao, Layagou, Liaowangtai, Luojiatian Tun, Panguwang, Shangan Tun, Tangying Village, Xiapeng Tun, Xiayandong Tun, Yigou.

Distribution in China: Widely distributed in most provinces of China.

▶ A. 种群；B. 植物体（背面观）；C. 弹丝和孢子；D. 植物体一段带雌苞（背面观）；E. 植物体一段带雄苞（背面观）；F. 侧叶；G. 雌苞纵切面；H. 茎横切面；I. 叶中部细胞，示角质层疣；J. 叶中部细胞，示油体。（凭证标本：韦玉梅等 201111-5）

A. Population; B. Plant (dorsal view); C. Elater and spores; D. Portion of plant with a gynoecium; E. Portion of plant with an androecium; F. Leaves; G. Longitudinal section of a gynoecium; H. Transverse section of stem; I. Median cells of leaf showing verrucose cuticle; J. Median cells of leaf showing oil bodies. (All from *Wei et al. 201111-5*)

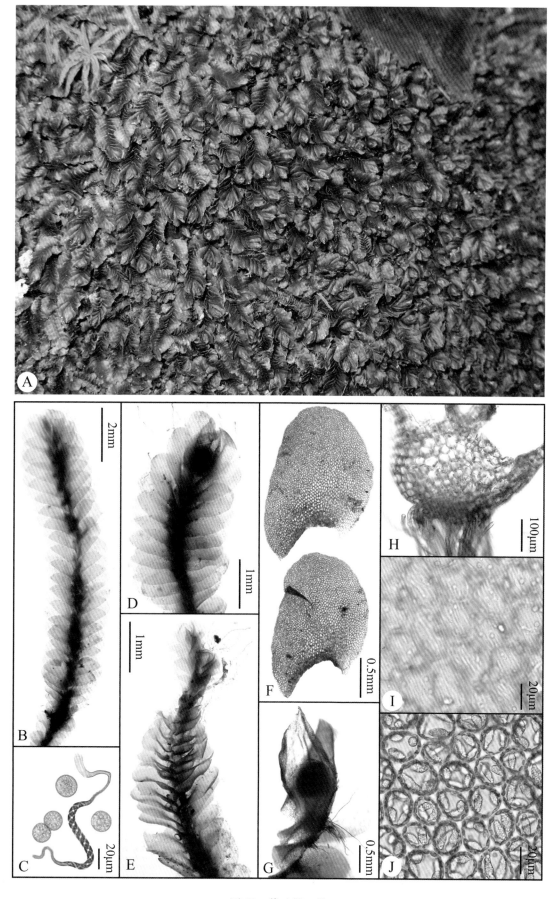

图 32 截叶管口苔

Fig. 32 *Solenostoma truncatum* (Nees) R. M. Schust. ex Váňa & D. G. Long

刺叶护蒴苔

Calypogeia arguta Nees et Mont., Naturgesch. Eur. Leberm. 3: 24. 1838.

植物体白绿色至浅绿色，带叶宽 0.75—0.95 mm，分枝少，间生型。茎横切面 5—7 个细胞厚，皮部细胞几乎不分化，内外细胞均薄壁。假根生于腹叶基部。侧叶远生，椭圆状卵形，顶端具 2 个齿，呈浅 2 裂状，齿 2—4 个细胞长，基部 1—2 个细胞宽，边缘全缘，腹侧基部下延。侧叶细胞薄壁，角质层具细疣，三角体小，不明显，中部球状加厚缺。油体聚合型，无色，每个细胞 2—5 个。腹叶远生，与茎等宽或略小于茎宽，顶端 2 裂几达基部，两侧边缘各具 1 个短齿，基部略下延。雄苞和雌苞未见。芽胞未见。

生境：土生。海拔 518—1862 m。
雅长分布：草王山，一沟。
中国分布：全国大部分省区均有分布。

Plants whitish green to light green when fresh, 0.75–0.95 mm wide, branches few, lateral-intercalary. Stem in transverse section 5–7 cells thick, cortex hardly differentiated, thin-walled. Rhizoids at base of underleaves. Leaves distant, oblong-ovate, apex with 2 teeth, forming shallowly bifid, teeth 2–4 cells long, 1–2 cells wide at base, margin entire, ventral base decurrent. Leaf cells thin-walled, trigones small, indistinct, intermediate thickenings absent. Cuticle verrucose. Oil bodies segmented, colorless, 2–5 per median cell. Underleaves distant, as wide as the stem or slightly smaller than stem width, mostly bilobed to the base, lateral margins usually with 1 short tooth, base slightly decurrent. Androecia and gynoecia not seen. Gemmae not seen.

Habitat: On soil. Elev. 518–1862 m.
Distribution in Yachang: Caowangshan, Yigou.
Distribution in China: Widely distributed in most provinces of China.

▶ A. 种群；B. 植物体一段；C. 侧叶；D. 茎横切面；E—F. 腹叶；G. 叶中部细胞，示角质层疣；H. 叶中部细胞，示油体。（凭证标本：*韦玉梅等 201108-33*）

A. Population; B. Portion of plant; C. Leaves; D. Transverse section of stem; E–F. Underleaves; G. Median cells of leaf showing verrucose cuticle; H. Median cells of leaf showing oil bodies. (All from *Wei et al. 201108-33*)

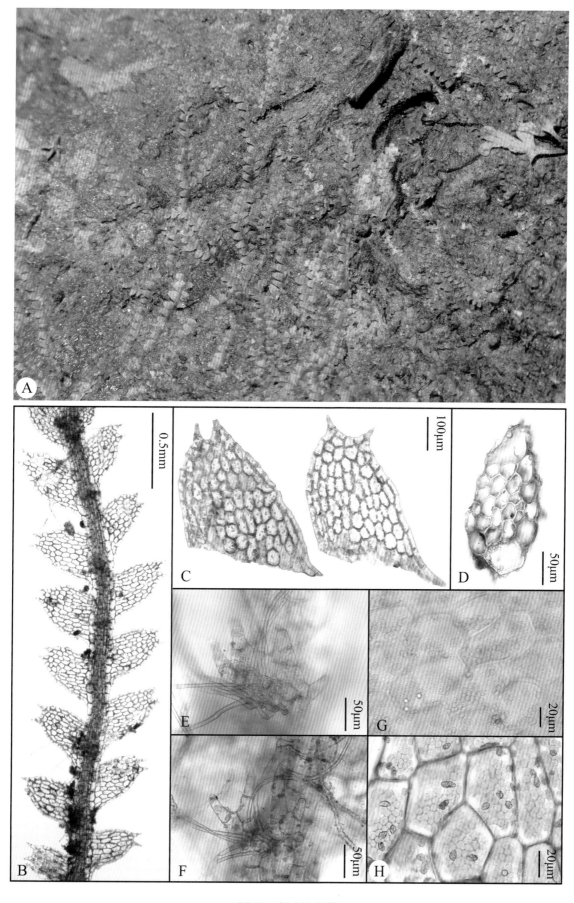

图 33 刺叶护蒴苔

Fig. 33 *Calypogeia arguta* Nees & Mont. ex Nees

全缘护蒴苔

Calypogeia japonica Steph., Sp. Hepat. (Stephani) 6: 448. 1924.

植物体浅绿色至绿色，带叶宽 1.20—1.58 mm，分枝少，间生型。茎横切面 9—10 个细胞厚，皮部细胞几乎不分化，内外细胞均薄壁。假根生于腹叶基部。侧叶疏松覆瓦状排列至毗邻，三角状卵形或卵圆形，顶端圆钝至钝尖，边缘全缘，腹侧基部下延。侧叶细胞薄壁，角质层平滑，三角体小，不明显，中部球状加厚缺。油体眼球型，无色，每个细胞 2—4 个。腹叶远生，为茎宽的 1.0—1.5 倍，顶端 2 裂约达 1/2，两侧边缘全缘，基部下延。雄苞和雌苞未见。芽胞 1—2 个细胞组成，常生于直立鞭状枝或茎枝的顶端。

生境： 土生。海拔 1339—1823 m。

雅长分布： 草王山，九十九堡，下棚屯。

中国分布： 安徽，福建，广东，贵州。首次记录于广西。

Plants light green to green when fresh, 1.20−1.58 mm wide, branches few, lateral-intercalary. Stem in transverse section 9−10 cells thick, cortex hardly differentiated, thin-walled. Rhizoids at base of underleaves. Leaves loosely imbricate to contiguous, triangular-ovate or rounded-ovate, apex rounded-obtuse to obtuse-acute, margin entire, ventral base decurrent. Leaf cells thin-walled, trigones small, indistinct, intermediate thickenings absent. Cuticle smooth. Oil bodies eyeball-like, colorless, 2−4 per median cell. Underleaves distant, 1.0−1.5 times as wide as the stem, bilobed to ca. 1/2 underleaf length, lateral margins usually with 1 short tooth, base decurrent. Androecia and gynoecia not seen. Gemmae 1−2-celled, produced at the tip of upright flagelliform branch or stem.

Habitat: On soil. Elev. 1339−1823 m.

Distribution in Yachang: Caowangshan, Jiushijiubao, Xiapeng Tun.

Distribution in China: Anhui, Fujian, Guangdong, Guizhou. New to Guangxi.

▶ A. 种群；B. 植物体一段；C. 芽胞；D. 植株体先端带顶端芽胞簇；E. 侧叶；F. 叶中部细胞，示油体；G. 茎横切面；H. 茎一段，示腹叶。（凭证标本：*韦玉梅等 201108-5*）

A. Population; B. Portion of plant; C. Gemmae; D. Apical part of plant with gemmae cluster at apex; E. Leaves; F. Median cells of leaf showing oil bodies; G. Transverse section of stem; H. Portion of stem showing underleaves. (All from *Wei et al. 201108-5*)

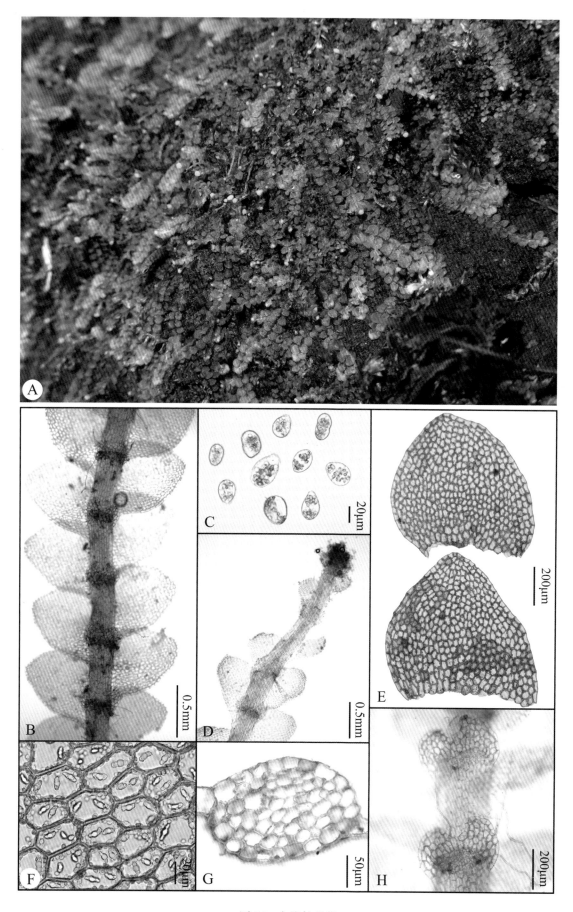

图 34 全缘护蒴苔

Fig. 34 *Calypogeia japonica* Steph.

双齿护蒴苔

Calypogeia tosana (Steph.) Steph., Sp. Hepat. (Stephani) 3: 410. 1908.

植物体浅绿色至棕绿色，带叶宽 1.88—2.74 mm，分枝少，间生型。茎横切面 8—10 个细胞厚，皮部细胞几乎不分化，内外细胞均薄壁。假根生于腹叶基部。侧叶疏松覆瓦状排列，三角状卵形，顶端具 2 个齿，呈浅 2 裂状，齿 1—2 个细胞长，基部 1—2 个细胞宽，边缘全缘，腹侧基部下延。侧叶细胞薄壁，角质层平滑或具细疣，三角体小，不明显，中部球状加厚缺。油体聚合型，无色，每个细胞 4—8 个。腹叶远生，为茎宽的 1—2 倍，顶端 2 裂 1/2—2/3，两侧边缘各具 1 个钝齿，基部下延。雌雄异株。雄苞生于侧短枝，雄苞叶 3—4 对。雌苞未见。芽胞 1—2 个细胞组成，常生于直立鞭状枝的顶端。

生境：石生，土生。海拔 1136—1747 m。

雅长分布：草王山，黄猄洞天坑，旁墙屯。

中国分布：全国大部分省区均有分布。

Plants light green to browish green when fresh, 1.88–2.74 mm wide, branches few, lateral-intercalary. Stem in transverse section 8–10 cells thick, cortex hardly differentiated, thin-walled. Rhizoids at base of underleaves. Leaves loosely imbricate, triangular-ovate, apex with 2 teeth, forming shallowly bifid, teeth 1–2 cells long, 1–2 cells wide at base, margin entire, ventral base decurrent. Leaf cells thin-walled, trigones small, indistinct, intermediate thickenings absent. Cuticle smooth or verrucose. Oil bodies segmented, colorless, 4–8 per median cell. Underleaves distant, 1–2 times as wide as the stem, bilobed to 1/2–2/3 underleaf length, lateral margins usually with 1 obtuse tooth, base decurrent. Androecia on short lateral branches, bracts in 3–4 pairs. Gynoecia not seen. Gemmae 1–2-celled, produced at the tip of upright flagelliform branch.

Habitat: On rocks and soil. Elev. 1136–1747 m.

Distribution in Yachang: Caowangshan, Huangjingdong Tiankeng, Pangqiang Tun.

Distribution in China: Widely distributed in most provinces of China.

▶ A. 种群；B. 植物体一段；C. 植株体先端；D. 侧叶；E. 茎一段，示腹叶；F. 茎横切面；G. 叶中部细胞，示油体。（凭证标本：*韦玉梅等 191012-69*）

A. Population; B. Portion of plant; C. Apical part of plant; D. Leaves; E. Portion of stem showing underleaves; F. Transverse section of stem; G. Median cells of leaf showing oil bodies. (All from *Wei et al. 191012-69*)

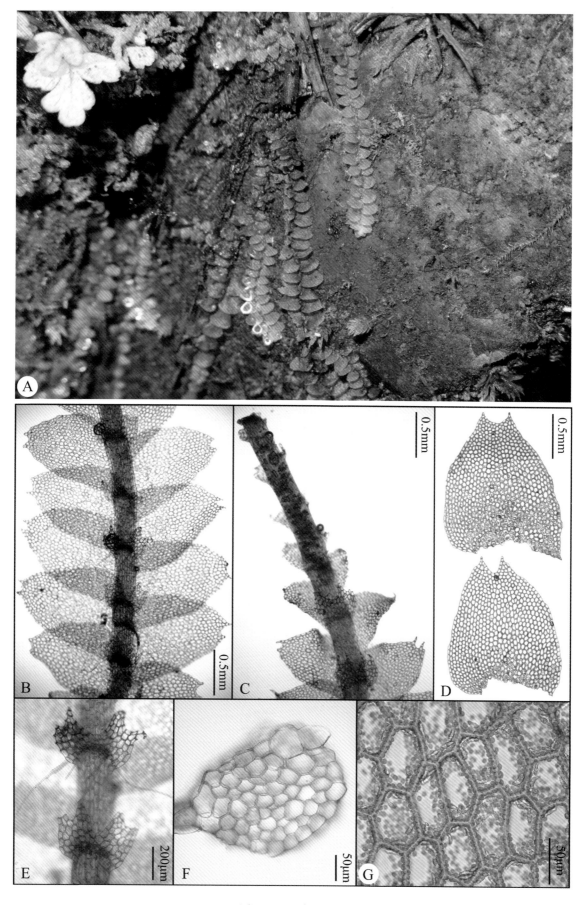

图 35　双齿护蒴苔
Fig. 35　*Calypogeia tosana* (Steph.) Steph.

东亚对耳苔（雌株）

Syzygiella nipponica (S. Hatt.) K. Feldberg, Váňa, Hentschel et Heinrichs, Cryptog. Bryol. 31(2): 145. 2010.

Jamesoniella nipponica S. Hatt., J. Jap. Bot. 19: 350. 1943.

植物体浅绿色至绿色，带叶宽 1.65—2.34 mm，分枝少，顶生型或间生型。茎横切面 10—11 个细胞厚，皮部细胞几乎不分化，内外细胞略厚壁。假根生于茎腹面近叶基处。侧叶覆瓦状排列，卵圆形或椭圆状卵形，顶端圆或圆钝，边缘全缘，背侧基部稍下延。侧叶细胞薄壁，角质层具颗粒状的或条纹状的疣，三角体大，简单三角形或肿胀形，中部球状加厚缺。油体聚合型，每个细胞 5—10 个。腹叶常缺失，或仅见于雌苞附近以及分枝基部。雌雄异株。雄苞未见。雌苞顶生或间生，雌苞叶 3 对，最内层雌苞叶具长裂片，最内层雌苞腹叶大，边缘具长裂片。蒴萼圆筒形，伸出苞叶，上部具多条不规则纵褶，口部略收缩，边缘具长纤毛状齿。

生境： 石生。海拔 1834 m。

雅长分布： 草王山。

中国分布： 安徽，福建，甘肃，贵州，河南，湖北，湖南，江西，陕西，四川，台湾，云南，浙江。

备注： 广西新记录属。

Plants light green to green when fresh, 1.65−2.34 mm wide, branches few, terminal or ventral-intercalary. Stem in transverse section 10−11 cells thick, cortex hardly differentiated, slightly thick-walled. Rhizoids near the leaf bases in ventral side of stem. Leaves imbricate, rounded-ovate or oblong-ovate, apex rounded to rounded-obtuse, margin entire, dorsal base slightly decurrent. Leaf cells thin-walled, trigones large, simple-triangulate or nodulate, intermediate thickenings absent. Cuticle granulate- or striate-verrucose. Oil bodies segmented, 5−10 per median cell. Underleaves usually absent, if present, confined to below gynoecia and near the bases of terminal branches. Dioicous. Androecia not seen. Gynoecia terminal or intercalary, bracts in 3 pairs, innermost bracts irregularly lobed, innermost bracteole large, irregularly lobed. Perianths cylindrical, emergent from bracts, irregularly pluriplicate near the mouth, mouth slightly constrated, marginal teeth long ciliate.

Habitat: On rocks. Elev. 1834 m.

Distribution in Yachang: Caowangshan.

Distribution in China: Anhui, Fujian, Gansu, Guizhou, Henan, Hubei, Hunan, Jiangxi, Shaanxi, Sichuan, Taiwan, Yunnan, Zhejiang.

Note: New genus record to Guangxi.

▶ A. 种群；B. 植物体带蒴萼（背面观）；C. 植物体一段（背面观）；D. 侧叶；E. 蒴萼口部的齿；F. 蒴萼；G. 最内层雌苞腹叶；H. 最内层雌苞叶；I. 茎横切面；J. 叶中部细胞，示角质层疣；K. 叶中部细胞，示油体。（凭证标本：*韦玉梅等 191013-131*）

A. Population; B. Plant with a perianth; C. Portion of plant (dorsal view); D. Leaves; E. Teeth of perianth mouth; F. Perianth; G. Innermost female bracteole; H. Innermost female bracts; I. Transverse section of stem; J. Median cells of leaf showing verrucose cuticle; K. Median cells of leaf showing oil bodies. (All from *Wei et al. 191013-131*)

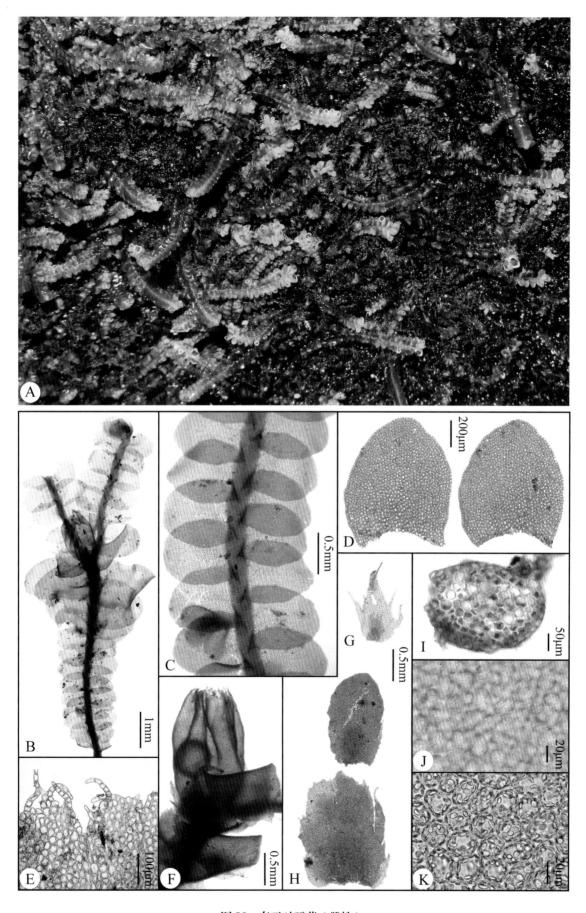

图 36 东亚对耳苔（雌株）

Fig. 36 *Syzygiella nipponica* (S. Hatt.) K. Feldberg

东亚对耳苔（雄株）

Syzygiella nipponica (S. Hatt.) K. Feldberg, Váňa, Hentschel et Heinrichs, Cryptog. Bryol. 31(2): 145. 2010.

Jamesoniella nipponica S. Hatt., J. Jap. Bot. 19: 350. 1943.

植物体绿色至棕红色，带叶宽 1.42—2.40 mm，分枝少，顶生型或间生型。茎横切面 7—10 个细胞厚，皮部细胞几乎不分化，内外细胞略厚壁。假根生于茎腹面近叶基处。侧叶通常远生，椭圆状卵形或近方形，顶端圆、平截或内凹，边缘全缘，背侧基部稍下延。侧叶细胞薄壁，角质层具疣或平滑，三角体小至大，简单三角形或肿胀形，中部球状加厚缺。油体聚合型，每个细胞 7—14 个。腹叶常缺失，或仅见于雌苞附近以及分枝基部。雌雄异株。雄苞间生，雄苞叶 3—5 对。雌苞和蒴萼未见。

生境： 树基，腐基质生。海拔 1878 m。

雅长分布： 草王山。

中国分布： 安徽，福建，甘肃，贵州，河南，湖北，湖南，江西，陕西，四川，台湾，云南，浙江。首次记录于广西。

Plants green to redish-brown when fresh, 1.42–2.40 mm wide, branches few, terminal or ventral-intercalary. Stem in transverse section 7–10 cells thick, cortex hardly differentiated, slightly thick-walled. Rhizoids near the leaf bases in ventral side of stem. Leaves usually remote, oblong-ovate to subquadrate, apex rounded, truncate to retuse, margin entire, dorsal base slightly decurrent. Leaf cells thin-walled, trigones small to large, simple-triangulate or nodulate, intermediate thickenings absent. Cuticle verrucose to smooth. Oil bodies segmented, 7–14 per median cell. Underleaves usually absent, if present, confined to below gynoecia and near the bases of terminal branches. Dioicous. Androecia intercalary, bracts in 3–5 pairs. Gynoecia and perianths not seen.

Habitat: On tree base and humus. Elev. 1878 m.

Distribution in Yachang: Caowangshan.

Distribution in China: Anhui, Fujian, Gansu, Guizhou, Henan, Hubei, Hunan, Jiangxi, Shaanxi, Sichuan, Taiwan, Yunnan, Zhejiang. New to Guangxi.

▶ A. 种群；B. 植物体带雄苞（背面观）；C. 植物体（背面观）；D. 侧叶；E. 植物体一段（背面观）；F. 植物体一段，示顶生型分枝；G. 植物体一段，示间生型分枝；H. 茎横切面；I. 侧叶中部细胞，示油体；J. 植物体一段（背面观），示雄苞。（凭证标本：韦玉梅等 221112-55）

A. Population; B. Plants with androecia (dorsal view); C. Plants (dorsal view); D. Leaves; E: Portion of plant (dorsal view); F: Portion of plant showing termonal branch; G: Portion of plant showing intercalary branch; H: Transverse section of stem; I: Median cells of leaf showing oil bodies; J: Portion of plant (dorsal view) showing a androecium. (All from *Wei et al. 221112-55*)

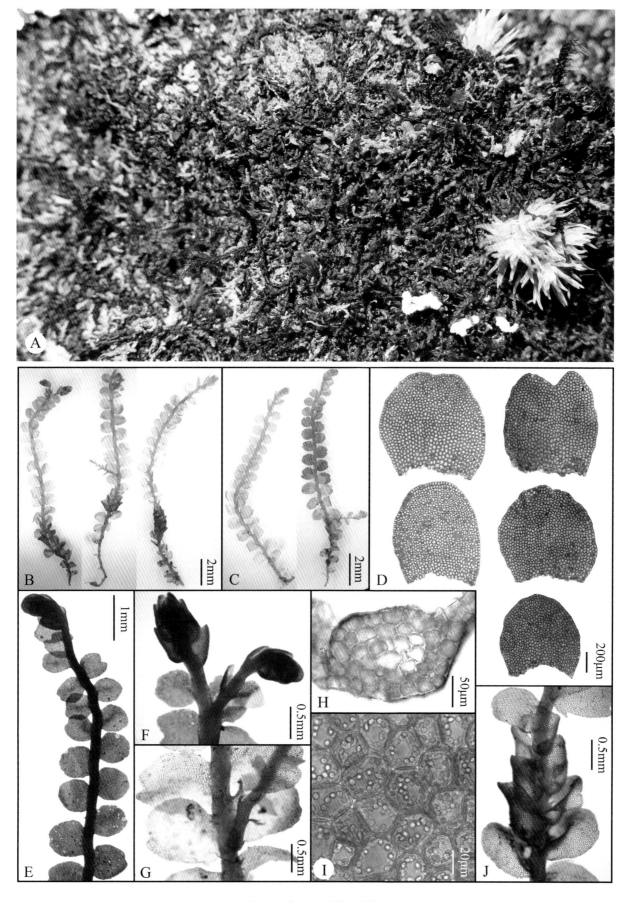

图 37　东亚对耳苔（雄株）
Fig. 37　*Syzygiella nipponica* (S. Hatt.) K. Feldberg

弯叶大萼苔（薄壁大萼苔）

Cephalozia hamatiloba Steph., Bull. Herb. Boissier (sér. 2) 8(6): 427 (303). 1908.

Cephalozia otaruensis Steph., Sp. Hepat. 6: 434. 1906.

植物体浅绿色，带叶宽 0.55—0.74 mm，不规则分枝，常具直立鞭状枝，分枝间生型。茎横切面约 8 个表皮细胞和 7 个内部细胞，表皮细胞大，薄壁，内部细胞小，稍厚壁。假根生于茎腹面。侧叶远生，卵形，顶端 2 裂至叶长的 1/2—2/3，裂瓣狭三角形，3—5 个细胞长，基部 2—4 个细胞宽，边缘全缘，背侧基部稍下延。侧叶细胞薄壁，角质层平滑，三角体无，中部球状加厚缺。油体缺。腹叶通常缺失，仅存在于雌苞上。雌雄异株。雄苞未见。雌苞侧生或顶生，最内层雌苞叶 2 裂，边缘全缘或具齿，最内层雌苞腹叶与苞叶相似，2 裂，有时一侧边缘具有 1 个较大的裂片。蒴萼长圆筒形，具 3 条长纵褶，口部略收缩，边缘具 1—2 个长细胞组成的齿。芽胞单细胞，生于直立鞭状枝或茎枝顶端。

生境： 土生环境常见，石生、腐木环境也有分布。海拔 1078—1772 m。

雅长分布： 草王山，黄狼洞天坑，盘古王，旁墙屯。

中国分布： 澳门，重庆，福建，广东，广西，贵州，湖北，湖南，江西，四川，香港，云南，浙江。

Plants light green when fresh, 0.55–0.74 mm wide, irregularly branched, flagelliform axes usually present, branches ventral-intercalary. Stem in transverse section with ca. 8 thin-walled epidermal cells and ca. 7 thick-walled medullary cell, epidermal cells larger than medullary cell. Rhizoids scattered on ventral side of stem. Leaves distant, ovate, apex bilobed to 1/2–2/3 leaf length, lobes narrow triangular, 3–5 cells long, 2–4 cells wide at base, margin entire, dorsal base slightly decurrent. Leaf cells thin-walled, without trigones, intermediate thickenings absent. Cuticle smooth. Oil bodies absent. Underleaves absent, if present, confined to below gynoecia. Dioicous. Androecia not seen. Gynoecia terminal or lateral, innermost bracts bilobed, margins entire or toothed, innermost bracteole similar to the bracts, bilobed. Perianths long-cylindrical, 3 plicate, mouth weakly constricted, marginal teeth 1–2 cells long. Gemmae 1-celled, produced at the tip of upright flagelliform branches or stem.

Habitat: Often on soil, sometimes on rocks and rotten logs. Elev. 1078–1772 m.

Distribution in Yachang: Caowangshan, Huangjingdong Tiankeng, Panguwang, Pangqiang Tun.

Distribution in China: Macao, Chongqing, Fujian, Guangdong, Guangxi, Guizhou, Hubei, Hunan, Jiangxi, Sichuan, Hongkong, Yunnan, Zhejiang.

▶ A. 种群；B. 植物体（背面观）；C. 植物体一段（背面观）；D. 植物体先端带芽胞；E. 植物体一段，示侧叶（背面观）；F. 最内层雌苞叶和雌苞腹叶；G. 植物体一段带蒴萼；H. 茎横切面；I. 蒴萼口部的齿。（凭证标本：*韦玉梅等 201110-19*）

A. Population; B. Plant (dorsal view); C. Portion of plant (dorsal view); D. Apical part of plant with gemmae; E. Portion of plant showing leaves (dorsal view); F. Innermost female bracts and bracteole; G. Portion of plant with perianth; H. Transverse section of stem; I. Teeth of perianth mouth. (All from *Wei et al. 201110-19*)

图 38 弯叶大萼苔

Fig. 38 *Cephalozia hamatiloba* Steph.

拳叶苔

Nowellia curvifolia (Dicks.) Mitt., Nat. Hist. Azores: 321. 1870.

植物体浅绿色至黄绿色，带叶宽 0.35—0.64 mm，不规则分枝，间生型或顶生型。茎横切面约 8 个表皮细胞和 5 个内部细胞，薄壁。假根散生于茎腹面。侧叶覆瓦状排列，斜卵圆形，顶端 2 裂，裂瓣狭三角形具毛状尖，边缘全缘，背缘拱起，腹缘内折并强烈鼓起形成囊状。侧叶细胞厚壁，角质层平滑，三角体无，中部球状加厚缺。油体缺或呈油滴状。腹叶仅存在于雌苞上。雌雄异株。雄苞未见。雌苞侧生或顶生，雌苞叶通常 3 对，基部无囊状内卷，最内层雌苞叶 2 裂，边缘具不规则齿。最内层雌苞腹叶大，2 裂或 3 裂。蒴萼长圆筒形，具 3—4 条长皱褶，口部近平截，边缘具 1—3 个细胞长的齿。

生境： 石生，腐木。海拔 1152—1450 m。

雅长分布： 盘古王，旁墙屯，下棚屯。

中国分布： 全国大部分省区均有分布。

Plants light green to yellowish green when fresh, 0.35−0.64 mm wide, irregularly branched, branches ventral-intercalary or lateral-terminal. Stem in transverse section with ca. 8 epidermal cells and ca. 5 medullary cell, cells thin-walled. Rhizoids scattered on ventral side of stem. Leaves imbricate, obliquely ovate, apex bilobes, lobes narrowly triangular, 4−6 cells uniseriate towards apex, margin entire, dorsal margin arched, ventral margin broadly folded inward to form a ventral sac. Leaf cells thick-walled, without trigones, intermediate thickenings absent. Cuticle smooth. Oil bodies absent or droplets. Underleaves absent, if present, confined to below gynoecia. Dioicous. Androecia not seen. Gynoecia terminal or lateral, bracts in 3 pairs, without ventral sac, innermost bracts bilobed, margins irregularly toothed, innermost bracteole large, bi- or tri-lobed. Perianths long-cylindrical, 3−4 plicate, mouth subtruncate, marginal teeth 1−3 cells long.

Habitat: On rocks and rotten logs. Elev. 1152−1450 m.

Distribution in Yachang: Panguwang, Pangqiang Tun, Xiapeng Tun.

Distribution in China: Widely distributed in most provinces of China.

▶ A. 种群；B. 植物体一段（背面观）；C. 植物体一段带蒴萼；D. 蒴萼口部；E. 侧叶；F. 茎横切面；G. 蒴萼横切面；H. 最内层雌苞叶和雌苞腹叶。（凭证标本：*韦玉梅等 201110-18*）

A. Population; B. Portion of plant (dorsal view); C. Portion of plant with a perianth; D. Perianth mouth; E. Leaves; F. Transverse section of stem; G. Transverse section of perianth; H. Innermost female bracts and bracteole. (All from *Wei et al. 201110-18*)

图 39 拳叶苔

Fig. 39 *Nowellia curvifolia* (Dicks) Mitt.

合叶裂齿苔

Odontoschisma denudatum (Mart.) Dumort., Recueil Observ. Jungerm.: 19. 1835.

植物体浅绿色，带叶宽 0.85—1.25 mm，常具匍匐枝和鞭状枝，分枝间生型。茎横切面约 7 个细胞厚，表皮细胞厚壁，略小于内部细胞，内部细胞薄壁。假根散生于茎腹面。侧叶覆瓦状排列至毗邻，近圆形，顶端圆或圆截形，边缘全缘，背弯，背侧基部稍下延。侧叶细胞厚壁，角质层平滑或具疣，三角体大，肿胀形，中部球状加厚缺。油体聚合型，每个细胞 2—5 个。腹叶通常缺失。雄苞和雌苞未见。芽胞 1—2 个细胞组成，生于直立鞭状枝的顶端。

生境： 石生，土生，腐木。海拔 1125—1756 m。

雅长分布： 盘古王，旁墙屯，下岩洞屯。

中国分布： 安徽，重庆，福建，广东，广西，贵州，海南，湖北，湖南，江西，台湾，云南，浙江。

Plants light green when fresh, 0.85–1.25 mm wide, irregularly branched, stolons and flagelliform axes usually present, branches ventral-intercalary. Stem in transverse section ca. 7 cells thick, epidermal cells thick-walled, slightly smaller than thin-walled medullary cell. Rhizoids scattered on ventral side of stem. Leaves imbricate to contiguous, suborbicular, apex rounded or rounded-truncate, margin entire, curved toward dorsal side, dorsal base slightly decurrent. Leaf cells thick-walled, trigones large, nodulate, intermediate thickenings absent. Cuticle verrucose to almost smooth. Oil bodies segmented, 2–5 per median cell. Underleaves absent. Androecia and gynoecia not seen. Gemmae 1–2-celled, produced at the tip of upright flagelliform branch.

Habitat: On rocks, soil and rotten logs. Elev. 1125–1756 m.

Distribution in Yachang: Panguwang, Pangqiang Tun, Xiayandong Tun.

Distribution in China: Anhui, Chongqing, Fujian, Guangdong, Guangxi, Guizhou, Hainan, Hubei, Hunan, Jiangxi, Taiwan, Yunnan, Zhejiang.

▶ A. 种群；B. 植物体一段（背面观）；C. 侧叶；D. 茎横切面；E. 植物体一段示直立的鞭状枝；F. 芽胞；G. 植物体一段；H. 叶中部细胞，示油体。（凭证标本：*韦玉梅等 191019-371B*）

A. Population; B. Portion of plant (dorsal view); C. Leaves; D. Transverse section of stem; E. Portion of plant showing upright flagelliform branch (ventral view); F. Gemmae; G. Portion of plant; H. Median cells of leaf showing oil bodies. (All from *Wei et al. 191019-371B*)

图 40 合叶裂齿苔

Fig. 40 *Odontoschisma denudatum* (Mart.) Dumort.

粗齿拟大萼苔

Cephaloziella dentata (Raddi) Steph., Bull. Herb. Boissier 5(2): 78. 1897.

植物体干时暗棕色，带叶宽 0.25—0.34 mm，分枝少，间生型。茎横切面约 7 个细胞厚，表皮细胞稍厚壁，与内部细胞近于等大。侧叶横生在茎上，远离，顶端 2 裂至叶长的 1/2—3/5，裂瓣三角形，边缘具单细胞齿突或粗齿，基部不下延。侧叶细胞厚壁，角质层平滑，三角体无，中部球状加厚缺。油体未见。腹叶缺。雌雄异株。雄苞生于侧短枝，雄苞叶 2—3 对。精子器单生于雄苞叶中。雌苞未见。

生境： 土生。海拔 1156—1765 m。

雅长分布： 草王山，瞭望台，下岩洞屯。

中国分布： 广东，贵州，海南，湖南，江西。首次记录于广西。

Plants dark brown in herbarium specimens, 0.25−0.34 mm wide, branches few, ventral-intercalary. Stem in transverse section ca. 7 cells thick, epidermal cells slighly thick-walled, as large as medullary cells. Rhizoids scattered on ventral side of stem. Leaves subtransverse, distant, apex bilobed to 1/2−3/5 leaf length, lobes triangular, margin with 1-celled teeth, base not decurrent. Leaf cells thick-walled, without trigones, intermediate thickenings absent. Cuticle smooth. Oil bodies not seen. Underleaves absent. Dioicous. Androecia on short lateral branches, bracts in 2−3 pairs; 1 antheridium per bract. Gynoecia not seen.

Habitat: On soil. Elev. 1156−1765 m.

Distribution in Yachang: Caowangshan, Liaowangtai, Xiayandong Tun.

Distribution in China: Guangdong, Guizhou, Hainan, Hunan, Jiangxi. New to Guangxi.

▶ A. 种群；B. 植物体一段（背面观）；C. 植物体一段（腹面观）；D−E. 植物体一段（背面观）；F. 植物体一段带雄苞（背面观）；G. 侧叶；H. 茎横切面。（凭证标本：*唐启明 & 张仕艳 20181002-300B*）

A. Population; B. Portion of plant (dorsal view); C. Portion of plant (ventral view); D−E. Portions of plants (dorsal view); F. Portion of plant with an androecium (dorsal view); G. Leaves; H. Transverse section of stem. (All from *Tang & Zhang 20181002-300B*)

图 41 粗齿拟大萼苔

Fig. 41 *Cephaloziella dentata* (Raddi) Steph.

小叶拟大萼苔

Cephaloziella microphylla (Steph.) Douin, Mém. Soc. Bot. France 29: 59. 1920.

植物体绿色，带叶宽 0.10—0.15 mm，不规则分枝，间生型。茎横切面约 5 个细胞厚，表皮细胞与内部细胞近于等大，均薄壁。侧叶横生在茎上，远离，顶端 2 裂几达叶基，裂瓣三角形，不对称，背侧裂瓣明显小，背表面常具 1—2 个细胞长的齿突，边缘具不规则粗齿，基部不下延。侧叶细胞薄壁，角质层具细疣，三角体无，中部球状加厚缺。油体不明显的聚合型，每个细胞 2—5 个。腹叶缺。雌雄同株。雄苞顶生，雄苞叶 3—6 对。精子器单生于雄苞叶中。雌苞顶生，最内层雌苞叶和雌苞腹叶中下部合生，上部 2 裂，边缘具单细胞粗齿。蒴萼圆柱形，具 4—5 条纵褶，口部平截，边缘具单细胞齿突。

生境：土生环境常见，潮湿土生、钙华基质、树基环境也有分布。海拔 1048—1424 m。

雅长分布：黄猄洞天坑，里郎天坑，瞭望台，下岩洞屯，中井天坑。

中国分布：安徽，澳门，福建，广东，广西，贵州，海南，湖北，湖南，内蒙古，四川，台湾，香港，云南，浙江。

Plants green when fresh, 0.10−0.15 mm wide, branches few, ventral-intercalary. Stem in transverse section ca. 5 cells thick, epidermal cells as large as medullary cells, both cells thin-walled. Rhizoids scattered on ventral side of stem. Leaves subtransverse, distant, apex mostly bilobed to leaf base, lobes triangular, asymmetrial, ventral lobe larger than dorsal ones, dorsal surface usually with 1−2-celled teeth, margin irregularly toothed, base not decurrent. Leaf cells thin-walled, without trigones, intermediate thickenings absent. Cuticle verrucose. Oil bodies indistinctly segmented, 2−5 per cell. Underleaves absent. Monoicous. Androecia terminal, bracts in 3−6 pairs; 1 antheridium per bract. Gynoecia terminal, innermost bracts and bracteole connate for about half their lenght, bilobed, margins with 1-celled teeth. Perianths cylindrical, 4−5 plicate, mouth truncate, crenulate with projecting cells at margin.

Habitat: Often on soil, sometimes on wet soil, calcareous substrates and tree bases. Elev. 1048−1424 m.

Distribution in Yachang: Huangjingdong Tiankeng, Lilang Tiankeng, Liaowangtai, Xiayandong Tun, Zhongjing Tiankeng.

Distribution in China: Anhui, Macao, Fujian, Guangdong, Guangxi, Guizhou, Hainan, Hubei, Hunan, Inner Mongolia, Sichuan, Taiwan, Hongkong, Yunnan, Zhejiang.

▶ A. 种群；B. 植物体一段（背面观）；C. 植物体一段带雌苞和雄苞（腹面观）；D. 植物体一段（背面观）；E. 最内层雌苞叶和雌苞腹叶；F. 蒴萼横切面；G. 蒴萼口部；H. 茎横切面；I. 侧叶（背面观）；J. 侧叶；K. 叶细胞，示油体。（凭证标本：A、H−K 拍摄自韦玉梅等 201104-15，其余拍摄自唐启明等 20210717-5）

A. Population; B. Portion of plant (dorsal view); C. Portion of plant with gynoecia and an androecium (ventral view); D. Portion of plant (dorsal view); E. Innermost female bracts and bracteole; F. Transverse section of perianth; G. Perianth mouth; H. Transverse section of stem; I. Leaves (dorsal view); J. Leaf; K. Leaf cells showing oil bodies. (A, H−K from *Wei et al. 201104-15*, the others from *Tang et al. 20210717-5*)

图 42　小叶拟大萼苔

Fig. 42　*Cephaloziella microphylla* (Steph.) Douin

鳞叶筒萼苔（鳞叶拟大萼苔）

Cylindrocolea kiaeri (Austin) Váňa, Phytotaxa 112(1): 2. 2013.
Cephaloziella kiaeri (Austin) Douin, Mem. Bot. Soc. France 29: 68. 1920.

植物体绿色，带叶宽 0.15—0.28 mm，不规则分枝，分枝顶端常呈鞭状，分枝间生型。茎横切面 6—7 个细胞厚，表皮细胞稍厚壁，与内部细胞近于等大。假根散生于茎腹面。侧叶横生在茎上，远离，顶端 2 裂至叶长的 1/2—2/3，裂瓣三角形，5—7 个细胞长，基部 4—6 个细胞宽，边缘全缘，基部不下延。侧叶细胞厚壁，角质层具疣，三角体无，中部球状加厚缺。油体未见。腹叶缺。雄苞和雌苞未见。芽胞 1—2 个细胞组成，梭形或椭圆形，生于茎枝顶端。

生境：土生。海拔 1254—1316 m。

雅长分布：下岩洞屯，中井天坑，盘古王。

中国分布：安徽，福建，甘肃，广西，贵州，海南，湖南，江苏，江西，辽宁，山东，云南，浙江。

Plants green when fresh, 0.15–0.28 mm wide, irregularly branched, flagelliform axes usually present, branches ventral-intercalary. Stem in transverse section 6–7 cells thick, epidermal cells thick-walled, as large as medullary cells. Rhizoids scattered on ventral side of stem. Leaves subtransverse, distant, apex bilobed to 1/2–2/3 leaf length, lobes triangular, 5–7 cells long, 4–6 cells wide at base, margin entire, dorsal base not decurrent. Leaf cells thick-walled, without trigones, intermediate thickenings absent. Cuticle verrucose. Oil bodies not seen. Underleaves absent. Androecia and gynoecia not seen. Gemmae 1–2-celled, fusiform or ellipsoid, produced at the tip of stems.

Habitat: On soil. Elev. 1254–1316 m.

Distribution in Yachang: Xiayandong Tun, Zhongjing Tiankeng, Panguwang.

Distribution in China: Anhui, Fujian, Gansu, Guangxi, Guizhou, Hainan, Hunan, Jiangsu, Jiangxi, Liaoning, Shandong, Yunnan, Zhejiang.

▶ A. 种群，B–D. 植物体一段（背面观）；E. 茎横切面；F. 芽胞；G. 侧叶（背面观）。（凭证标本：*唐启明 & 张仕艳 20181001-190*）

A. Population; B–D. Portions of plants (dorsal view); E. Transverse section of stem; F. Gemmae; G. Leaves (dorsal view). (All from *Tang & Zhang 20181001-190*)

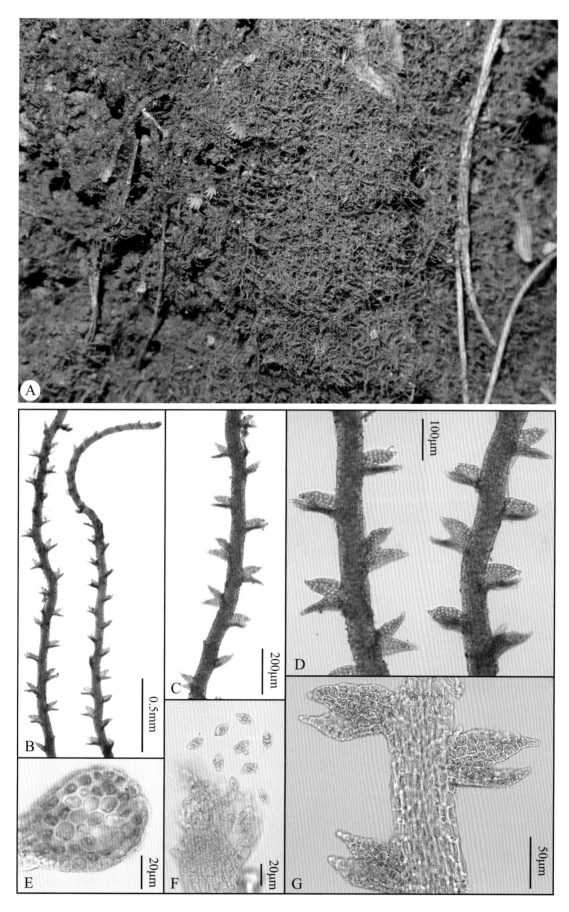

图 43 鳞叶筒萼苔

Fig. 43 *Cylindrocolea kiaeri* (Austin) Váňa

甲克苔

Jackiella javanica Schiffn., Hep. Fl. Buitenzorg: 212. 1900.

植物体浅绿色至棕绿色，带叶宽 1.15—1.52 mm，分枝少，在茎腹面分出，间生型。茎横切面 7—8 个细胞厚，表皮细胞和内部细胞几乎等大，均厚壁。侧叶覆瓦状排列，阔卵形至三角状卵形，顶端圆，边缘全缘，腹侧基部心形，背侧基部楔形。侧叶细胞稍厚壁，角质层平滑，三角体大，肿胀形，中部球状加厚缺。油体聚合型，每个细胞 2—3 个。腹叶退化，丝状，2—3 个细胞，顶端具有一个透明细胞。雄苞和雌苞未见。芽胞 1—2 个细胞，生于叶顶端。

生境： 土生。海拔 1325—1823 m。

雅长分布： 草王山，九十九堡，盘古王，旁墙屯，下棚屯。

中国分布： 福建，广东，贵州，海南，湖南，台湾，香港，云南。首次记录于广西。

备注： 广西新记录科和属。

Plants light green to brownish green when fresh, 1.15−1.52 mm wide, rarely branched, branches ventral-intercalary. Stem in transverse section 7−8 cells thick, epidermal cells equal to medullary cells in size, both cells thick-walled. Leaves imbricate, broadly ovate to triangular-ovate, apex rounded, margin entire, ventral base cordate, dorsal base cuneate. Leaf cells slightly thick-walled, trigones large, nodulate, intermediate thickenings absent. Cuticle smooth. Oil bodies segmented, 2−3 per median cell. Underleaves vestigial, filiform, 2−3-celled, with an apical hyaline papilla. Androecia and gynoecia not seen. Gemmae 1−2-celled, usually produced from apical leaf margin.

Habitat: On soil. Elev. 1325−1823 m.

Distribution in Yachang: Caowangshan, Jiushijiubao, Panguwang, Pangqiang Tun, Xiapeng Tun.

Distribution in China: Fujian, Guangdong, Guizhou, Hainan, Hunan, Taiwan, Hongkong, Yunnan. New to Guangxi.

Note: New family and genus records for Guangxi.

▶ A. 种群；B. 植物体一段（背面观）；C. 植物体一段（腹面观）；D. 侧叶；E. 腹叶；F. 侧叶顶端具芽胞；G. 叶中部细胞，示油体。（凭证标本：韦玉梅等 201110-1）

A. Population; B. Portion of plant (dorsal view); C. Portion of plant (ventral view); D. Leaves; E. Underleaf; F. Apex of leaf with gemmae; G. Median cells of leaf showing oil bodies. (All from *Wei et al. 201110-1*)

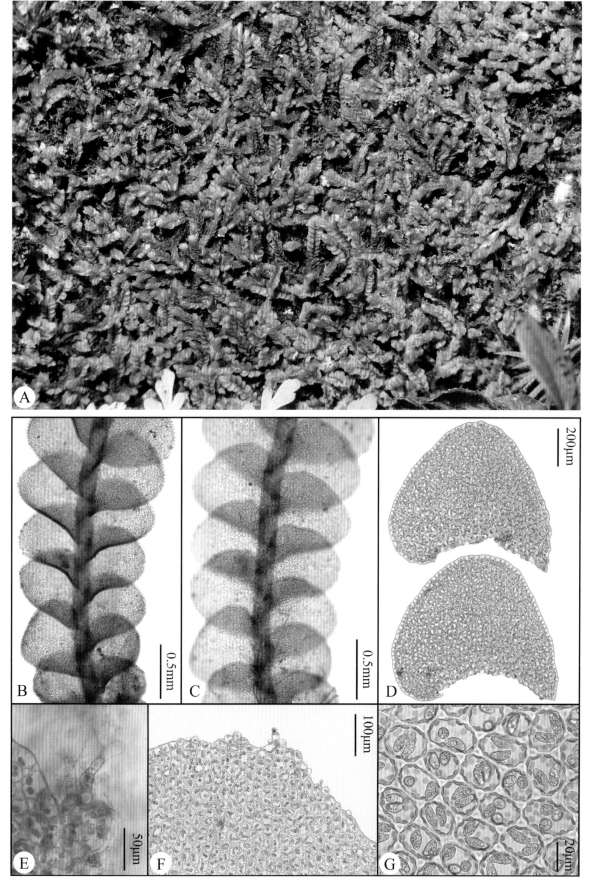

图 44 甲克苔

Fig. 44 *Jackiella javanica* Schiffn.

柯氏合叶苔

Scapania koponenii Potemkin, Ann. Bot. Fenn. 37(1): 41. 2000.

植物体深绿色，带叶宽 2.4—3.3 mm，分枝少，从茎腹面生出，间生型。茎横切面 16—18 个细胞厚，皮部细胞 2—3 层，腹面一部分皮部细胞薄壁，其余厚壁，内部细胞薄壁。侧叶覆瓦状排列至毗邻，背瓣卵形或圆方形，长为腹瓣的 1/3—1/2，顶端细尖，边缘具不规则细齿，齿 2—4 个细胞长，基部 1—3 个细胞宽，基部短下延；背脊直或稍内曲，为腹瓣长的 1/4—2/5；腹瓣椭圆状卵形，顶端细尖，边缘具不规则刺状齿，齿 2—4 个细胞长，基部 2—4 个细胞宽，基部长下延。侧叶细胞薄壁，角质层具密集的粗疣，三角体小，简单三角形，中部球状加厚缺。油体聚合型，每个细胞 3—6 个。腹叶缺。雄苞和雌苞未见。芽胞椭圆形，由（1—）2 个细胞构成，生于茎先端叶腹瓣顶部。

生境： 土生。海拔 1415 m。
雅长分布： 九十九堡。
中国分布： 福建，广东，广西，贵州，湖北，湖南，江西，四川，云南，浙江。

Plants dark green when fresh, 2.4−3.3 mm wide, rarely branched, branches ventral-intercalary. Stem in transverse section 16−18 cells thick, cortical cells of 2−3 layers, thick-walled, except several tiers of ventral cells thin-walled, medullary cells thin-walled. Leaves imbricate to contiguous, dorsal lobe ovate or rounded-rectangular, 1/3−1/2 as long as the ventral lobe, apex apiculate, margin irregularly dentate-toothed, teeth 2−4 cells long, 1−3 cells wide at base, base short decurrent; keel mostly straight or slightly sinuate, 1/4−2/5 as long as the ventral lobe; ventral lobe oblong-ovate, apex apiculate, margin irregularly spinose-toothed, teeth 2−4 cells long, 2−4 cells wide at base, base long decurrent. Leaf cells thin-walled, trigones small, simple-triangulate, intermediate thickenings absent. Cuticle densely and coarsely papillose. Oil bodies segmented, 3−6 per median cell. Underleaves absent. Androecia and gynoecia not seen. Gemmae ellipsoid, (1−)2-celled, usually produced from apex of leaf ventral lobe.

Habitat: On soil. Elev. 1415 m.
Distribution in Yachang: Jiushijiubao.
Distribution in China: Fujian, Guangdong, Guangxi, Guizhou, Hubei, Hunan, Jiangxi, Sichuan, Yunnan, Zhejiang.

▶ A. 种群；B. 植物体（背面观）；C. 侧叶；D. 侧叶横切面；E. 芽胞；F. 植物体一段（腹面观）；G. 植物体一段（背面观）；H. 腹瓣顶端，示边缘齿；I. 腹瓣中部细胞，示油体；J. 腹瓣中部细胞，示角质层的疣。（凭证标本：*韦玉梅等 201111-6*）

A. Population; B. Plant (dorsal view); C. Leaves; D. Transverse section of leaf; E. Gemmae; F. Portion of plant (ventral view); G. Portion of plant (dorsal view); H. Apex of leaf ventral lobe showing marginal teeth; I. Median cells of leaf ventral lobe showing oil bodies; J. Median cells of leaf ventral lobe showing papillose cuticle. (All from *Wei et al. 201111-6*)

图 45　柯氏合叶苔

Fig. 45　*Scapania koponenii* Potemkin

小睫毛苔

Blepharostoma minus Horik., Hikobia 1: 104. 1951 [1952].

植物体浅绿色，带叶宽 0.25—0.55 mm，不规则分枝，耳叶苔型。茎横切面 8—9 个表皮细胞和 3 个内部细胞。假根束状生于腹叶基部。侧叶横生，3—4 裂达基部，基部 3—4 个细胞宽，1 个细胞高，裂瓣丝状，单列细胞，8—15 个细胞长。侧叶细胞方形或长方形，薄壁，角质层具细条纹状疣。油体均一型，每个细胞 2—4 个。腹叶 2 裂，与侧叶同形，裂瓣 5—9 个细胞长。雄苞和雌苞未见。

生境： 石生。海拔 1553—1755 m。

雅长分布： 盘古王，塘英村。

中国分布： 安徽，重庆，福建，甘肃，广西，贵州，湖北，江西，陕西，山东，四川，西藏，云南，浙江。

Plants light green when fresh, 0.25−0.55 mm wide, irregularly branched, branching of the *Frullania*-type. Stem in transverse section with 8−9 epidermal cells and 3 medullary cell. Rhizoids in tufts from underleaf bases. Leaves transverse, 3−4-lobed, divided up to the base, basal part 3−4 cells wide, 1 cell high, lobes filiform, 8−15 celled, uniseriate. Leaf cell quadrate or rectangular, thin-walled. Cuticle striate-verrucose. Oil bodies homogenous, 2−4 per cell. Underleaves identical in morphology to leaves, 2-lobed, lobes 5−9-celled. Androecia and gynoecia not seen.

Habitat: On rocks. Elev. 1553−1755 m.

Distribution in Yachang: Panguwang, Tangying Village.

Distribution in China: Anhui, Chongqing, Fujian, Gansu, Guangxi, Guizhou, Hubei, Jiangxi, Shaanxi, Shandong, Sichuan, Tibet, Yunnan, Zhejiang.

A. 植物体（背面观）；B. 植物体一段（背面观）；C. 植物体一段，示腹叶；D. 叶中部细胞；E. 茎横切面；F. 侧叶。（凭证标本：*韦玉梅等 201110-37B*）

A. Plant (dorsal view); B. Portion of plant (dorsal view); C. Portion of plant showing underleaves; D. Median cells of leaf; E. Transverse section of stem; F. Leaves. (All from *Wei et al. 201110-37B*)

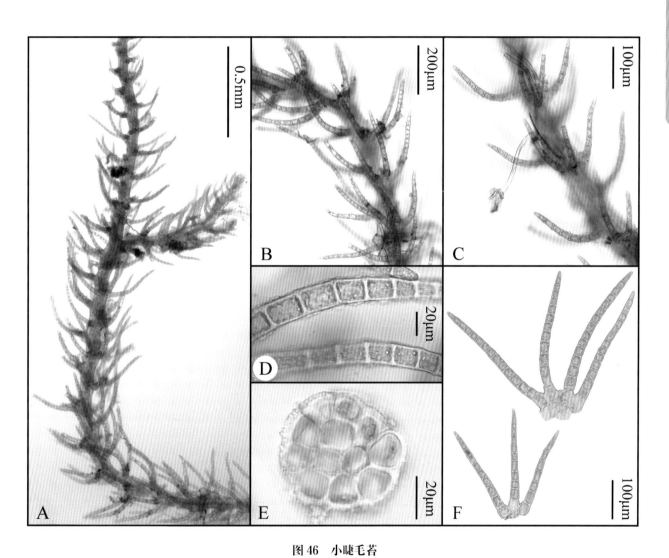

图 46　小睫毛苔
Fig. 46　*Blepharostoma minus* Horik.

卵叶鞭苔

Bazzania angustistipula N. Kitag., J. Hattori Bot. Lab. 30: 268, 1967.

植物体暗绿色至暗棕色，带叶宽 0.86—1.39 mm，叉状分枝，耳叶苔型，鞭状枝多，分枝鞭苔型。茎横切面 6—7 个细胞厚，表皮细胞和内部细胞几乎等大，均厚壁。侧叶远生至疏松覆瓦状排列，易脱落，卵状三角形，顶端渐尖或 2 裂，边缘全缘。侧叶细胞薄壁，角质层平滑，三角体小，简单三角形，中部球状加厚缺。油体均一型，每个细胞 2—3 个。腹叶远生，宽为茎的 1.5—2.5 倍，顶端浅凹至 2 裂，有时圆，边缘全缘，基部楔形。腹叶细胞均不透明，同叶细胞。雄苞和雌苞未见。

生境： 倒木生。海拔 1902 m。

雅长分布： 草王山。

中国分布： 广西，贵州，四川，云南。

Plants dark green to dark brown when fresh, 0.86–1.39 mm wide, dichotomously branched, branching of the *Frullania*-type, ventral flagella frequent, branching of the Bazzania-type. Stem in transverse section 6–7 cells thick, epidermal cells equal to medullary cells in size, both cells thick-walled. Leaves remote to loosely imbricate, caducous, triangular-ovate, apex acuminate or narrowly 2-lobed, margin entire. Leaf cells thin-walled, trigones small, simple-triangulate, intermediate thickenings absent. Cuticle smooth. Oil bodies homogeneous, 2–3 per median cell. Underleaves distant, 1.5–2.5 times as wide as the stem, apex retuse to bilobed, sometimes rounded, margin entire, base cuneate. Underleaf cells not hyaline, similar to those of the leaf. Androecia and gynoecia not seen.

Habitat: On fallen trees. Elev.1902 m.

Distribution in Yachang: Caowangshan.

Distribution in China: Guangxi, Guizhou, Sichuan, Yunnan.

▶ A. 种群；B. 植物体；C–D. 植物体一段；E. 茎一段；F. 侧叶；G. 腹叶；H. 茎横切面；I. 叶中部细胞，示油体。（凭证标本：*韦玉梅等 221112-68*）

A. Population; B. Plant; C–D. Portions of plants; E. Portion of stem; F. Leaves; G. Underleaves; H. Transverse section of stem; I. Median cells of leaf showing oil bodies. (All from *Wei et al. 221112-68*)

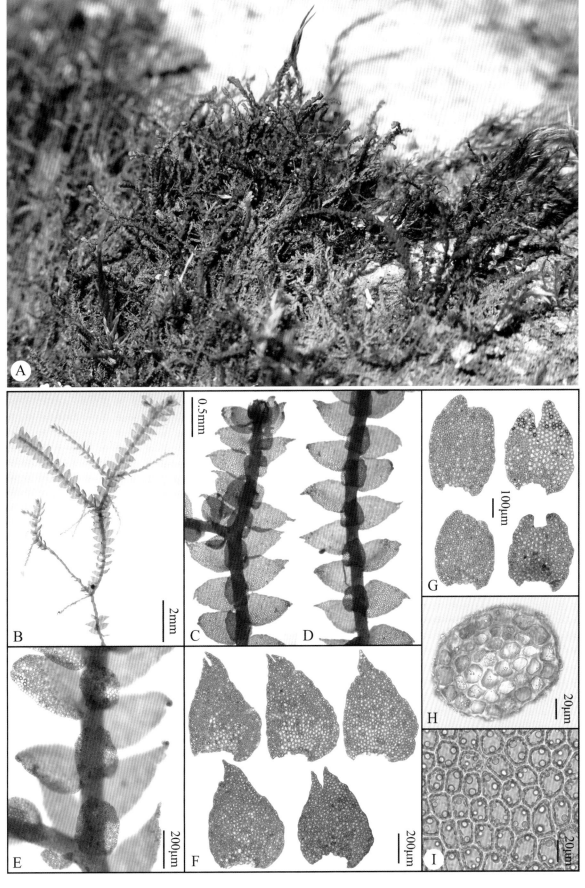

图 47 卵叶鞭苔

Fig. 47 *Bazzania angustistipula* N. Kitag.

三裂鞭苔

Bazzania tridens (Reinw., Blume et Nees) Trevis., Mem. Reale Ist. Lombardo Sci. (Ser. 3), C. Sci. Mat. 4(13): 415. 1877.

植物体浅绿色至绿色，带叶宽 2.65—3.50 mm，叉状分枝，耳叶苔型，鞭状枝多，分枝鞭苔型。茎横切面 14—16 个细胞厚，皮部细胞 2—3 层，厚壁，内部细胞薄壁。侧叶覆瓦状排列，长椭圆状卵形，呈镰刀形弯曲，顶端斜截形，具 3 个齿。侧叶细胞薄壁至稍厚壁，角质层平滑（边缘有时具细疣），三角体小至大，简单三角形，中部球状加厚缺。油体均一型，每个细胞 3—6 个。腹叶毗邻至远生，宽为茎的 1.5—2.0 倍，顶端浅波状凹陷或偶具小齿，边缘全缘，基部楔形。腹叶细胞除基部几列绿色细胞外，其余细胞薄壁透明。雄苞和雌苞未见。

生境： 石生，土生。海拔 1138—1553 m。

雅长分布： 旁墙屯，塘英村。

中国分布： 全国范围广布。

Plants light green to green when fresh, 2.65−3.50 mm wide, dichotomously branched, branching of the *Frullania*-type, ventral flagella frequent, branching of the Bazzania-type. Stem in transverse section 14−16 cells thick, cortical cells of 2−3 layers, thick-walled, medullary cells thin-walled. Leaves imbricate, oblong-ovate, falcate, apex obliquely truncate, with 3 teeth. Leaf cells thin-walled to slightly thick-walled, trigones small to large, simple-triangulate, intermediate thickenings absent. Cuticle smooth (except leaf margin sometimes verrucose). Oil bodies homogeneous, 3−6 per median cell. Underleaves contiguous to distant, 1.5−2.0 times as wide as the stem, apex repand or occasionally with small teeth, base cuneate. Underleaf cells hyaline except for a few chlorophyllous cells in the base. Androecia and gynoecia not seen.

Habitat: On rocks and soil. Elev. 1138−1553 m.

Distribution in Yachang: Pangqiang Tun, Tangying Village.

Distribution in China: Widely distributed in China.

▶ A. 种群；B. 植物体；C. 植物体一段；D. 侧叶；E. 茎一段，示腹叶；F. 叶中部细胞，示油体；G. 腹叶；H. 茎横切面。（凭证标本：*韦玉梅等 191019-369*）

A. Population; B. Plant; C. Portion of plant; D. Leaves; E. Portion of stem showing underleaves; F. Median cells of leaf showing oil bodies; G. Underleaves; H. Transverse section of stem. (*All from Wei et al. 191019-369*)

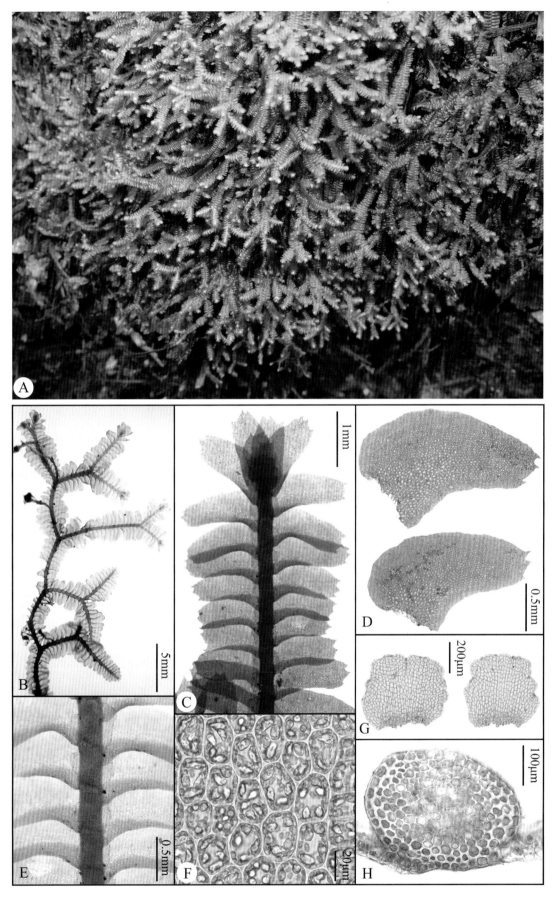

图 48 三裂鞭苔

Fig. 48 *Bazzania tridens* (Reinw., Blume & Nees) Trevis.

指叶苔

Lepidozia reptans (L.) Dumort., Recueil Observ. Jungerm.: 19. 1835.

植物体绿色至褐绿色，带叶宽 0.55—0.85 mm，羽状分枝，分枝顶生型或间生型，常具鞭状枝。茎横切面 6—7 个细胞厚，表皮细胞略大于内部细胞。茎侧叶远生至毗邻，近方形，内凹，3—4 裂达叶长的 1/3—1/2，裂瓣三角形，基部宽 4—8 个细胞，先端具 2—3 个单列细胞，盘状基部高 5—8 个细胞。枝侧叶通常覆瓦状排列，比茎侧叶稍小，常 2—3 裂。侧叶细胞稍厚壁，角质层平滑，三角体无，中部球状加厚缺。油体均一型，每个细胞 9—20 个。茎腹叶远生，与茎粗等宽，4 裂（偶 3 裂）达腹叶长的 1/4—2/5，裂瓣短，先端钝。枝腹叶毗邻至远生，比茎腹叶稍小，常 2—3 裂。雌雄异株。雄苞未见。雌苞生于腹侧短枝上。

生境： 腐木生。海拔 1878 m。

雅长分布： 草王山。

中国分布： 全国范围广布。

Plants green to brownish-green when fresh, 0.55−0.85 mm wide, pinnate branched, lateral-terminal and ventral-intercalary, often flagelliform. Stem in transverse section 6−7 cells thick, epidermal cells slightly larger than medullary cells. Stem leaves remote to contigues, usually convex, subquadrate, 3−4-lobed to 1/3−1/2, lobes triangular, 4−8 cells at base, with 2−3 uniseriate cells at tips, disc 5−8 cells high. Branch leaves usually imbricate, smaller than stem ones, usually 2−3-lobed. Leaf cells slightly thick-walled, trigones small, simple-triangulate, intermediate thickenings absent. Cuticle smooth. Oil bodies homogeneous, 9−20 per median cell. Stem underleaves distant, as wide as the stem, (3−)4-lobed to 1/4−2/5, lobes short, obtuse at apex. Branch underleaves contigous to distant, smaller than stem ones, usually 2−3-lobed. Dioicous. Androecia not seen. Gynoecia on short ventral branches.

Habitat: On rotten log. Elev. 1878 m.

Distribution in Yachang: Caowangshan.

Distribution in China: Widely distributed in China.

▶ A. 种群；B. 植物体；C. 分枝一段；D. 主茎一段；E. 主茎一段，示腹叶；F. 侧叶；G. 分枝一段，示腹叶；H. 腹叶；I. 茎横切面。（凭证标本：韦玉梅等 *221112-40*）

A. Population; B. Plant; C. Portion of branch; D. Portion of main stem; E. Portion of main stem showing underleaves; F. Stem leaves; G. Portion of branch showing underleaves; H. Stem underleaves; I. Transverse section of stem. (All from *Wei et al. 221112-40*)

图 49 指叶苔

Fig. 49 *Lepidozia reptans* (L.) Dumort.

长角剪叶苔

Herbertus dicranus (Taylor) Trevis., Mem. Reale Ist. Lombardo Sci. (Ser. 3), C. Sci. Mat. 4(13): 397. 1877.

植物体暗绿色至棕褐色，体型多变，可从纤细到粗壮，不规则分枝，分枝常从腹面伸出，顶生型。茎横切面15—16个细胞厚，皮部细胞2—3层，明显厚壁，内部细胞稍厚壁。侧叶覆瓦状排列，2 裂约达叶长的3/5，近对称，裂瓣狭披针形，直或略向一侧偏曲，先端具1—4个单列细胞，基部卵形，边缘常全缘，具黏液瘤。侧叶细胞厚壁，角质层具疣，三角体大，肿胀形，中部球状加厚大。假肋明显，由叶片中间的加长细胞构成，于基盘中部或中下部分叉，达裂瓣中上部。油体均一型，数量变化大。腹叶与侧叶同形，略小。雄苞和雌苞未见。

生境：树干，树基生。海拔 1861—1878 m。

雅长分布：草王山。

中国分布：广泛分布于除东北以外的我国大部分省区。

Plants green to brownish-green when fresh, highly variable in size, from thin and slender to robust, irregularly branched, ventral-intercalary. Stem in transverse section 15−16 cells thick, cortical cells of 2−3 layers, thick-walled, medullary cells less thick-walled. Leaves imbricate, bifid usually to ca. 3/5, subsymmetrical, lobes narrowlly laceolate, spreading or falcate-secund, with 1−4 uniseriate cells at tips, basal disc ovate, margin subentire, usually with several slime papillae. Leaf cells thick-walled, trigones large, nodulate, intermediate thickenings large. Vitta conspicuous, composed of elongate cells, extending to near the lobe apices, bifurcating uasually at the mid of basal disc or the lower part. Cuticle papillose. Oil bodies homogeneous, variable in number. Underleaves simillar to leaves in shape, but smaller. Androecia and gynoecia not seen.

Habitat: On tree trunks and tree bases. Elev. 1861−1878 m.

Distribution in Yachang: Caowangshan.

Distribution in China: Widely distributed in most provinces of China except Northeast Region.

▶ A. 种群；B. 植物体；C. 植物体一段；D. 侧叶；E. 腹叶；F. 叶基盘；G. 叶尖；H. 叶基盘边缘，示黏液瘤；I. 茎横切面；J. 叶边缘细胞和加长的假肋细胞。（凭证标本：A 拍摄自韦玉梅等 *221112-52*，其余来自韦玉梅等 *221112-60*）

A. Population; B. Plant; C. Portion of plant; D. Leaves; E. Underleaves; F. Basal disc of the leaf; G. Apex of leaves; H. Margin of basal disc showing slime papillae; I. Transverse section of stem; J. Marginal and elongate vitta cells. (A from *Wei et al. 221112-52*, the others from *Wei et al. 221112-60*)

图 50 长角剪叶苔

Fig. 50 *Herbertus dicranus* (Gottsche, Lindenb. et Nees) Trevis.

树羽苔（羽状羽苔）

Chiastocaulon dendroides (Nees) Carl, Flora 126: 59. 1931.

Plagiochila dendroides (Nees) Lindenb., Sp. Hepat. (Lindenberg) (fasc. 5): 146. 1843.

植物体深绿色，带叶宽 1.20—1.75 mm，具横茎，树状分枝，上部分枝顶生型，下部间生型，分枝顶部有时鞭状，向地性鞭状枝常可见。茎横切面 15—17 个细胞厚，皮部细胞 2—3 层，明显厚壁，内部细胞薄壁。假根分布于横茎上。侧叶毗邻至远生，斜生，但在靠近主茎下部近横生，椭圆状卵形至长方形，顶端通常短 2 裂，有时中间夹生 1—2（—3）个小齿，背侧边缘轻微拱起，稍内折，常全缘，偶在近顶部具 1—2 个小齿，基部稍下延，腹侧边缘稍拱起，具 1—4 个短齿，基部稍下延。侧叶细胞稍厚壁，角质层平滑，三角体小，简单三角形，中部球状加厚偶尔可见。油体聚合型，每个细胞 4—7 个。腹叶退化，丝状。无鳞毛。雄苞顶生或间生，雄苞叶 3—8 对。雌苞未见。

生境：树干生。海拔 1861 m。

凭证标本：草王山。

雅长分布：安徽，广东，贵州，海南，湖北，台湾，西藏，云南。首次记录于广西。

备注：广西新记录属。

Plants dark green when fresh, 1.20–1.75 mm wide, with creeping rhizome, dendroid, branches lateral-intercalary and ventral-intercalary in lower part of shoot, mainly terminal in upper part, flagelliform branches and geotropic flagella present. Stem in transverse section 15–17 cells thick, cortical cells of 2–3 layers, thick-walled, medullary cells thin-walled. Rhizoids mainly on rhizomes. Leaves contiguous to remote, usually obliquely inserted, but transversely inserted on the low part of main stem, oblong-ovate to rectangular, apex short bilobed, sometimes with 1–2(–3) small teeth in between, dorsal margin weakly arched, slightly incurved, almost entire, occasionally with 1–2 teeth near apex, base short-decurrent, ventral margin weakly arched, with 1–4 short teeth, base weakly decurrent. Leaf cells slightly thick-walled, trigones small, simple-triangulate, intermediate thickenings occasionally present. Cuticle smooth. Oil bodies segmented, 4–7 per median cell. Underleaves vestigial, filiform. Paraphyllia absent. Androecia terminal or intercalary, bracts in 3–8 pairs. Gynoecia not seen.

Habitat: On tree trunk. Elev. 1861 m.

Distribution in Yachang: Caowangshan.

Distribution in China: Anhui, Guangdong, Guizhou, Hainan, Hubei, Taiwan, Tibet, Yunnan. New to Guangxi.

Note: New genus record to Guangxi.

▶ A. 种群；B. 植物体；C. 分枝一段；D. 枝叶；E. 茎叶；F. 植物体一段（背面观）；G. 植物体一段（腹面观）；H. 茎横切面；I. 主茎一段；J. 叶中部细胞，示油体。（凭证标本：*韦玉梅等 221112-61*）

A. Population; B. Plant; C. Portion of branch; D. branch leaves; E. Stem leaves; F. Portion of branch (dorsal view); G. Portion of branch (ventral view); H. Transverse section of stem; I. Portion of main stem; J. Median cells of leaf showing oil bodies. (All from *Wei et al. 221112-61*)

图 51 树羽苔

Fig. 51 *Chiastocaulon dendroides* (Nees) Carl

埃氏羽苔

Plagiochila akiyamae Inoue, Bull. Natl. Sci. Mus. Tokyo, B 12(3): 73. 1986.

植物体深绿色，带叶宽 1.8—3.0 mm，具横茎，稀疏分枝或分枝多，上部分枝顶生型，基部间生型。茎横切面 12—13 个细胞厚，皮部细胞 2—3 层，厚壁，内部细胞薄壁。假根分布于横茎上。侧叶毗邻，卵形至椭圆状卵形，顶端截形，具 2—3 个齿，背侧边缘直，稍内折，常全缘，偶在近顶部具 1—2 个齿，基部长下延，腹侧边缘拱起，下半部常内折，具 3—5 个齿，基部稍下延；全叶具 4—9 个齿，齿分布于近顶端或中上部，2—4（—7）个细胞长，2—4 个细胞宽。侧叶细胞薄壁，角质层平滑，三角体小至大，简单三角形，中部球状加厚少见。油体未见。腹叶退化，丝状。无鳞毛。雄苞和雌苞未见。无性芽有时可见生于叶腹面。

生境： 石生。海拔 1305 m。
雅长分布： 大宴坪天坑漏斗。
中国分布： 广西，海南，云南。

Plants dark green when fresh, 1.8–3.0 mm wide, with creeping rhizome, sparingly or frequently branched, branches lateral-intercalary in lower part of shoot, always terminal in upper part. Stem in transverse section 12–13 cells thick, cortical cells of 2–3 layers, thick-walled, medullary cells thin-walled. Rhizoids mainly on rhizomes. Leaves contiguous, ovate to oblong-ovate, apex truncate, with 2–3 teeth, dorsal margin straight, slightly incurved, almost entire, occasionally with 1–2 teeth near apex, base long-decurrent, ventral margin arched, incurved at the lower half, with 3–5 teeth, base weakly decurrent; teeth almost restricted to the upper half, 4–9 per leaf, 2–4 (–7) cells long, 2–4 cells wide. Leaf cells thin-walled, trigones small to large, simple-triangulate, intermediate thickenings scarce. Cuticle smooth. Oil bodies not seen. Underleaves vestigial, filiform. Paraphyllia absent. Androecia and gynoecia not seen. Propagules sometimes occurring on ventral leaf surface.

Habitat: On rocks. Elev. 1305 m.
Distribution in Yachang: Dayanping Tiankeng.
Distribution in China: Guangxi, Hainan, Yunnan.

▶ A. 种群；B. 植物体（背面观）；C. 侧叶；D. 植物体一段（背面观）；E. 植物体一段（腹面观）；F. 叶中部细胞；G. 侧叶带无性芽；H. 茎横切面。（凭证标本：*唐启明 & 韦玉梅 20191017-256*）

A. Population; B. Plant (dorsal view); C. Leaves; D. Portion of plant (dorsal view); E. Portion of plant (ventral view); F. Median cells of leaf; G. Leaf with propagules; H. Transverse section of stem. (All from *Tang & Wei 20191017-256*)

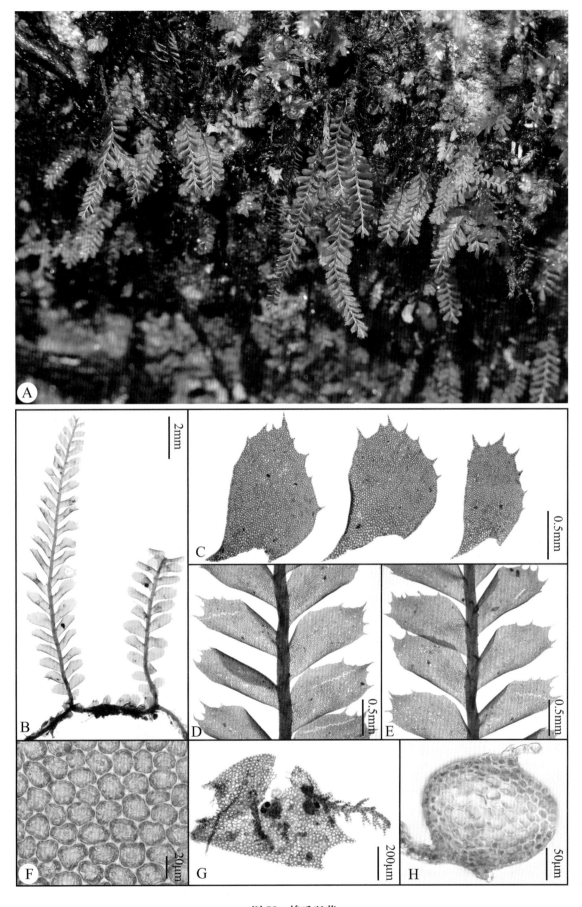

图 52 埃氏羽苔

Fig. 52 *Plagiochila akiyamae* Inoue

阿萨羽苔

Plagiochila assamica Steph., Sp. Hepat. (Stephani) 6: 125. 1917.

植物体绿色至褐绿色，带叶宽 1.6—3.6 mm，具横茎，树状分枝，上部分枝顶生型为主，下部间生型，分枝顶部有时鞭状。茎横切面 25—27 个细胞厚，皮部细胞 3—5 层，明显厚壁，内部细胞稍厚壁。假根分布在横茎上。侧叶毗邻至远生，长方形至椭圆状卵形，顶端常截形，具 3—4 个齿，背侧边缘直或轻微向内弯曲，稍内折，全缘或在近顶部具 1—3 个齿，基部长下延，腹侧边缘直或轻微拱起，有时稍内折，具稀疏的齿，基部稍下延；全叶具 6—9 个齿，3—12 个细胞长，3—7 个细胞宽。侧叶细胞薄壁，角质层平滑，三角体小，不明显，中部球状加厚缺。油体聚合型，每个细胞 3—6 个。腹叶退化，丝状。鳞毛刺毛状，密布于主茎背面，腹面几无。雄苞和雌苞未见。

生境：石生。海拔 1893 m。

雅长分布：草王山。

中国分布：广西，云南。

Plants green to browish green when fresh, 1.6−3.6 mm wide, with creeping rhizome, dendroid, branches lateral-intercalary in lower part of shoot, predominantly terminal in upper part, tips of branches sometimes becoming minute-leaved flagelliform. Stem in transverse section 25−27 cells thick, cortical cells of 3−5 layers, very thick-walled, medullary cells slightly thick-walled. Rhizoids mainly on rhizomes. Leaves contigous to remote, rectangular to oblong-ovate, apex truncate, with 3−4 teeth, dorsal margin straight or slightly sinuate, incurved, entire or with 1−3 teeth near apex, base long-decurrent, ventral margin straight or slightly arched, sometimes incurved, with sparsely toothed, base slightly decurrent; teeth 6−9 per leaf, 3−12 cells long, 3−7 cells wide. Leaf cells thin-walled, trigones small, indistinct, intermediate thickenings absent. Cuticle smooth. Oil bodies segmented, 3−6 per median cell. Underleaves vestigial, filiform. Paraphyllia spinelike, numerous on dorsal stem surface, ventral stem surface almost smooth. Androecia and gynoecia not seen.

Habitat: On rocks. Elev. 1893 m.

Distribution in Yachang: Caowangshan.

Distribution in China: Guangxi, Yunnan.

▶ A. 种群；B. 植物体（背面观）；C. 侧叶；D. 茎横切面；E. 叶中部细胞，示油体；F. 植物体一段（背面观）；G. 植物体一段（腹面观）；H. 茎一段带鳞毛（背面观）；I. 茎一段（腹面观）。（凭证标本：*唐启明等 20210719-15*）

A. Population; B. Plant (dorsal view); C. Leaves; D. Transverse section of stem; E. Median cells of leaf showing oil bodies; F. Portion of plant (dorsal view); G. Portion of plant (ventral view); H. Portion of stem with paraphyllia (dorsal view); I. Portion of stem (ventral view). (All from *Tang et al. 20210719-15*)

图 53 阿萨羽苔

Fig. 53 *Plagiochila assamica* Steph.

中华羽苔

Plagiochila chinensis Steph., Mém. Soc. Nat. Sci. Nat. Math. Cherbourg 29: 223. 1894.

植物体绿色，带叶宽 1.8—2.6 mm，横茎未见，稀疏分枝，分枝间生型。茎横切面 12—14 个细胞厚，皮部细胞 2—3 层，厚壁，内部细胞薄壁。假根主要分布于植物体基部。侧叶覆瓦状排列至毗邻，卵形至椭圆状卵形，顶端圆形或截形，具 5—8 个齿，背侧边缘直或轻微拱起，内折，近顶部具 2—4 个齿，基部下延，腹侧边缘拱起，有时稍内折，具 7—10 个齿，基部稍下延；全叶具 17—21 个刺状齿，2—5 个细胞长，1—3 个细胞宽。侧叶细胞稍厚壁，角质层平滑，三角体小，简单三角形，中部球状加厚缺。油体未见。腹叶退化，丝状。无鳞毛。雄苞间生，雄苞叶 4—10 对。雌苞未见。

生境：石生。海拔 1823 m。

雅长分布：草王山。

中国分布：重庆，福建，甘肃，广西，贵州，河北，湖北，湖南，江西，陕西，四川，台湾，西藏，香港，云南，浙江。

Plants dark green when fresh, 1.8−2.6 mm wide, creeping rhizome not seen, sparingly branched, branches exclusively lateral-intercalary. Stem in transverse section 12−14 cells thick, cortical cells of 2−3 layers, thick-walled, medullary cells thin-walled. Rhizoids on basal part of plant. Leaves imbricate to contiguous, ovate to oblong-ovate, apex rounded or truncate, with 5−8 teeth, dorsal margin straight or slightly arched, incurved, with 2−4 teeth near apex, base decurrent, ventral margin arched, sometimes incurved, with 7−10 teeth, base slightly decurrent; teeth 17−21 per leaf, spinose, 2−5 cells long, 1−3 cells wide. Leaf cells slightly thick-walled, trigones small, simple-triangulate, intermediate thickenings absent. Cuticle smooth. Oil bodies not seen. Underleaves vestigial, filiform. Paraphyllia absent. Androecia intercalary, bracts in 4−10 pairs. Gynoecia not seen.

Habitat: On rocks. Elev. 1823 m.

Distribution in Yachang: Caowangshan.

Distribution in China: Chongqing, Fujian, Gansu, Guangxi, Guizhou, Hebei, Hubei, Hunan, Jiangxi, Shaanxi, Sichuan, Taiwan, Tibet, Hongkong, Yunnan, Zhejiang.

A. 种群；B. 植物体带雄苞（背面观）；C. 侧叶；D. 植物体一段（背面观）；E. 植物体一段（腹面观）；F. 茎横切面；G. 叶中部细胞。（凭证标本：唐启明 & 韦玉梅 20191013-119）

A. Population; B. Plant with androecia (dorsal view); C. Leaves; D. Portion of plant (dorsal view); E. Portion of plant (ventral view); F: Transverse section of stem; G: Median cells of leaf. (All from *Tang & Wei 20191013-119*)

图 54 中华羽苔

Fig. 54 *Plagiochila chinensis* Steph.

树生羽苔

Plagiochila corticola Steph., Mém. Soc. Nat. Sci. Nat. Math. Cherbourg 29: 224. 1894.

植物体绿色，带叶宽 0.52—1.38 mm，横茎未见，分枝少，间生型。茎横切面约 8 个细胞厚，皮部细胞 1—2 层，稍厚壁，内部细胞薄壁。假根多见于幼枝。侧叶远生至毗邻，卵形，顶端圆形或截形，具 2 个锐或钝的齿，背侧边缘轻微拱起，全缘或近顶部具 1 个齿，基部下延，腹侧边缘拱起，具 1—3 个齿，基部下延不明显；全叶具 2—5 个齿，大小变化大。侧叶细胞薄壁，角质层平滑，三角体小，不明显，中部球状加厚缺。油体聚合型，每个细胞 2—5 个。腹叶退化，丝状。无鳞毛。雄苞和雌苞未见。

生境： 石生。海拔 1240 m。

雅长分布： 蓝家湾天坑。

中国分布： 福建，广西，贵州，四川，西藏，云南，浙江。

Plants green when fresh, 0.52–1.38 mm wide, creeping rhizome not seen, rarely branched, branches exclusively lateral-intercalary. Stem in transverse section ca. 8 cells thick, cortical cells of 1–2 layers, slightly thick-walled, medullary cells thin-walled. Rhizoids usually on young shoots. Leaves remote to contigous, ovate, apex rounded or truncate, with 2 spinose or obtuse teeth, dorsal margin slightly arched, almost entire or with 1 tooth near apex, base decurrent, ventral margin arched, with 1–3 teeth, base not decurrent; teeth 2–5 per leaf, variable in size. Leaf cells thin-walled, trigones small, indistinct, intermediate thickenings absent. Cuticle smooth. Oil bodies segmented, 2–5 per median cell. Underleaves vestigial, filiform. Paraphyllia absent. Androecia and gynoecia not seen.

Habitat: On rocks. Elev. 1240 m.

Distribution in Yachang: Lanjiawan Tiankeng.

Distribution in China: Fujian, Guangxi, Guizhou, Sichuan, Tibet, Yunnan, Zhejiang.

▶ A. 种群；B. 植物体（背面观）；C. 植物体一段（背面观）；D. 腹叶；E. 茎横切面；F. 叶中部细胞，示油体。（凭证标本：*韦玉梅等 191019-374*）

A. Population; B. Plant (dorsal view); C. Portions of plants (dorsal view); D. Underleaf; E. Transverse section of stem; F. Median cells of leaf showing oil bodies. (All from *Wei et al. 191019-374*)

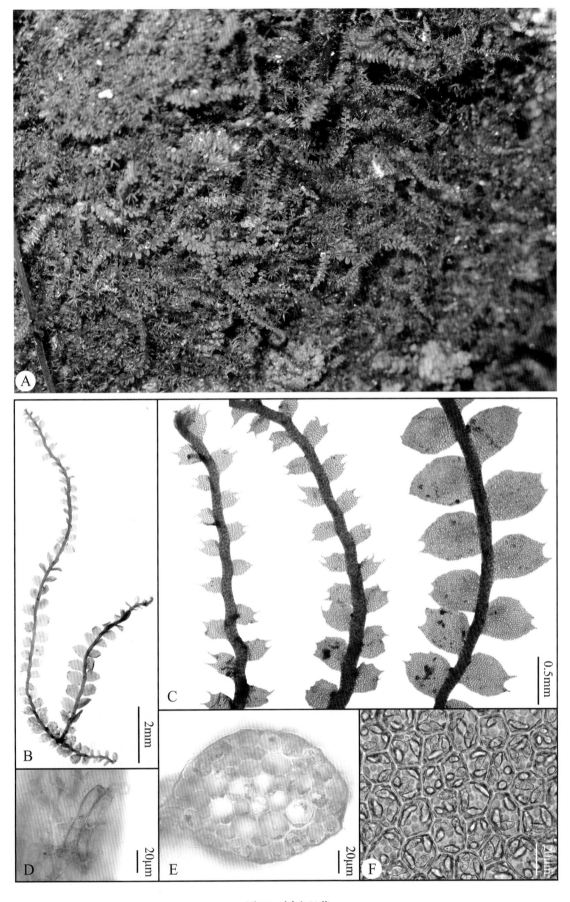

图 55 树生羽苔

Fig. 55 *Plagiochila corticola* Steph.

德氏羽苔

Plagiochila delavayi Steph., Mém. Soc. Nat. Sci. Nat. Math. Cherbourg 29: 224. 1894.

植物体干时暗棕色，带叶宽 1.8—3.9 mm，横茎未见，稀疏分枝，间生型，偶尔具有向地性鞭状枝。茎横切面 16—19 个细胞厚，皮部细胞 2—3 层，厚壁，内部细胞薄壁。假根少。侧叶疏松覆瓦状排列至毗邻，阔卵形至椭圆状卵形，顶端圆，具密集的齿，背侧边缘直或轻微拱起，强烈内卷，中上部开始具齿，基部长下延，腹侧边缘拱起，具密集的齿，基部轻微下延；全叶具 21—30 个齿，齿多为细齿状，1—3（—4）个细胞长，1—2（—3）个细胞宽。侧叶细胞厚壁，角质层平滑，三角体小至大，简单三角形，中部球状加厚缺。油体未见。腹叶退化，丝状。无鳞毛。雄苞顶生或间生，雄苞叶 4—10 对。雌苞未见。

生境：石生。海拔 1343 m。

雅长分布：逻家田屯。

中国分布：重庆，甘肃，贵州，陕西，山西，四川，台湾，西藏，云南。首次记录于广西。

Plants dark brown in herbarium specimens, 1.8–3.9 mm wide, creeping rhizome not seen, sparingly branched, branches exclusively lateral-intercalary, sometimes with geotropic flagella. Stem in transverse section 16–19 cells thick, cortical cells of 2–3 layers, thick-walled, medullary cells thin-walled. Rhizoids scarce. Leaves loosely imbricate to contigous, broadly ovate to oblong-ovate, apex rounded, densely toothed, dorsal margin straight or slightly arched, strongly involute, toothed from the middle part, base long decurrent, ventral margin arched, densely toothed, base slightly decurrent; teeth 21–30 per leaf, dentate, 1–3(–4) cells long, 1–2(–3) cells wide. Leaf cells thick-walled, trigones small to large, simple-triangulate, intermediate thickenings absent. Cuticle smooth. Oil bodies not seen. Underleaves vestigial, filiform. Paraphyllia absent. Androecia terminal or intercalary, bracts in 4–10 pairs. Gynoecia not seen.

Habitat: On rocks. Elev. 1343 m.

Distribution in Yachang: Luojiatian Tun.

Distribution in China: Chongqing, Gansu, Guizhou, Shaanxi, Shanxi, Sichuan, Taiwan, Tibet, Yunnan. New to Guangxi.

▶ A. 植物体带雄苞（背面观）；B. 侧叶；C. 茎横切面；D. 植物体一段（背面观）；E. 植物体一段（腹面观）；F. 叶中部细胞。（凭证标本：*唐启明等 20190521-424*）

A. Plant with an androecium (dorsal view); B. Leaves; C. Transverse section of stem; D. Portion of plant (dorsal view); E. Portion of plant (ventral view); F. Median cells of leaf. (All from *Tang et al. 20190521-424*)

图 56 德氏羽苔

Fig. 56 *Plagiochila delavayi* Steph.

裂叶羽苔

Plagiochila furcifolia Mitt., Trans. Linn. Soc. London, Bot. 3(3): 194. 1891.

植物体绿色至黄绿色，带叶宽 1.7—3.5 mm，具横茎，分枝多，分枝以顶生型为主，偶尔间生型。茎横切面 12—13 个细胞厚，皮部细胞 2—3 层，明显厚壁，内部细胞稍厚壁。假根主要分布于横茎上。侧叶覆瓦状排列至毗邻，易碎，椭圆状卵形，顶端 2 裂达叶长的 1/5—3/5，有时背侧边缘和腹侧边缘近于平行，背侧边缘直，轻微内弯，全缘，基部下延，腹侧边缘稍拱起或直，全缘或偶尔具 1—4 个齿，基部轻微下延。侧叶细胞薄壁，角质层平滑，三角体小，不明显，中部球状加厚缺。油体聚合型，每个细胞 5—8 个。腹叶退化，丝状。无鳞毛。雄苞和雌苞未见。无性芽常见于破碎的叶组织上。

生境： 树基。海拔 1421 m。

雅长分布： 中井天坑。

中国分布： 福建，广东，广西，贵州，海南，湖南，江西，云南，浙江。

Plants green to yellowish green when fresh, 1.7–3.5 mm wide, with creeping rhizome, frequently branched, branches predominantly terminal, occasionally lateral-intercalary. Stem in transverse section 12–13 cells thick, cortical cells of 2–3 layers, thick-walled, medullary cells slightly thick-walled. Rhizoids mainly on rhizomes. Leaves imbricate to contiguous, fragmenting, oblong-ovate, apex bilobed to 1/5–3/5 underleaf length, dorsal and ventral margins of some leaves almost parallel, dorsal margin straight, slightly incurved, entire, base decurrent, ventral margin slightly arched or straight, entire or with 1–4 teeth, base weakly decurrent. Leaf cells thin-walled, trigones small, indistinct, intermediate thickenings absent. Cuticle smooth. Oil bodies segmented, 5–8 per median cell. Underleaves vestigial, filiform. Paraphyllia absent. Androecia and gynoecia not seen. Propagules usually occurring on fragmenting leaves.

Habitat: On tree bases. Elev. 1421 m.

Specimens examined: Zhongjing Tiankeng.

Distribution in China: Fujian, Guangdong, Guangxi, Guizhou, Hainan, Hunan, Jiangxi, Yunnan, Zhejiang.

▶ A. 种群；B. 植物体（背面观）；C. 侧叶；D. 植物体一段（背面观）；E. 无性芽；F. 叶中部细胞，示油体；G. 腹叶；H. 茎横切面。（凭证标本：唐启明 & 张仕燕 20181007-568）

A. Population; B. Plant (dorsal view); C. Leaves; D. Portion of plant (dorsal view); E. Propagules; F. Median cells of leaf showing oil bodies; G. Underleaves; H. Transverse section of stem. (All from *Tang & Zhang 20181007-568*)

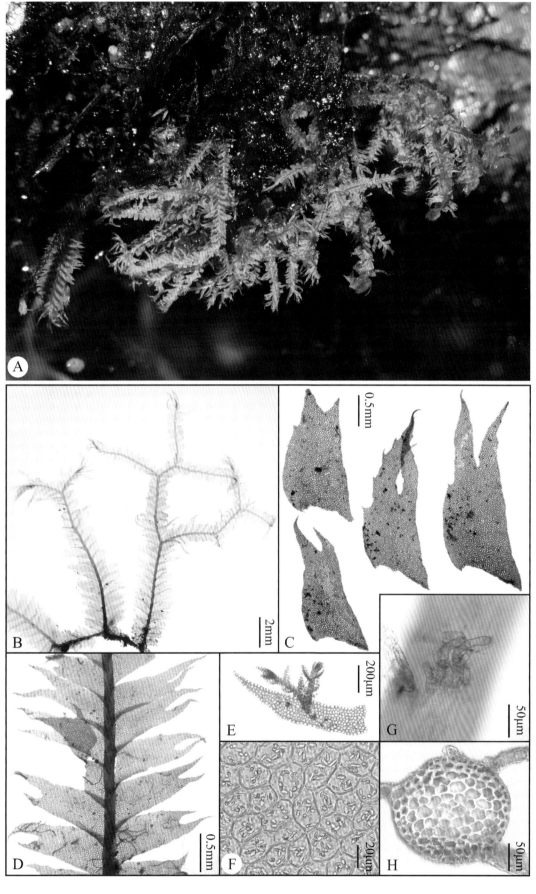

图 57 裂叶羽苔

Fig. 57 *Plagiochila furcifolia* Mitt.

裸茎羽苔

Plagiochila gymnoclada Sande Lac., Plagiochila Sandei: 6. 1856.

植物体干时棕黄色，带叶宽 1.5—3.1 mm，横茎未见，分枝少，间生型，有时具向地性鞭状枝。茎横切面约 12 个细胞厚，皮部细胞 1—2 层，厚壁，内部细胞薄壁。假根少。侧叶覆瓦状排列至毗邻，卵形至椭圆状卵形，顶端圆或截形，具 6—9 个齿，背侧边缘直，稍内弯，近顶部具 3—8 个齿，基部下延，腹侧边缘拱起，具密集的齿，近基部稍膨大，基部下延；全叶具 27—39 个齿，齿刺状。侧叶细胞厚壁，角质层平滑，三角体大，肿胀形，中部球状加厚缺。油体未见。腹叶退化，丝状。无鳞毛。雄苞和雌苞未见。

生境：石生，土生。海拔 1124—1747 m。

雅长分布：草王山，蓝家湾天坑。

中国分布：福建，广西，贵州，湖北，江西，四川，台湾，云南。

Plants brownish yellow in herbarium specimens, 1.5−3.1 mm wide, creeping rhizome not seen, sparingly branched, branches exclusively lateral-intercalary, sometimes with geotropic flagella. Stem in transverse section ca. 12 cells thick, cortical cells of 1−2 layers, thick-walled, medullary cells thin-walled. Rhizoids scarce. Leaves imbricate to contigous, ovate to oblong-ovate, apex rounded or truncate, with 6−9 teeth, dorsal margin straight, slightly incurved, with 3−8 teeth near apex, base decurrent, ventral margin arched, densely toothed, base decurrent, sometimes ampliate; teeth 27−39 per leaf, spinose. Leaf cells thick-walled, trigones large, nodulate, intermediate thickenings absent. Cuticle smooth. Oil bodies not seen. Underleaves vestigial, filiform. Paraphyllia absent. Androecia and gynoecia not seen.

Habitat: On rocks and soil. Elev. 1124−1747 m.

Distribution in Yachang: Caowangshan, Lanjiawan Tiankeng.

Distribution in China: Fujian, Guangxi, Guizhou, Hubei, Jiangxi, Sichuan, Taiwan, Yunnan.

▶ A. 植物体带向地性鞭状枝（背面观）；B. 植物体（背面观）；C. 侧叶；D. 植物体一段（背面观）；E. 植物体一段（腹面观）；F. 茎横切面；G. 叶中部细胞。（凭证标本：*唐启明 & 韦玉梅 20191019-384*）

A. Plant with geotropic flagella (dorsal view); B. Plant (dorsal view); C. Leaves; D. Portion of plant (dorsal view); E. Portion of plant (ventral view); F. Transverse section of stem; G. Median cells of leaf. (All from *Tang & Wei 20191019-384*)

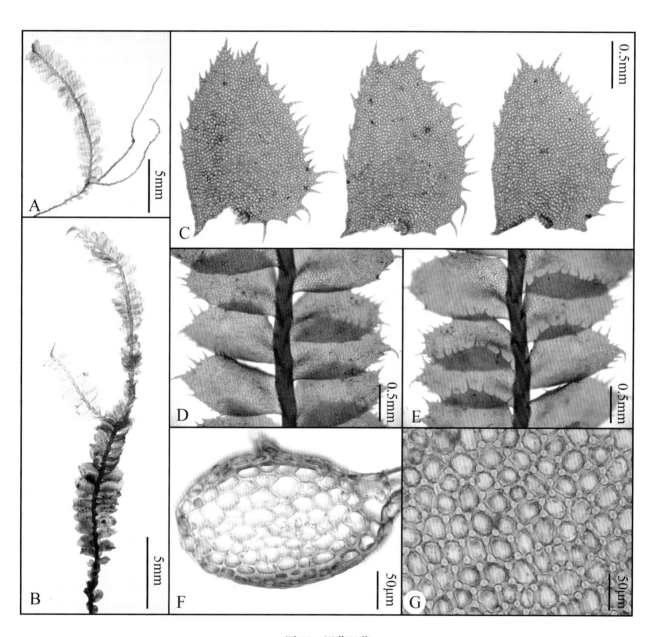

图 58　裸茎羽苔

Fig. 58　*Plagiochila gymnoclada* Sande Lac.

容氏羽苔

Plagiochila junghuhniana Sande Lac., Ned. Kruidk. Arch. 3: 416. 1854.

植物体浅绿色至褐绿色，带叶宽 1.9—3.0 mm，横茎未见，二叉分枝，顶生型。茎横切面 14—16 个细胞厚，皮部细胞 2—3 层，厚壁，内部细胞薄壁。假根少。侧叶覆瓦状排列，易碎，椭圆状卵形，顶端 2 裂，中间偶尔具 1 个小齿，背侧边缘略直，内弯，全缘或近顶部具 1 个齿，基部长下延，腹侧边缘拱起，具 3—5 个刺状或钝齿，近基部稍膨大，基部下延。侧叶细胞薄壁，角质层平滑，三角体小至大，简单三角形，中部球状加厚缺。油体聚合型，每个细胞 5—10 个。腹叶退化，丝状。无鳞毛。雄苞未见。雌苞顶生，具 1—2 个新生枝。蒴萼钟形，脊具有齿的翼，口部弧形，具纤毛状齿。无性芽生于叶腹面。

生境： 石生，树基。海拔 1125—1218 m。

雅长分布： 黄猄洞天坑，逻家田屯。

中国分布： 福建，广西，贵州，海南，江西，台湾。

Plants light green to browish green when fresh, 1.9–3.0 mm wide, creeping rhizome not seen, dichtomously branched, branches terminal. Stem in transverse section 14–16 cells thick, cortical cells of 2–3 layers, thick-walled, medullary cells thin-walled. Rhizoids scarce. Leaves imbricate to contigous, fragmenting, oblong-ovate, apex bilobed, occasionally with a small tooth in between, dorsal margin almost straight, slightly incurved, entire or with 1 tooth near apex, base long decurrent, ventral margin arched, with 3–5 spinose or obtuse teeth, base decurrent, slightly ampliate. Leaf cells thin-walled, trigones small to large, simple-triangulate, intermediate thickenings absent. Cuticle smooth. Oil bodies segmented, 5–10 per median cell. Underleaves vestigial, filiform. Paraphyllia absent. Androecia not seen. Gynoecia terminal, with 1–2 lateral-intercalary innovations. Perianths campanulate, dorsal keel with a dentate wing, teeth of mouth irregularly ciliate. Propagules usually occurring on ventral leaf surface.

Habitat: On rocks and tree bases. Elev. 1125–1218 m.

Distribution in Yachang: Huangjingdong Tiankeng, Luojiatian Tun.

Distribution in China: Fujian, Guangxi, Guizhou, Hainan, Jiangxi, Taiwan.

A. 种群；B. 植物体（背面观）；C. 植物体一段（背面观）；D. 侧叶；E. 蒴萼；F. 茎横切面；G. 蒴萼口部的齿；H. 叶中部细胞，示油体；I. 侧叶带无性芽；J. 植物体一段（背面观）；K. 植物体一段（腹面观）。（凭证标本：韦玉梅等 *191020-414*）

A. Population; B. Plant (dorsal view); C. Portion of plant (dorsal view); D. Leaves; E. Perianth; F. Transverse section of stem; G. Teeth of perianth mouth; H. Median cells of leaf lobe showing oil bodies; I. Leaf with propagules; J. Portion of plant (dorsal view); K. Portion of plant (ventral view). (All from *Wei et al. 191020-414*)

图 59 容氏羽苔

Fig. 59 *Plagiochila junghuhniana* Sande Lac.

加萨羽苔

Plagiochila khasiana Mitt., J. Proc. Linn. Soc., Bot. 5 (18): 95, 1860 [1861].

植物体绿色至褐绿色，带叶宽 3.5—4.5 mm，有横茎，分枝多，顶生型和间生型。茎横切面 15—18 个细胞厚，皮部细胞 3—4 层，厚壁，内部细胞薄壁。假根主要分布于横茎上。侧叶覆瓦状排列，易碎，椭圆状卵形，顶端狭圆截形，具 2—4 个刺状齿，背侧边缘直，稍内弯，近顶部具 2—3 个齿，基部长下延，腹侧边缘拱起，具不规则齿，近基部膨大，基部下延；全叶具 7—13 个齿，1—7 个细胞长，1—3 个细胞宽，最末端细胞呈长刺状。侧叶细胞薄壁，角质层平滑，三角体大，肿胀形，中部球状加厚在叶基部偶尔可见。油体聚合型，每个细胞 3—9 个。腹叶退化，丝状。鳞毛生于茎背面叶基处，片状。雄苞未见。雌苞顶生，常具 2 个新生枝。蒴萼钟形，脊平滑无翼，口部稍弧形，具不规则齿。无性芽常生于叶腹面。

生境：倒木生。海拔 1902 m。

雅长分布：草王山。

中国分布：安徽，福建，甘肃，广东，贵州，陕西，台湾，云南。首次记录于广西。

Plants green to browish green when fresh, 3.5–4.5 mm wide, with creeping rhizome, frequently branched, branches terminal and lateral-intercalary. Stem in transverse section 15–18 cells thick, cortical cells of 3–4 layers, thick-walled, medullary cells thin-walled. Rhizoids mainly on rhizomes. Leaves imbricate, fragmenting, oblong-ovate, apex narrowly rounded-truncate, with 2–4 spinose teeth, dorsal margin straight, slightly incurved, with 2–3 teeth near apex, base long decurrent, ventral margin arched, irregularly toothed, base decurrent, ampliate; teeth 7–13 per leaf, 1–7 cells long, 1–3 cells wide, terminal cell elongate, spinose. Leaf cells thin-walled, trigones large, nodulate, intermediate thickenings occasionally present at base. Cuticle smooth. Oil bodies segmented, 3–9 per median cell. Underleaves vestigial, filiform. Paraphyllia lamelliform, at each dorsal leaf base. Androecia not seen. Gynoecia terminal, usually with 2 lateral-intercalary innovations. Perianths campanulate, dorsal keel without wing, teeth of mouth irregularly dentate. Propagules usually occurring on ventral leaf surface.

Habitat: On fallen tree. Elev. 1902 m.

Distribution in Yachang: Caowangshan.

Distribution in China: Anhui, Fujian, Gansu, Guangdong, Guizhou, Shaanxi, Taiwan, Yunnan. New to Guangxi.

▶ A. 种群；B. 植物体（背面观）；C. 植物体一段（腹面观）；D. 茎一段（背面观）；E. 侧叶；F. 鳞毛；G. 茎横切面；H. 蒴萼；I. 侧叶带无性芽；J. 叶顶端的齿；K. 叶中部细胞，示油体。（凭证标本：*韦玉梅等 221112-69*）

A. Population; B. Plant (dorsal view); C. Portion of plant (ventral view); D. Portion of stem (dorsal view); E. Leaves; F. Paraphyllium; G. Transverse section of stem; H. Perianth; I. Leaf with propagules; J. Apical teeth of leaf; K. Median cells of leaf lobe showing oil bodies. (All from *Wei et al. 221112-69*)

图 60　加萨羽苔
Fig. 60　*Plagiochila khasiana* Mitt.

昆明羽苔

Plagiochila kunmingensis Piippo, Ann. Bot. Fenn. 34(4): 281. 1997.

植物体绿色至褐绿色，带叶宽 1.7—3.6 mm，具横茎，分枝多，以顶生型为主，偶尔间生型。茎横切面 16—18 个细胞厚，皮部细胞 3—4 层，厚壁，内部细胞薄壁。假根主要分布于横茎上。侧叶覆瓦状排列，易碎，卵形至椭圆状卵形，顶端阔圆形或截形，具 3—5 个齿，背侧边缘直，稍内弯，近顶部具 1—2 个齿，基部长下延，腹侧边缘拱起，具 5—11 个齿，近基部稍膨大至膨大，基部轻微下延；全叶具 9—16 个齿，3—7 个细胞长，2—6 个细胞宽。侧叶细胞薄壁，角质层平滑，三角体小至大，简单三角形，中部球状加厚缺。油体聚合型，每个细胞 5—9 个。腹叶退化，丝状。无鳞毛。雄苞未见。雌苞顶生，常具 1 个新生枝。蒴萼钟形，脊具或不具翼，口部弧形，具长刺状或细齿状齿。无性芽常生于叶腹面。

生境： 石生环境常见，倒木、树干和树基环境也有分布。海拔 1048—1750 m。

雅长分布： 草王山，大棚屯，吊井天坑，黄猄洞天坑，逻家田屯，盘古王，下棚屯，香当天坑。

中国分布： 云南。首次记录于广西。

Plants green to browish green when fresh, 1.7–3.6 mm wide, with creeping rhizome, frequently branched, branches predominantly terminal, occasionally lateral-intercalary. Stem in transverse section 16–18 cells thick, cortical cells of 3–4 layers, thick-walled, medullary cells thin-walled. Rhizoids mainly on rhizomes. Leaves imbricate, fragmenting, ovate to oblong-ovate, apex broadly rounded or truncate, with 3–5 teeth, dorsal margin straight, slightly incurved, with 1–2 teeth near apex, base long decurrent, ventral margin arched, with 5–11 teeth, base decurrent, slightly ampliate to ampliate; teeth 9–16 per leaf, 3–7 cells long, 2–6 cells wide. Leaf cells thin-walled, trigones small to large, simple-triangulate, intermediate thickenings absent. Cuticle smooth. Oil bodies segmented, 5–9 per median cell. Underleaves vestigial, filiform. Paraphyllia absent. Androecia not seen. Gynoecia terminal, usually with 1 lateral-intercalary innovation. Perianths campanulate, dorsal keel with or without wing, teeth of mouth irregularly spinose or dentate. Propagules usually occurring on ventral leaf surface.

Habitat: Often on rocks, sometimes on fallen trees, tree trunks and tree bases. Elev. 1048–1750 m.

Distribution in Yachang: Caowangshan, Dapeng Tun, Diaojing Tiankeng, Huangjingdong Tiankeng, Luojiatian Tun, Panguwang, Xiapeng Tun, Xiangdang Tiankeng.

Distribution in China: Yunnan. New to Guangxi.

▶ A. 种群；B. 植物体（背面观）；C. 侧叶；D. 植物体一段（腹面观）；E. 植物体一段（背面观）；F. 茎横切面；G. 蒴萼；H. 叶中部细胞，示油体。（凭证标本：*韦玉梅等 201110-46*）

A. Population; B. Plant (dorsal view); C. Leaves; D. Portion of plant (ventral view); E. Portion of plant (dorsal view); F. Transverse section of stem; G. Perianth; H. Median cells of leaf showing oil bodies. (All from *Wei et al. 201110-46*)

图 61 昆明羽苔

Fig. 61 *Plagiochila kunmingensis* Piippo

尼泊尔羽苔

Plagiochila nepalensis Lindenb., Sp. Hepat. (Lindenberg) 2-4: 93. 1840.

植物体绿色，带叶宽 3.0—5.0 mm，具横茎，分枝多，以顶生型为主。茎横切面约 16 个细胞厚，皮部细胞 2—3 层，厚壁，内部细胞薄壁。假根主要分布于横茎上。侧叶毗邻至远生，椭圆状卵形，顶端稍圆，常具 2 个较明显的大齿，背侧边缘直或轻微拱起，稍内弯，全缘或有时近顶部具 1 个齿，基部下延，腹侧边缘拱起，具 1—5 个齿，近基部稍膨大至膨大，基部轻微下延；全叶具 3—7 个齿，齿多为刺状。侧叶细胞薄壁至稍厚壁，角质层平滑，三角体小至大，简单三角形或肿胀形，中部球状加厚缺。油体聚合型，每个细胞 5—9 个。腹叶退化，丝状。无鳞毛。雄苞未见。雌苞顶生，具 1—2 个新生枝。蒴萼钟形，背脊具明显的翼，口部弧形，具刺状长齿。无性芽有时可见生于叶腹面。

生境： 石生环境常见，腐木和树基环境也有分布。海拔 1048—1675 m。

雅长分布： 大宴坪天坑漏斗，黄猄洞天坑，蓝家湾天坑，老屋基天坑，盘古王，香当天坑，霄罗湾洞穴。

中国分布： 安徽，福建，甘肃，广西，贵州，湖北，湖南，江西，陕西，四川，台湾，西藏，云南，浙江。

Plants green when fresh, 3.0–5.0 mm wide, with creeping rhizome, frequently branched, branches predominantly terminal, occasionally lateral-intercalary. Stem in transverse section ca. 16 cells thick, cortical cells of 2–3 layers, thick-walled, medullary cells thin-walled. Rhizoids mainly on rhizomes. Leaves contigous to remote, oblong-ovate, apex subrounded, usually with 2 larger teeth, dorsal margin straight, slightly incurved, entire or with 1 tooth near apex, base long decurrent, ventral margin arched, with 1–5 teeth, base weakly decurrent, slightly ampliate to ampliate; teeth 3–7 per leaf, usually spinose. Leaf cells thin-walled to slightly thick-walled, trigones small to large, simple-triangulate or nodulate, intermediate thickenings absent. Cuticle smooth. Oil bodies segmented, 5–9 per median cell. Underleaves vestigial, filiform. Paraphyllia absent. Androecia not seen. Gynoecia terminal, with 1–2 lateral-intercalary innovations. Perianths campanulate, dorsal keel with wing, teeth of mouth irregularly long-spinose. Propagules usually occurring on ventral leaf surface.

Habitat: Often on rocks, sometimes on rotten logs and tree bases. Elev. 1048–1675 m.

Distribution in Yachang: Dayanping Tiankeng, Huangjingdong Tiankeng, Lanjiawan Tiankeng, Laowuji Tiankeng, Panguwang, Xiangdang Tiankeng, Xiaoluowan Cave.

Distribution in China: Anhui, Fujian, Gansu, Guangxi, Guizhou, Hubei, Hunan, Jiangxi, Shaanxi, Sichuan, Taiwan, Tibet, Yunnan, Zhejiang.

A. 种群；B. 植物体（背面观）；C. 侧叶；D. 植物体一段（背面观）；E. 植物体一段（腹面观）；F. 蒴萼；G. 茎横切面；H. 叶腹面的无性芽；I. 叶中部细胞，示油体。（凭证标本：*韦玉梅等 191012-77*）

A. Population; B. Plant (dorsal view); C. Leaves; D. Portion of plant (dorsal view); E. Portion of plant (ventral view); F. Perianth; G. Transverse section of stem; H. Propagules on ventral leaf surface; I. Median cells of leaf showing oil bodies. (All from *Wei et al. 191012-77*)

图 62　尼泊尔羽苔

Fig. 62　*Plagiochila nepalensis* Lindenb.

卵叶羽苔

Plagiochila ovalifolia Mitt., Trans. Linn. Soc. London, Bot. 3(3): 193. 1891.

植物体深绿色，带叶宽 2.1—4.5 mm，横茎未见，稀疏分枝，间生型。茎横切面 15—18 个细胞厚，皮部细胞 2—4 层，厚壁，内部细胞薄壁。假根少。侧叶覆瓦状排列，阔卵形，顶端圆，具 6—12 个齿，背侧边缘直，强烈内卷，近顶部具 3—8 个齿，基部长下延，腹侧边缘拱起，具密集的齿，近基部膨大，基部轻微下延；全叶具 25—38 个齿，齿刺状。侧叶细胞薄壁，角质层平滑，三角体小至大，简单三角形，中部球状加厚缺。油体聚合型，每个细胞 3—6 个。腹叶退化，丝状。无鳞毛。雄苞间生，雄苞叶 6—9 对。雌苞顶生，不具或具 1—2 个新生枝。蒴萼圆柱形，脊无翼，口部具不规则纤毛状的齿。

生境： 石生。海拔 1124—1823 m。

雅长分布： 蓝家湾天坑，盘古王。

中国分布： 全国范围广布。

Plants dark green when fresh, 2.1–4.5 mm wide, creeping rhizome not seen, sparingly branched, branches exclusively lateral-intercalary. Stem in transverse section 15–18 cells thick, cortical cells of 2–4 layers, thick-walled, medullary cells thin-walled. Rhizoids scarce. Leaves imbricate, broadly ovate, apex rounded, with 6–12 teeth, dorsal margin straight, strongly involute, with 3–8 teeth near apex, base long decurrent, ventral margin arched, with densely teeth, base weakly decurrent, ampliate; teeth 25–38 per leaf, spinose. Leaf cells thin-walled, trigones small to large, simple-triangulate, intermediate thickenings absent. Cuticle smooth. Oil bodies segmented, 3–6 per median cell. Underleaves vestigial, filiform. Paraphyllia absent. Androecia intercalary, bracts in 6–9 pairs. Gynoecia terminal, without or with 1–2 lateral-intercalary innovations. Perianths long-cylindric, keels without wing, teeth of mouth irregularly long-spinose.

Habitat: On rocks. Elev. 1124–1823 m.

Distribution in Yachang: Lanjiawan Tiankeng, Panguwang.

Distribution in China: Widely distributed in China.

▶ A. 种群；B. 植物体带雄苞（背面观）；C. 植物体一段带蒴萼（背面观）；D. 侧叶；E. 茎一段（腹面观）；F. 茎横切面；G. 蒴萼口部；H. 叶中部细胞，示油体。（凭证标本：*韦玉梅等 201110-42*）

A. Population; B. Plant with androecia (dorsal view); C. Portion of plant with perianth (dorsal view); D. Leaves; E. Portion of stem (ventral view); F. Transverse section of stem; G. Perianth mouth; H. Median cells of leaf showing oil bodies. (All from *Wei et al. 201110-42*)

图 63 卵叶羽苔

Fig. 63 *Plagiochila ovalifolia* Mitt.

圆头羽苔

Plagiochila parvifolia Lindenb., Sp. Hepat. (Lindenberg) 1: 28. 1839.

植物体绿色至褐绿色，带叶宽 1.5—3.6 mm，具横茎，分枝多，以顶生型为主。茎横切面 15—16 个细胞厚，皮部细胞 2—3 层，厚壁，内部细胞薄壁。假根主要分布于横茎。侧叶覆瓦状排列，易碎，三角形至椭圆状卵形，顶端圆或截形，具 4—5 个齿，背侧边缘镰状弯曲，内弯，全缘，基部长下延，腹侧边缘略直或波曲，具稀疏的齿，近基部明显膨大，基部下延；全叶具 5—8 个齿，齿 2—4 个细胞长，1—3 个细胞宽。侧叶细胞稍厚壁，角质层平滑，三角体大，简单三角形或肿胀形，中部球状加厚偶尔可见。油体聚合型，每个细胞 4—7 个。腹叶退化，但通常大，片状，边缘具纤毛。无鳞毛。雄苞和雌苞未见。无性芽常见于破碎的叶组织上。

生境： 石生，倒木，腐木，树基。海拔 1115—1440 m。

雅长分布： 黄猄洞天坑，蓝家湾天坑，盘古王，下棚屯，悬崖天坑。

中国分布： 安徽，福建，甘肃，广东，广西，贵州，海南，湖南，陕西，四川，台湾，西藏，香港，云南，浙江。

Plants green to brownish green when fresh, 1.5−3.6 mm wide, with creeping rhizome, frequently branched, branches predominantly terminal. Stem in transverse section 15−16 cells thick, cortical cells of 2−3 layers, thick-walled, medullary cells thin-walled. Rhizoids mainly on rhizomes. Leaves imbricate, fragmenting, triangular to oblong-ovate, apex rounded or truncate, with 4−5 teeth, dorsal margin falcate, incurved, entire, base long decurrent, ventral margin straight or undulating, with sparingly teeth, base decurrent, strongly ampliate; teeth 5−8 per leaf, 2−4 cells long, 1−3 cells wide. Leaf cells slightly thick-walled, trigones large, simple-triangulate or nodulate, intermediate thickenings occasionally present. Cuticle smooth. Oil bodies segmented, 4−7 per median cell. Underleaves vestigial, but large, lamelliform, with ciliate margin. Paraphyllia absent. Androecia and gynoecia not seen. Propagules usually occurring on fragmenting leaves.

Habitat: On rocks, fallen trees, rotten logs and tree bases. Elev. 1115−1440 m.

Distribution in Yachang: Huangjingdong Tiankeng, Lanjiawan Tiankeng, Panguwang, Xiapeng Tun, Xuanya Tiankeng.

Distribution in China: Anhui, Fujian, Gansu, Guangdong, Guangxi, Guizhou, Hainan, Hunan, Shaanxi, Sichuan, Taiwan, Tibet, Hongkong, Yunnan, Zhejiang.

▶ A. 种群；B. 植物体（背面观）；C. 植物体一段（背面观）；D. 侧叶；E. 茎横切面；F. 叶中部细胞，示油体；G. 茎一段（背面观）；H. 茎一段（腹面观）；I. 腹叶。（凭证标本：*韦玉梅等 191019-396*）

A. Population; B. Plant (dorsal view); C. Portion of plant (dorsal view); D. Leaves; E. Transverse section of stem; F. Median cells of leaf showing oil bodies; G. Portion of stem (dorsal view); H. Portion of stem (ventral view); I. Underleaves. (All from *Wei et al. 191019-396*)

图 64　圆头羽苔

Fig. 64　*Plagiochila parvifolia* Lindenb.

刺叶羽苔

Plagiochila sciophila Nees ex Lindenb., Sp. Hepat. (Lindenberg) (fasc. 2-4): 100. 1840.

植物体绿色，带叶宽 2.5—5.5 mm，具横茎，分枝多，间生型。茎横切面 23—25 个细胞厚，皮部细胞 2—3 层，厚壁，内部细胞薄壁。假根主要分布于横茎。侧叶覆瓦状排列，有时易脱落，椭圆状卵形，顶端截形，具 2—4 个齿，背侧边缘直，稍内弯，全缘，基部短下延，不达茎中间位置，腹侧边缘略直或稍拱起，具稀疏的齿，基部几乎不下延；全叶具 4—6 个齿，长齿状，在不同的群落里，齿的数目变化大，少的 4 个，多的可达 15 个。侧叶细胞薄壁，角质层平滑，三角体小，不明显，中部球状加厚缺。油体聚合型，每个细胞 4—9 个。腹叶退化，丝状。无鳞毛。雄苞和雌苞未见。无性芽未见。

生境：石生环境常见，土生、钙华基质、腐木、树干、树基、岩面薄土生环境也有分布。海拔 503—1834 m。

雅长分布：草王山，大宴坪竖井，大宴坪天坑漏斗，吊井天坑，黄猄洞天坑，拉洞天坑，拉雅沟，蓝家湾天坑，里郎天坑，逻家田屯，盘古王，深洞，下棚屯，香当天坑，一沟，中井屯，中井屯竖井。

中国分布：全国大部分省区均有分布。

Plants green when fresh, 2.5–5.5 mm wide, with creeping rhizome, frequently branched, branches lateral-intercalary. Stem in transverse section 23–25 cells thick, cortical cells of 2–3 layers, thick-walled, medullary cells thin-walled. Rhizoids mainly on rhizomes. Leaves imbricate, sometimes caducous, oblong-ovate, apex truncate, with 2–4 teeth, dorsal margin straight, slightly incurved, entire, base short decurrent, not reach stem midline, ventral margin straight or slightly arched, sparingly toothed, base hardly decurrent; teeth 4–6 per leaf, long spinose, number of teeth in different populations variable, from as few as 4 to as many as 15 per leaf. Leaf cells thin-walled, trigones small, indistinct, intermediate thickenings absent. Cuticle smooth. Oil bodies segmented, 4–9 per median cell. Underleaves vestigial, filiform. Paraphyllia absent. Androecia and gynoecia not seen. Propagules not seen.

Habitat: Often on rocks, sometimes on soil, calcareous substrates, rotten logs, tree trunks, tree bases and on rocks with a thin layer of soil. Elev. 503–1834 m.

Distribution in Yachang: Caowangshan, Dayanping Tun, Dayanping Tiankeng, Diaojing Tiankeng, Huangjingdong Tiankeng, Ladong Tiankeng, Layagou, Lanjiawan Tiankeng, Lilang Tiankeng, Luojiatian Tun, Panguwang, Shendong, Xiapeng Tun, Xiangdang Tiankeng, Yigou, Zhongjing Tun.

Distribution in China: Widely distributed in most provinces of China.

▶ A. 种群；B. 植物体（背面观）；C. 植物体一段（背面观）；D. 侧叶；E. 腹叶；F. 叶中部细胞，示油体；G. 茎横切面。（凭证标本：*韦玉梅等 201105-46*）

A. Population; B. Plant (dorsal view); C. Portion of plant (dorsal view); D. Leaves; E. Underleaf; F. Median cells of leaf showing oil bodies; G. Transverse section of stem. (All from *Wei et al. 201105-46*)

图 65 刺叶羽苔

Fig. 65 *Plagiochila sciophila* Nees ex Lindenb.

大耳羽苔

Plagiochila subtropica Steph., Bull. Soc. Roy. Bot. Belgique 38(1): 46. 1899.

植物体绿色至棕绿色，带叶宽 2.1—4.0 mm，具横茎，分枝多，以顶生型为主。茎横切面 20—23 个细胞厚，皮部细胞 2—3 层，厚壁，内部细胞薄壁。假根主要分布于横茎。侧叶覆瓦状排列，易碎，三角状卵形，顶端圆或截形，具 3—4 个齿，背侧边缘镰刀形弯曲，稍内弯，中部以上具 2—4 个齿，基部长下延，腹侧边缘拱起，具稀疏的齿，近基部明显膨大，基部下延；全叶具 11—17 个齿，中上部齿多为长纤毛状，中下部多为长刺状。侧叶细胞稍厚壁，角质层平滑，三角体小至大，简单三角形或肿胀形，中部球状加厚少见。油体聚合型，每个细胞 3—6 个。腹叶退化，丝状。无鳞毛。雄苞和雌苞未见。无性芽常见于破碎的叶组织上。

生境： 石生，树干，树基。海拔 1312—1759 m。

雅长分布： 草王山。

中国分布： 广西，云南。

Plants green to brownish green when fresh, 2.1−4.0 mm wide, with creeping rhizome, frequently branched, branches predominantly terminal. Stem in transverse section 20−23 cells thick, cortical cells of 2−3 layers, thick-walled, medullary cells thin-walled. Rhizoids mainly on rhizomes. Leaves imbricate, fragmenting, triangular-ovate, apex rounded or truncate, with 3−4 teeth, dorsal margin falcate, slightly incurved, with 2−4 teeth from middle to apex, base long decurrent, ventral margin arched, sparingly toothed throughout the margin, base decurrent, strongly ampliate; teeth 11−17 per leaf, long ciliate in the upper part, long spinose in the lower part. Leaf cells slightly thick-walled, trigones small to large, simple-triangulate or nodulate, intermediate thickenings scarce. Cuticle smooth. Oil bodies segmented, 3−6 per median cell. Underleaves vestigial, filiform. Paraphyllia absent. Androecia and gynoecia not seen. Propagules usually occurring on fragmenting leaves.

Habitat: On rocks, tree trunks and tree bases. Elev. 1312−1759 m.

Distribution in Yachang: Caowangshan.

Distribution in China: Guangxi, Yunnan.

▶ A. 种群；B. 植物体（背面观）；C. 植物体一段（背面观）；D. 侧叶；E. 茎一段（背面观）；F. 茎一段（腹面观）；G. 茎横切面；H. 叶中部细胞，示油体。（凭证标本：*韦玉梅等 191013-108*）

A. Population; B. Plant (dorsal view); C. Portion of plant (dorsal view); D. Leaves; E. Portion of stem (dorsal view); F. Portion of stem (ventral view); G. Transverse section of stem; H. Median cells of leaf showing oil bodies. (All from *Wei et al. 191013-108*)

图 66 大耳羽苔

Fig. 66 *Plagiochila subtropica* Steph.

短齿羽苔

Plagiochila vexans Schiffn. ex Steph., Sp. Hepat. 6: 237. 1921.

植物体绿色至棕绿色，带叶宽 1.3—3.6 mm，横茎未见，稀疏分枝，间生型，有时具向地性鞭状枝。茎横切面 16—19 个细胞厚，皮部细胞 3—4 层，厚壁，内部细胞薄壁。假根少。侧叶覆瓦状排列，易碎，椭圆状卵形，顶端圆，具齿，背侧边缘直或稍拱起，稍内弯，近顶部具短齿，基部长下延，腹侧边缘拱起，具密集的齿，基部明显下延；全叶具 24—30 个齿，齿刺状，1—4 个细胞长，1—3 个细胞宽。侧叶细胞稍厚壁，角质层平滑，三角体小至大，简单三角形或肿胀形，中部球状加厚偶尔可见，基部细胞加长形成明显的假肋区。油体聚合型，每个细胞 3—7 个。腹叶退化，丝状。鳞毛未见。雄苞未见。雌苞顶生，具 1—2 个新生枝。蒴萼圆柱形，脊无翅，口部具不规则刺状长齿。无性芽常见于叶腹面。

生境： 石生。海拔 1893 m。

雅长分布： 草王山。

中国分布： 安徽，重庆，福建，广西，贵州，江西，四川，台湾，西藏，云南，浙江。

Plants green to brownish green when fresh, 1.3–3.6 mm wide, creeping rhizome not seen, sparingly branched, branches exclusively lateral-intercalary, sometimes with geotropic flagella. Stem in transverse section 16–19 cells thick, cortical cells of 3–4 layers, thick-walled, medullary cells thin-walled. Rhizoids scarce. Leaves imbricate, fragmenting, oblong-ovate, apex rounded, toothed, dorsal margin straight or weakly arched, slightly incurved, with short teeth near apex, base long decurrent, ventral margin arched, densely toothed throughout the margin, base distinctly decurrent; teeth 24–30 per leaf, spinose, 1–4 cells long, 1–3 cells wide. Leaf cells slightly thick-walled, trigones small to large, simple-triangulate or nodulate, intermediate thickenings occasionally present, basal cells elongate forming a distinct vitta area. Cuticle smooth. Oil bodies segmented, 3–7 per median cell. Underleaves vestigial, filiform. Paraphyllia not seen. Androecia not seen. Gynoecia terminal, with 1–2 innovations. Perianths cylindric, keels unwing, mouth irregularly spinosely toothed. Propagules usually occurring on ventral leaf surface.

Habitat: On rocks. Elev. 1893 m.

Distribution in Yachang: Caowangshan.

Distribution in China: Anhui, Chongqing, Fujian, Guangxi, Guizhou, Jiangxi, Sichuan, Taiwan, Tibet, Yunnan, Zhejiang.

▶ A. 种群；B. 植物体（背面观）；C. 侧叶；D. 蒴萼口部的齿；E. 茎横切面；F. 植物体一段（背面观）；G. 植物体一段（腹面观）；H. 蒴萼；I. 叶基部细胞；J. 叶中部细胞，示油体。（凭证标本：*唐启明等* 20210719-14）

A. Population; B. Plant (dorsal view); C. Leaves; D. Teeth of perianth mouth; E. Transverse section of stem; F. Portion of plant (dorsal view); G. Portion of plant (ventral view); H. Perianth, I. Basal cells of leaf; J. Median cells of leaf showing oil bodies. (All from *Tang et al.* 20210719-14)

图 67　短齿羽苔

Fig. 67　*Plagiochila vexans* Schiffn. ex Steph.

韦氏羽苔

Plagiochila wightii Nees ex Lindenb., Sp. Hepat. (Lindenberg) (fasc. 2-4): 43. 1840.

植物体浅绿色，带叶宽 2.8—3.8 mm，具横茎，分枝少，间生型或顶生型。茎横切面约 15 个细胞厚，皮部细胞 2—3 层，厚壁，内部细胞薄壁。假根主要分布于横茎上。侧叶覆瓦状排列，易碎，椭圆状卵形，顶端截形或斜截形，具 4—6 个齿，背侧边缘略直，稍内弯，全缘或近顶部具 2—4 齿，基部长下延，腹侧边缘拱起，具密集或稀疏的齿，近基部明显膨大，基部轻微下延；全叶齿的数目变化大，12—29 个不等，刺状，2—6 个细胞长，1—3 个细胞宽。侧叶细胞薄壁，角质层平滑，三角体小，简单三角形，中部球状加厚缺。油体均一型，每个细胞 6—10 个。腹叶退化，丝状。鳞毛未见。雄苞和雌苞未见。无性芽常见于叶腹面。

生境： 潮湿石生，石生，倒木，腐木，树干，树基。海拔 1083—1342 m。

雅长分布： 大宴坪天坑漏斗，黄猄洞天坑，蓝家湾天坑，逻家田屯，下棚屯，中井屯。

中国分布： 安徽，广西，四川，云南。

Plants light green when fresh, 2.8–3.8 mm wide, with creeping rhizome, rarely branched, branches terminal or lateral-intercalary. Stem in transverse section ca. 15 cells thick, cortical cells of 2–3 layers, thick-walled, medullary cells thin-walled. Rhizoids mainly on rhizomes. Leaves imbricate, fragmenting, oblong-ovate, apex truncate or oblique truncate, with 4–6 teeth, dorsal margin mostly straight, slightly incurved, entire or with 2–4 teeth near apex, base long decurrent, ventral margin arched, densely toothed throughout the margin, base weakly decurrent, ampliate; teeth variable in number, 12–29 per leaf, spinose, 2–6 cells long, 1–3 cells wide. Leaf cells thin-walled, trigones small, simple-triangulate, intermediate thickenings absent. Cuticle smooth. Oil bodies homogeneous, 6–10 per median cell. Underleaves vestigial, filiform. Paraphyllia not seen. Androecia and gynoecia not seen. Propagules usually occurring on ventral leaf surface.

Habitat: On (wet) rocks, fallen trees, rotten logs, tree trunks and tree bases. Elev. 1083–1342 m.

Distribution in Yachang: Dayanping Tiankeng, Huangjingdong Tiankeng, Lanjiawan Tiankeng, Luojiatian Tun, Xiapeng Tun, Zhongjing Tun.

Distribution in China: Anhui, Guangxi, Sichuan, Yunnan.

▶ A. 种群；B. 植物体（背面观）；C. 植物体带无性芽（腹面观）；D. 侧叶；E. 茎一段（背面观）；F. 茎一段（腹面观）；G. 叶中部细胞，示油体；H. 茎横切面；I. 无性芽。（凭证标本：*韦玉梅等 201105-52*）

A. Population; B. Plant (dorsal view); C. Plant with propagules (ventral view); D. Leaves; E. Portion of stem (dorsal view); F. Portion of stem (ventral view); G. Median cells of leaf showing oil bodies; H. Transverse section of stem; I. Propagules. (All from *Wei et al. 201105-52*)

图 68 韦氏羽苔

Fig. 68 *Plagiochila wightii* Nees ex Lindenb.

裂萼苔

Chiloscyphus polyanthos (L.) Corda, Naturalientausch 12 [Opiz, Beitr. Naturgesch.]: 651. 1829.

植物体干时暗棕色，带叶宽 1.85—2.25 mm，不规则分枝，耳叶苔型或间生型。茎横切面 9—11 个细胞厚，内外细胞不明显分化，薄壁。假根成束生于腹叶基部。侧叶覆瓦状排列，椭圆形状卵形，顶端圆或近平截，边缘全缘，背侧基部略下延。侧叶细胞薄壁，角质层平滑，三角体小，不明显，中部球状加厚缺。油体未见。腹叶远生，与茎等宽或略大于茎，顶端 2 裂 1/2—2/3，两侧边缘各具 1 个小齿，基部不与侧叶基部相连。雄苞和雌苞未见。

生境： 石生。海拔 1533 m。
雅长分布： 逻家田屯。
中国分布： 全国范围广布。

Plants dark brown in herbarium specimens, 1.85–2.25 mm wide, irregularly branched, branches *Frullania*-type or intercalary. Stem in transverse section 9–11 cells thick, thin-walled, without differentiated cortex. Rhizoids in tufts from underleaf bases. Leaves imbricate, oblong-ovate, apex rounded to subtruncate, margin entire, dorsal base slightly decurrent. Leaf cells thin-walled, trigones small, indistinct, intermediate thickenings absent. Cuticle smooth. Oil bodies not seen. Underleaves distant, as wide as the stem or slightly wider, bilobed to 1/2–2/3 underleaf length, lateral margins usually with 1 small tooth, base not connate with leaf bases. Androecia and gynoecia not seen.

Habitat: On rock. Elev. 1533 m.
Distribution in Yachang: Luojiatian Tun.
Distribution in China: Widely distributed in China.

▶ A. 植物体（背面观）；B. 侧叶；C. 腹叶；D. 侧叶顶部和中部细胞。（凭证标本：*唐启明等 20190521-428*）

A. Plant (dorsal view); B. Leaves; C. Underleaves; D. Leaf apical and median cells. (All from *Tang et al. 20190521-428*)

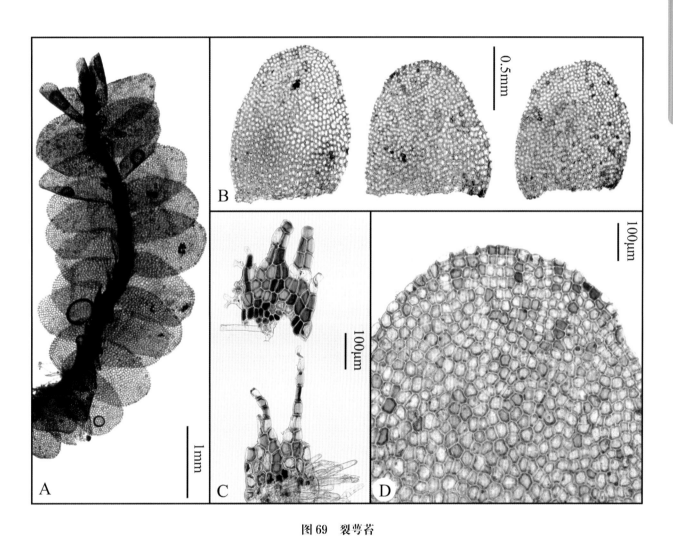

图 69 裂萼苔

Fig. 69 *Chiloscyphus polyanthos* (L.) Corda

四齿异萼苔

Heteroscyphus argutus (Reinw., Blume et Nees) Schiffn., Österr. Bot. Z. 60(5): 172. 1910.

植物体浅绿色，带叶宽 1.85—3.10 mm，不规则分枝，间生型。茎横切面 10—11 个细胞厚，内外细胞不明显分化，薄壁。假根生于腹叶基部。侧叶毗邻至疏松覆瓦状排列，圆长方形，顶端平截或近圆，具 4—8 个齿，齿 2—4 个细胞长，基部 1—2 个细胞宽，侧缘全缘，腹侧和背侧基部均略下延。侧叶细胞薄壁，角质层平滑，三角体小，不明显，中部球状加厚缺。油体聚合型，每个细胞 3—8 个。腹叶远生，宽为茎的 1—2 倍，顶端深 2 裂几乎达基部，两侧边缘各具 1 个长齿，通常基部一侧与侧叶基部相连，偶尔两侧相连。雌雄异株。雄苞生于侧短枝上，雄苞叶 4—10 对，雄苞腹叶生于整个雄苞。雌苞于侧短枝上。蒴萼钟形，无脊，口部具 3 片大裂瓣，大裂瓣通常分裂成小的裂瓣，小裂瓣边缘具长刺状齿。孢蒴球形，成熟时 4 瓣开裂。孢子球形，表面具蠕虫状纹饰。弹丝 2 螺旋加厚。

生境： 石生、土生、腐木生境常见，树干和树基环境也有分布。海拔 474—1531 m。

雅长分布： 吊井天坑，黄猄洞天坑，拉雅沟，蓝家湾天坑，盘古王，深洞，二沟，一沟，中井天坑，中井屯竖井。

中国分布： 除了东北以外的全国大部分省区均有分布。

Plants light green when fresh, 1.85–3.10 mm wide, irregularly branched, branches intercalary. Stem in transverse section 10–11 cells thick, thin-walled, without differentiated cortex. Rhizoids in tufts from underleaf bases. Leaves contiguous to loosely imbricate, rounded-rectangular, apex truncate to subrouned, 4–8-toothed, teeth 2–4 cells long, 1–2 cells wide at base, lateral margin entire, ventral and dorsal base slightly decurrent. Leaf cells thin-walled, trigones small, indistinct, intermediate thickenings absent. Cuticle smooth. Oil bodies segmented, 3–8 per median cell. Underleaves distant, 1–2 times as wide as the stem, mostly bilobed to the base, lateral margins usually with 1 long tooth, base narrowly connate on one side of leaf base, occasionally both sides. Androecia on short lateral branches, bracts in 4–10 pairs, bracteoles present throughout the androecium. Gynoecia on short lateral branches. Perianths campanulate, without keels, mouth 3-lobed, lobes further irregularly lobed, margins with long spinose teeth. Capsules spherical, longitudinally dehiscing by 4 regular valves at maturity. Spores globose, surface vermiculate. Elaters with 2 spiral thickenings.

Habitat: Often on rocks, soil and rotten logs, sometimes on tree trunks and tree bases. Elev. 474–1531 m.

Distribution in Yachang: Diaojing Tiankeng, Huangjingdong Tiankeng, Layagou, Lanjiawan Tiankeng, Panguwang, Shendong, Ergou, Yigou, Zhongjing Tiankeng, Zhongjing Tun.

Distribution in China: Widely distributed in most provinces of China except Northeast Region.

▶ A. 种群；B. 植物体一段；C. 侧叶；D. 植物体一段带蒴萼；E. 植物体一段，示腹叶；F. 叶中部细胞，示油体；G. 茎横切面；H. 孢子；I. 叶顶端的齿；J. 弹丝。（凭证标本：*韦玉梅等 201104-54*）

A. Population; B. Portion of plant; C. Leaves; D. Portion of plant with a perianth; E. Portion of plant showing underleaves; F. Median cells of leaf showing oil bodies; G. Transverse section of stem; H. Spores; I. Teeth of leaf apex; J. Elaters. (All from *Wei et al. 201104-54*)

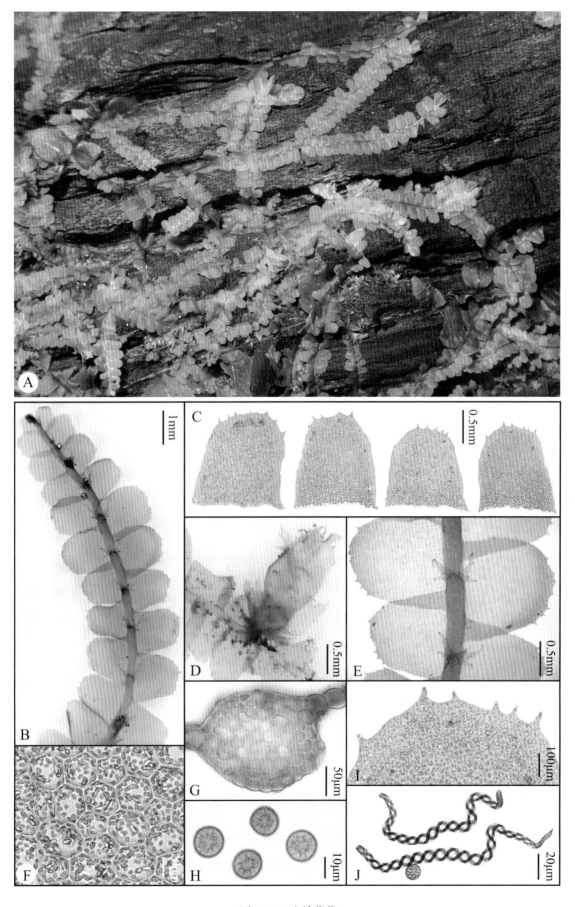

图 70 四齿异萼苔

Fig. 70 *Heteroscyphus argutus* (Reinw., Blume & Nees) Schiffn.

双齿异萼苔

Heteroscyphus coalitus (Hook.) Schiffn., Österr. Bot. Z. 60(5): 172. 1910.

植物体浅绿色，带叶宽 3.62—4.05 mm，不规则分枝，间生型。茎横切面 8—9 个细胞厚，内外细胞不明显分化，薄壁。假根生于腹叶基部。侧叶覆瓦状排列，长方形，顶端平截，具 2 个边缘齿，齿 3—5 个细胞长，基部 2—4 个细胞宽，侧缘全缘，腹侧和背侧基部均略下延。侧叶细胞薄壁，角质层平滑，三角体小，不明显，中部球状加厚缺。油体聚合型，每个细胞 2—5 个。腹叶远生，宽为茎的 1—2 倍，顶端 2 裂约达 2/3，两侧边缘各具 1 个长齿，基部两侧与侧叶基部相连。雄苞和雌苞未见。

生境： 石生环境常见，土生、腐木、岩面薄土生环境也有分布。海拔 828—1678 m。

雅长分布： 拉洞天坑，拉雅沟，蓝家湾天坑，李家坨屯，逻家田屯，盘古王，深洞，霄罗湾洞穴，中井屯竖井。

中国分布： 全国范围广布。

Plants light green when fresh, 3.62−4.05 mm wide, irregularly branched, branches intercalary. Stem in transverse section 8−9 cells thick, thin-walled, without differentiated cortex. Rhizoids in tufts from underleaf bases. Leaves imbricate, rectangular, apex truncate, with 2 marginal teeth, teeth 3−5 cells long, 2−4 cells wide at base, lateral margin entire, ventral and dorsal base slightly decurrent. Leaf cells thin-walled, trigones small, indistinct, intermediate thickenings absent. Cuticle smooth. Oil bodies segmented, 2−5 per median cell. Underleaves distant, 1−2 times as wide as the stem, bilobed to 2/3 underleaf length, lateral margins usually with 1 long tooth, base connate on both sides of leaf base. Androecia and gynoecia not seen.

Habitat: Often on rocks, sometimes on soil, rotten logs and on rocks with a thin layer of soil. Elev. 828−1678 m.

Distribution in Yachang: Ladong Tiankeng, Layagou, Lanjiawan Tiankeng, Lijiatuo Tun, Luojiatian Tun, Panguwang, Shendong, Xiaoluowan Cave, Zhongjing Tun.

Distribution in China: Widely distributed in China.

▶ A. 种群；B. 植物体；C. 植物体一段；D. 侧叶；E. 叶顶端；F. 茎一段，示腹叶；G. 茎横切面；H. 叶中部细胞，示油体。（凭证标本：韦玉梅等 *201110-14*）

A. Population; B. Plant; C. Portion of plant; D. Leaves; E. Apex of leaf; F. Portion of stem showing underleaves; G. Transverse section of stem; H. Median cells of leaf showing oil bodies. (All from *Wei et al. 201110-14*)

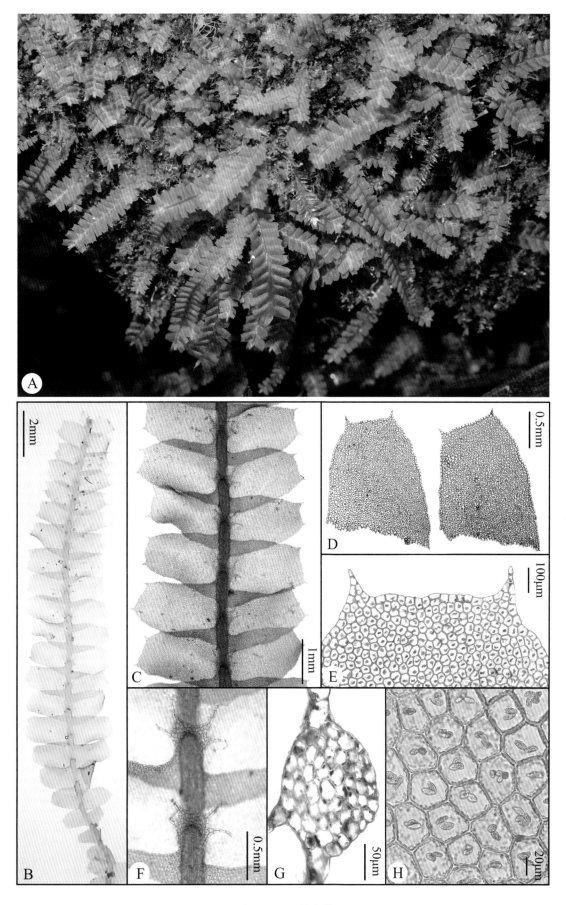

图 71 双齿异萼苔

Fig. 71 *Heteroscyphus coalitus* (Hook.) Schiffn.

平叶异萼苔

Heteroscyphus planus (Mitt.) Schiffn., Österr. Bot. Z. 60(5): 171. 1910.

植物体浅绿色，带叶宽 1.55—2.06 mm，不规则分枝，间生型。茎横切面 10—11 个细胞厚，内外细胞不明显分化，薄壁。假根生于腹叶基部。侧叶覆瓦状排列，近长方形，顶端平截，具 2—4 个不规则齿，齿 3—6 个细胞长，基部 2—4 个细胞宽，侧缘全缘，腹侧和背侧基部均略下延。侧叶细胞稍厚壁，角质层平滑，三角体小，不明显，中部球状加厚缺。油体聚合型。腹叶远生，宽为茎的 1.0—1.5 倍，顶端 2 裂约达 2/3，两侧边缘各具 1 个长或短齿，通常基部一侧与侧叶基部相连，偶尔两侧相连。雄苞和雌苞未见。

生境：土生。海拔 1136—1278 m。
雅长分布：黄猄洞天坑，中井屯。
中国分布：全国大部分省区均有分布。

Plants light green when fresh, 1.55−2.06 mm wide, irregularly branched, branches intercalary. Stem in transverse section 10−11 cells thick, thin-walled, without differentiated cortex. Rhizoids in tufts from underleaf bases. Leaves imbricate, subrectangular, apex truncate, with 2−4 irregular teeth, teeth 3−6 cells long, 2−4 cells wide at base, lateral margin entire, ventral and dorsal base slightly decurrent. Leaf cells slightly thick-walled, trigones small, indistinct, intermediate thickenings absent. Cuticle smooth. Oil bodies segmented. Underleaves distant, 1.0−1.5 times as wide as the stem, bilobed to 2/3 underleaf length, lateral margins usually with 1 long or short tooth, base narrowly connate on one side of leaf base, occasionally both sides. Androecia and gynoecia not seen.

Habitat: On soil. Elev. 1136−1278 m.
Distribution in Yachang: Huangjingdong Tiankeng, Zhongjing Tun.
Distribution in China: Widely distributed in most provinces of China.

▶ A. 种群；B. 植物体；C. 植物体一段；D. 侧叶；E. 茎一段，示腹叶；F. 茎横切面；G. 叶中部细胞。（凭证标本：*韦玉梅等 191012-61*）

A. Population; B. Plant; C. Portion of plant; D. Leaves; E. Portion of stem showing underleaves; F. Transverse section of stem; G. Median cells of leaf. (All from *Wei et al. 191012-61*)

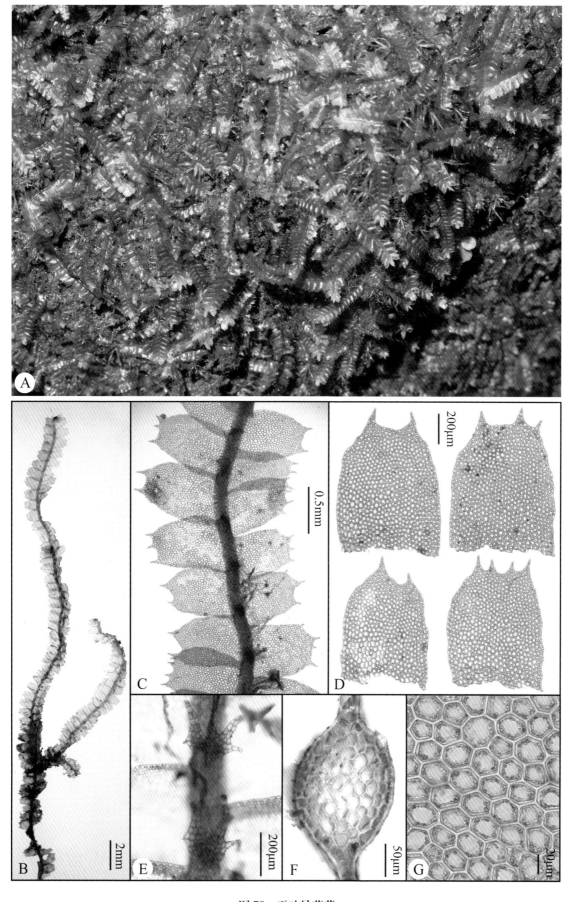

图 72 平叶异萼苔

Fig. 72 *Heteroscyphus planus* (Mitt.) Schiffn.

南亚异萼苔

Heteroscyphus zollingeri (Gottsche) Schiffn., Österr. Bot. Z. 60(5): 171. 1910.

植物体浅绿色，带叶宽 1.49—2.11 mm，不规则分枝，间生型。茎横切面 10—12 个细胞厚，内外细胞不明显分化，薄壁。假根生于腹叶基部。侧叶覆瓦状排列，近长方形，顶端近平截或圆，全缘或具 1—3 个小齿，齿 2—3 个细胞长，基部 1—2 个细胞宽，侧缘全缘，腹侧和背侧基部均略下延。侧叶细胞薄壁，角质层平滑，三角体小，不明显，中部球状加厚缺。油体未见。腹叶远生，宽为茎的 1.0—1.5 倍，顶端 2 裂几达基部，两侧边缘各具 1 个短齿，通常基部一侧与侧叶基部相连，偶尔两侧相连。雄苞和雌苞未见。

生境：石生，土生，腐木。海拔 1348 m。

雅长分布：逻家田屯。

中国分布：除了东北以外的全国大部分省区均有分布。

Plants light green when fresh, 1.49−2.11 mm wide, irregularly branched, branches intercalary. Stem in transverse section 10−12 cells thick, thin-walled, without differentiated cortex. Rhizoids in tufts from underleaf bases. Leaves imbricate, subrectangular, apex subtruncate or rounded, entire or with 1−3 small teeth, teeth 2−3 cells long, 1−2 cells wide at base, lateral margin entire, ventral and dorsal base slightly decurrent. Leaf cells thin-walled, trigones small, indistinct, intermediate thickenings absent. Cuticle smooth. Oil bodies not seen. Underleaves distant, 1.0−1.5 times as wide as the stem, mostly bilobed to the base, lateral margins usually with 1 short tooth, base narrowly connate on one side of leaf base, occasionally both sides. Androecia and gynoecia not seen.

Habitat: On rocks, soil and rotten logs. Elev. 1348 m.

Distribution in Yachang: Luojiatian Tun.

Distribution in China: Widely distributed in most provinces of China except Northeast Region.

▶ A. 种群；B. 植物体；C. 植物体一段；D. 侧叶；E. 茎一段，示腹叶；F. 茎横切面；G. 叶中部细胞。（凭证标本：*唐启明*等 *20210721-113*）

A. Population; B. Plant; C. Portion of plant; D. Leaves; E. Portion of stem showing underleaves; F. Transverse section of stem; G. Median cells of leaf. (All from *Tang et al. 20210721-113*)

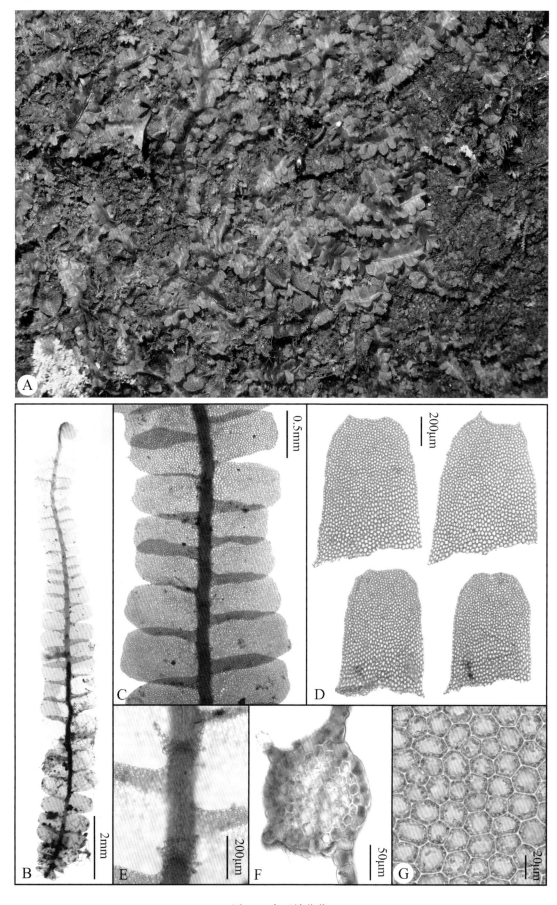

图 73 南亚异萼苔

Fig. 73 *Heteroscyphus zollingeri* (Gottsche) Schiffn.

尖叶齿萼苔（尖叶裂萼苔）

Lophocolea bidentata (L.) Dumort., Recueil Observ. Jungerm.: 17. 1835.

Chiloscyphus cuspidatus (Nees) J. J. Engel & R. M. Schust., Nova Hedwigia 39: 413. 1984.

植物体黄绿色至绿色，带叶宽 1.36—1.82 mm，不规则分枝，耳叶苔型或间生型。茎横切面 9—11 个细胞厚，内外细胞不明显分化，薄壁。假根成束生于腹叶基部。侧叶覆瓦状排列，椭圆状卵形或圆方形，顶端 2 裂约达叶长的 1/3，裂瓣长细尖，弯缺半月形，边缘全缘，背侧基部略下延。侧叶细胞薄壁，角质层平滑，三角体小，不明显，中部球状加厚缺。油体聚合型，每个细胞 3—7 个。腹叶远生，宽为茎的 1.5—2.0 倍，顶端 2 裂约达 2/3，两侧边缘各具 1 个长齿，基部不与侧叶基部相连。雌雄同株。雄苞顶生，雄苞叶 10—17 对，雄苞腹叶生于整个雄苞。雌苞顶生，雌苞叶与侧叶同形，略大。蒴萼三棱形，具 3 个脊，脊具窄的翅，口部具 3 阔裂瓣，裂瓣边缘具不规则纤毛状齿。

生境： 腐木和石生环境常见，潮湿土生、土生、树基环境也有分布。海拔 491—1829 m。

雅长分布： 草王山，吊井天坑，二沟，黄猄洞天坑，拉洞天坑，老屋基天坑，盘古王，旁墙屯，全达村，深洞，悬崖天坑，一沟。

中国分布： 重庆，甘肃，广东，广西，贵州，河北，湖北，吉林，江西，陕西，山西，四川，台湾，西藏，云南，浙江。

Plants yellowish green to green when fresh, 1.36–1.82 mm wide, irregularly branched, branches *Frullania*-type or intercalary. Stem in transverse section 9–11 cells thick, thin-walled, without differentiated cortex. Rhizoids in tufts from underleaf bases. Leaves imbricate, oblong-ovate to rounded-quadrate, apex bilobed to ca. 1/3 leaf length, lobes longly acuminate, sinus broadly lunate, margin entire, dorsal base slightly decurrent. Leaf cells thin-walled, trigones small, indistinct, intermediate thickenings absent. Cuticle smooth. Oil bodies segmented, 3–7 per median cell. Underleaves distant, 1.5–2.0 times as wide as the stem, bilobed to ca. 2/3 underleaf length, lateral margins usually with 1 long tooth, base not connate with leaf bases. Monoicous. Androecia terminal, bracts in 10–17 pairs, bracteoles present throughout the androecium. Gynoecia terminal, Bracts resembling the leaves, but larger. Perianths trigonous, 3-keeled, keels with narrow wings, mouth 3-lobed, lobes with irregularly ciliate teeth.

Habitat: Often on rotten logs and rocks, sometimes on (wet) soil and tree bases. Elev. 491–1829 m.

Distribution in Yachang: Caowangshan, Diaojing Tiankeng, Ergou, Huangjingdong Tiankeng, Ladong Tiankeng, Laowuji Tiankeng, Panguwang, Pangqiang Tun, Quanda Village, Shendong, Xuanya Tiankeng, Yigou.

Distribution in China: Chongqing, Gansu, Guangdong, Guangxi, Guizhou, Hebei, Hubei, Jilin, Jiangxi, Shaanxi, Shanxi, Sichuan, Taiwan, Tibet, Yunnan, Zhejiang.

▶ A 种群；B. 植物体；C. 植物体一段；D. 侧叶；E. 腹叶；F. 蒴萼横切面；G. 雌苞和雄苞；H. 叶中部细胞，示油体。（凭证标本：*韦玉梅等 201108-12*）

A. Population; B. Plant; C. Portion of plant; D. Leaves; E. Underleaves; F. Transverse section of perianth; G. Androeciaum and gynoecium; H. Median cells of leaf showing oil bodies. (All from *Wei et al. 201108-12*)

图 74 尖叶齿萼苔

Fig. 74 *Lophocolea bidentata* (L.) Dumort.

拟异叶齿萼苔（新拟中文名）

Lophocolea concreta Mont., Ann. Sci. Nat. Bot. (sér. 3)4: 350. 1845.

植物体浅绿色，带叶宽 1.00—1.82 mm，不规则分枝，耳叶苔型或间生型。茎横切面 7—8 个细胞厚，内外细胞不明显分化，薄壁。假根成束生于腹叶基部。侧叶覆瓦状排列，椭圆形或圆方形，顶端平截，不裂、微凹或浅 2 裂，湿时内弯，干时强烈内卷，边缘全缘或稍具细圆齿，背侧基部略下延。侧叶细胞薄壁，角质层平滑，三角体小至大，简单三角形或偶尔肿胀形，中部球状加厚缺。油体聚合型，每个细胞 2—5 个。腹叶远生，宽为茎的 1.5—2.0 倍，顶端 2 裂约达 2/3，两侧边缘各具 1 个长齿，基部不与侧叶基部相连。雌雄同株同苞。雄苞生于雌苞下，雄苞叶 2—4 对，雄苞腹叶生于整个雄苞。雌苞顶生，雌苞叶与侧叶同形，略大。蒴萼三棱形，具 3 个脊，脊具窄的翅，口部具 3 阔裂瓣，裂瓣边缘具不规则长裂片状齿。无性芽常生于叶边缘。

生境： 腐木和树基环境常见，土生环境也有分布。海拔 766—1361 m。

雅长分布： 白岩坨屯，大宴坪天坑漏斗，吊井天坑，黄猄洞天坑，拉雅沟，旁墙屯，悬崖天坑。

中国分布： 广西。

Plants yellowish green to green when fresh, 1.00–1.82 mm wide, irregularly branched, branches *Frullania*-type or intercalary. Stem in transverse section 7–8 cells thick, thin-walled, without differentiated cortex. Rhizoids in tufts from underleaf bases. Leaves imbricate, oblong to rounded-quadrate, apex truncate, entire, slightly retuse or weakly 2-lobed, incurved when moist, strongly involute when dry, margin entire or slightly crenate, dorsal base slightly decurrent. Leaf cells thin-walled, trigones small to large, simple-triangulate or nodulate, intermediate thickenings absent. Cuticle smooth. Oil bodies segmented, 2–5 per median cell. Underleaves distant, 1.5–2.0 times as wide as the stem, bilobed to ca. 2/3 underleaf length, lateral margins usually with 1 long tooth, base not connate with leaf bases. Paroicous. Androecia below the Gynoecium, bracts in 2–4 pairs, bracteoles present throughout the androecium. Gynoecia terminal, Bracts resembling the leaves, but larger. Perianths trigonous, 3-keeled, keels with narrow wings, mouth 3-lobed, lobes with irregularly laciniate teeth Propagules usually produced from leaf margins.

Habitat: Often on rotten logs and tree bases, sometimes on soil. Elev. 766–1361 m.

Distribution in Yachang: Baiyantuo Tun, Dayanping Tiankeng, Diaojing Tiankeng, Huangjingdong Tiankeng, Layagou, Pangqiang Tun, Xuanya Tiankeng.

Distribution in China: Guangxi.

▶ A. 种群；B. 植物体一段；C. 侧叶；D. 侧叶带无性芽；E. 植物体一段带雌苞和雄苞（背面观）；F. 蒴萼（展开）；G. 蒴萼横切面；H. 茎一段，示腹叶；I. 雌苞叶和雌苞腹叶；J. 叶中部细胞，示油体。（凭证标本：*韦玉梅等 201104-37*）

A. Population; B. Portion of plant; C. Leaves; D. Leaf with Propagules; E. Portion of plant with a gynoecium and an androecium; F. Perianth, open out; G. Transverse section of perianth; H. Portion of stem showing underleaves; I. Female bracts and bracteole; J. Median cells of leaf showing oil bodies. (All from *Wei et al. 201104-37*)

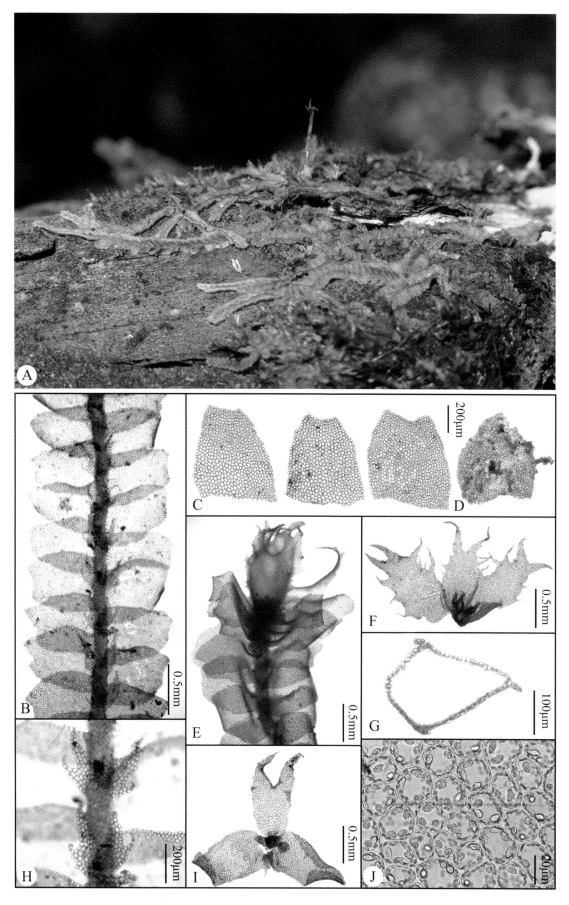

图 75 拟异叶齿萼苔

Fig. 75 *Lophocolea concreta* Mont.

疏叶齿萼苔（疏叶裂萼苔）

Lophocolea itoana Inoue, J. Jap. Bot. 31(11): 340. 1956.

Chiloscyphus itoanus (Inoue) J. J. Engel & R. M. Schust., Nova Hedwigia 39: 417. 1984.

植物体黄绿色至浅绿色，带叶宽 1.10—1.50 mm，不规则分枝，耳叶苔型或间生型。茎横切面约 5 个细胞厚，内外细胞不明显分化，薄壁。假根成束生于腹叶基部。侧叶毗邻至远生，椭圆状卵形或圆方形，顶端 2 裂约达叶长的 1/4，裂瓣三角形，弯缺圆钝，边缘全缘，背侧基部略下延。侧叶细胞薄壁，角质层平滑，三角体小，不明显，中部球状加厚缺。油体聚合型，每个细胞 2—5 个。腹叶远生，宽为茎的 1.0—1.5 倍，顶端 2 裂约达 2/3，两侧边缘各具 1 个长齿，基部不与侧叶基部相连或一侧与侧叶基部相连。雄苞和雌苞未见。芽胞生于侧叶裂瓣顶端。

生境： 腐木，钙华基质，枯枝。海拔 1235—1834 m。

雅长分布： 草王山，蓝家湾天坑，里郎天坑，逻家田屯，中井屯竖井。

中国分布： 福建，甘肃，广东，广西，贵州，湖南，吉林，江西，四川，云南。

Plants yellowish green to light green when fresh, 1.10−1.50 mm wide, irregularly branched, branches *Frullania*-type or intercalary. Stem in transverse section ca. 5 cells thick, thin-walled, without differentiated cortex. Rhizoids in tufts from underleaf bases. Leaves contigous to distant, oblong-ovate to rounded-quadrate, apex bilobed to ca. 1/4 leaf length, lobes triangular, sinus obtuse, margin entire, dorsal base slightly decurrent. Leaf cells thin-walled, trigones small, indistinct, intermediate thickenings absent. Cuticle smooth. Oil bodies segmented, 2−5 per median cell. Underleaves distant, 1.0−1.5 times as wide as the stem, bilobed to ca. 2/3 underleaf length, lateral margins usually with 1 long tooth, base not connate with leaf bases or narrowly connate on one side. Androecia and gynoecia not seen. Gemmae abundantly occurring on the apex of leaf lobes.

Habitat: On rotten logs, calcareous substrates and dead branches. Elev. 1235−1834 m.

Distribution in Yachang: Caowangshan, Lanjiawan Tiankeng, Lilang Tiankeng, Luojiatian Tun, Zhongjing Tun.

Distribution in China: Fujian, Gansu, Guangdong, Guangxi, Guizhou, Hunan, Jilin, Jiangxi, Sichuan, Yunnan.

A. 种群；B. 植物体一段；C. 侧叶；D. 叶中部细胞，示油体；E. 茎一段，示腹叶；F−G. 腹叶。（凭证标本：*韦玉梅等 191018-328*）

A. Population; B. Portion of plant; C. Leaves; D. Median cells of leaf showing oil bodies; E. Portion of stem showing underleaves; F−G. Underleaves. (All from *Wei et al. 191018-328*)

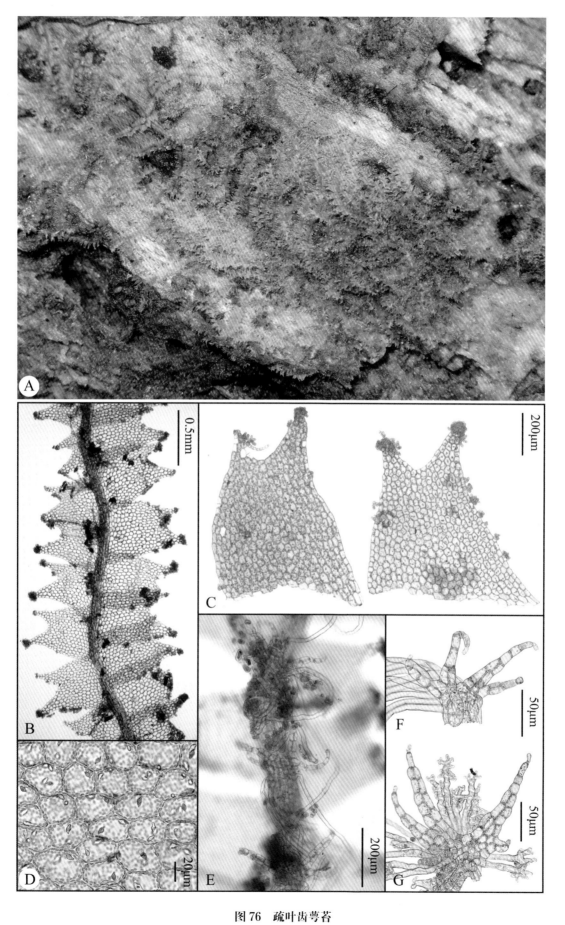

图 76 疏叶齿萼苔

Fig. 76 *Lophocolea itoana* Inoue

芽胞齿萼苔（芽胞裂萼苔）

Lophocolea minor Nees, Naturgesch. Eur. Leberm. 2: 330. 1836.

Chiloscyphus minor (Nees) J. J. Engel et R. M. Schust., Nova Hedwigia 39(3–4): 419. 1984.

植物体黄绿色至浅绿色，带叶宽 0.81—1.40 mm，不规则分枝，耳叶苔型或间生型。茎横切面 7—8 个细胞厚，内外细胞不明显分化，薄壁。假根成束生于腹叶基部。侧叶远生，偶尔毗邻，近方形或近长方形，顶端 2 裂约达叶长的 1/5，由于长有大量的芽胞，有时 2 裂不明显，裂瓣三角形，弯缺圆钝，边缘全缘，背侧基部略下延。侧叶细胞薄壁，角质层平滑，三角体小，不明显，中部球状加厚缺。油体聚合型，每个细胞 2—3 个。腹叶远生，宽为茎的 1.0—1.5 倍，顶端深 2 裂几乎达基部，两侧边缘各具 1 个长齿，基部不与侧叶基部相连。雌雄异株。雄苞未见。雌苞顶生，雌苞叶与侧叶同形，略大。蒴萼三棱形，具 3 个脊，脊具窄的翅，口部具 3 裂瓣，裂瓣边缘具不规则粗齿。孢蒴球形，成熟时 4 瓣开裂。孢子球形，表面具细疣。弹丝 2 螺旋加厚。芽胞常大量生于叶顶端以及蒴萼口部。

生境：土生和腐木环境常见，石生、树干、树基、岩面薄土生环境也有分布。海拔 590—1774 m。

雅长分布：广泛分布于保护区各处。

中国分布：全国范围广布。

Plants yellowish green to light green when fresh, 0.81−1.40 mm wide, irregularly branched, branches *Frullania*-type or intercalary. Stem in transverse section 7−8 cells thick, thin-walled, without differentiated cortex. Rhizoids in tufts from underleaf bases. Leaves distant, sometimes contigous, subquadrate to subrectangular, apex bilobed to ca. 1/5 leaf length, lobes triangular, sometimes indistinct due to formation of gemmae, sinus obtuse, margin entire, dorsal base slightly decurrent. Leaf cells thin-walled, trigones small, indistinct, intermediate thickenings absent. Cuticle smooth. Oil bodies segmented, 2−3 per median cell. Underleaves distant, 1.0−1.5 times as wide as the stem, deeply bilobed to the base, lateral margins usually with 1 long tooth, base not connate with leaf bases. Dioicous. Androecia not seen. Gynoecia terminal, Bracts resembling the leaves, but larger. Perianths trigonous, 3-keeled, keels with narrow wings, mouth 3-lobed, lobes with irregularly teeth. Capsules spherical, longitudinally dehiscing by 4 regular valves at maturity. Spores globose, surface verrucose. Elaters with 2 spiral thickenings. Gemmae abundantly occurring on leaf apex and perianth mouth.

Habitat: Often on soil and rotten logs, sometimes on rocks, tree trunks, tree bases and on rocks with a thin layer of soil. Elev. 590−1774 m.

Distribution in Yachang: Widely distributed in Yachang.

Distribution in China: Widely distributed in China.

▶ A. 种群；B. 植物体一段；C. 侧叶；D–E. 腹叶；F. 蒴萼；G. 叶中部细胞，示油体；H. 裂开的孢蒴；I. 孢子；J. 弹丝。（凭证标本：韦玉梅等 *191020-404*）

A. Population; B. Portion of plant; C. Leaves; D–E. Underleaves; F. Perianth; G. Median cells of leaf showing oil bodies; H. Dehisced capsule; I. Spores; J. Elaters. (All from *Wei et al. 191020-404*)

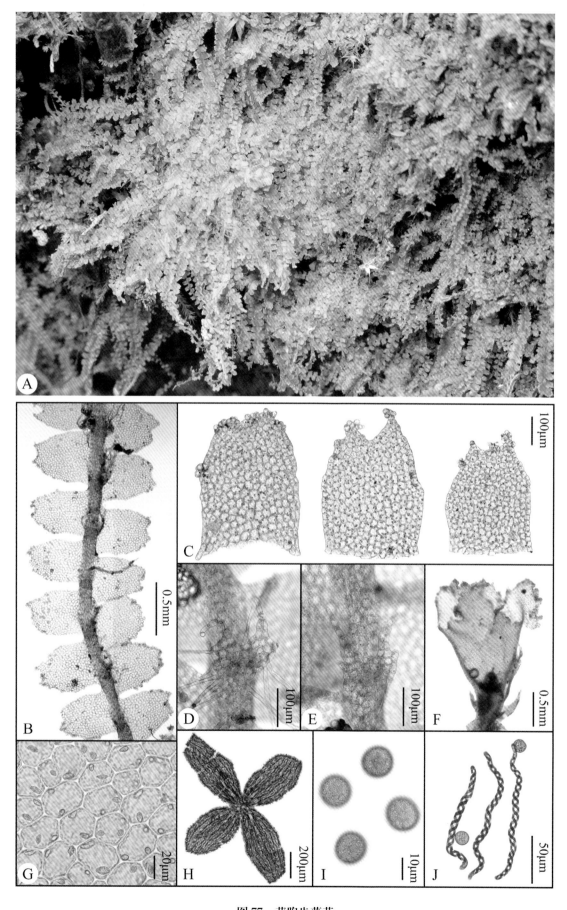

图 77 芽胞齿萼苔
Fig. 77 *Lophocolea minor* Nees

尖瓣光萼苔原亚种

Porella acutifolia (Lehm. et Lindenb.) Trevis. subsp. **acutifolia**, Mem. Reale Ist. Lombardo Sci. (Ser. 3), C. Sci. Mat. 4(13): 408. 1877.

植物体绿色至棕绿色，带叶宽 2.40—3.25 mm，羽状分枝，分枝耳叶苔型。茎横切面 16—18 个细胞厚，皮部细胞 2—4 层，厚壁，内部细胞薄壁。假根少，成束生于腹叶基部。侧叶覆瓦状排列，略倾斜伸展，背瓣椭圆状卵形，顶端具顶生的短齿，顶生齿两侧常具 3—5 个锐齿或钝齿，腹侧边缘近乎直，下半部内弯，上半部外弯，基部截形，背侧边缘稍拱起，近基部稍膨大，基部心形或稍耳状；腹瓣舌形，边缘全缘，或顶部偶尔具齿，与背瓣相连处具明显的短脊，近轴侧基部沿茎长下延，下延部分常具波状钝齿。侧叶细胞薄壁，但边缘常明显厚壁，角质层平滑，三角体大，简单三角形或肿胀形，中部球状加厚缺。油体均一型。腹叶毗邻，三角状卵形，宽为茎的 1.0—1.5 倍，顶端圆钝或截形，全缘或偶具齿，两侧边缘全缘，基部两侧沿茎长下延，下延部分全缘。雌雄异株。雄苞顶生，雄苞叶 3—8 对，雄苞腹叶生于雄苞全枝。雌苞未见。

生境： 石生。海拔 1342 m。

雅长分布： 下棚屯。

中国分布： 重庆，福建，甘肃，广西，贵州，湖北，湖南，江西，陕西，四川，西藏，云南，浙江。

Plants green to brownish green when fresh, 2.40–3.25 mm wide, pinnately branched, branching of the *Frullania*-type. Stem in transverse section 16–18 cells thick, cortical cells of 2–4 layers, thick-walled, medullary cells thin-walled. Rhizoids few, in tufts from underleaf bases. Leaves imbricate, obliquely spreading, lobe oblong-ovate, apex with a apical short tooth and 3–5 additional obtuse or acute teeth, dorsal margin straight, incurved at lower part, recurved at upper part, base truncate, ventral margin arched, slightly ampliate near base, base cordate to auriculate; lobule lingulate, margin entire, occasionally toothed at apex, attached to the lobe by a short but distinct keel, the proximal base long-decurrent, crispate in decurrent area. Leaf cells thin-walled, except thick-walled along margin, trigones large, simple-triangulate or nodulate, intermediate thickenings absent. Cuticle smooth. Oil bodies homogeneous. Underleaves contigous, triangular-ovate, 1.0–1.5 times as wide as the stem, apex rounded-obtuse to truncate, entire, occasionally toothed, lateral margins entire, base long-decurrent, entire in decurrent area. Dioicous. Androecia terminal, bracts in 3–8 pairs, bracteoles present throughout the androecium. Gynoecia not seen.

Habitat: On rocks. Elev. 1342 m.

Distribution in Yachang: Xiapeng Tun.

Distribution in China: Chongqing, Fujian, Gansu, Guangxi, Guizhou, Hubei, Hunan, Jiangxi, Shaanxi, Sichuan, Tibet, Yunnan, Zhejiang.

▶ A. 种群；B. 植物体；C. 植物体一段；D. 侧叶；E. 茎一段，示腹瓣和腹叶；F. 茎横切面；G. 腹瓣；H. 腹叶；I. 叶中部细胞，示油体；J. 腹瓣和背瓣连接处。（凭证标本：*韦玉梅等 191015-166*）

A. Population; B. Plant; C. Portion of plant; D. Leaves; E. Portion of stem, showing leaf lobules and underleaves; F. Transverse section of stem; G. Leaf lobules; H. Underleaves; I. Median cells of leaf lobe showing oil bodies; J. Connection between the leaf lobe and lobule. (All from *Wei et al. 191015-166*)

图 78a 尖瓣光萼苔原亚种

Fig. 78a *Porella acutifolia* (Lehm. et Lindenb.) Trevis. subsp. *acutifolia*

尖瓣光萼苔东亚亚种

Porella acutifolia subsp. **tosana** (Steph.) S. Hatt., J. Hattori Bot. Lab. 44: 100. 1978.

植物体绿色至深绿色，带叶宽 3.2—4.5 mm，羽状分枝，分枝耳叶苔型。茎横切面 15—17 个细胞厚，皮部细胞 1—2 层，厚壁，内部细胞薄壁。假根成束生于腹叶基部。侧叶覆瓦状排列至毗邻，稍倾斜或近水平伸展，背瓣椭圆状卵形，顶端渐尖，具顶生的长齿，顶生齿两侧常具 2—5 短锐齿，腹侧边缘略直，外弯或呈波曲状，基部截形，背侧边缘稍拱起，近基部稍膨大，基部耳状；腹瓣舌形，边缘全缘，或近顶部偶尔具少数齿，与背瓣相连处具明显的短脊，近轴侧基部沿茎长下延，下延部分常具波状钝齿。侧叶细胞薄壁，角质层平滑，三角体小至大，简单三角形，中部球状加厚缺。油体均一型。腹叶毗邻至远生，三角状卵形，宽为茎的 1.0—1.5 倍，顶端圆钝或截形，全缘或具齿，两侧边缘全缘，基部两侧沿茎长下延，下延部分全缘。雄苞和雌苞未见。

生境： 石生。海拔 1039—1342 m。

雅长分布： 下棚屯。

中国分布： 安徽，福建，广东，广西，贵州，海南，湖南，陕西，山东，四川，台湾，西藏，云南。

Plants green to dark green when fresh, 3.2–4.5 mm wide, irregularly pinnately branched, branching of the *Frullania*-type. Stem in transverse section 15–17 cells thick, cortical cells of 1–2 layers, thick-walled, medullary cells thin-walled. Rhizoids few, in tufts from underleaf bases. Leaves imbricate, obliquely or horizontally spreading, lobe oblong-ovate, apex with a apical long tooth and 2–5 additional short teeth, dorsal margin straight, recurved or crispate, base truncate, ventral margin arched, slightly ampliate near base, base auriculate; lobule lingulate, margin entire, or with few teeth near apex, attached to the lobe by a short but distinct keel, the proximal base long-decurrent, crispate in decurrent area. Leaf cells thin-walled, trigones small to large, simple-triangulate, intermediate thickenings absent. Cuticle smooth. Oil bodies homogeneous. Underleaves contigous to distant, triangular-ovate, 1.0–1.5 times as wide as the stem, apex rounded-obtuse to truncate, entire or toothed, lateral margins entire, base long-decurrent, entire in decurrent area. Androecia and gynoecia not seen.

Habitat: On rocks. Elev. 1039–1342 m.

Distribution in Yachang: Xiapeng Tun.

Distribution in China: Anhui, Fujian, Guangdong, Guangxi, Guizhou, Hainan, Hunan, Shaanxi, Shandong, Sichuan, Taiwan, Tibet, Yunnan.

▶ A. 种群；B. 植物体；C. 植物体一段；D. 侧叶；E. 茎一段，示腹瓣和腹叶；F. 茎横切面；G. 腹瓣；H. 叶中部细胞，示油体；I. 腹叶；J. 腹瓣和背瓣连接处。（凭证标本：*韦玉梅等 191015-172*）

A. Population; B. Plant; C. Portion of plant; D. Leaves; E. Portion of stem showing leaf lobules and underleaves; F. Transverse section of stem; G. Leaf lobules; H. Median cells of leaf lobe showing oil bodies; I. Underleaves; J. Connection between the leaf lobe and lobule. (All from *Wei et al. 191015-172*)

图 78b 尖瓣光萼苔东亚亚种

Fig. 78b *Porella acutifolia* subsp. *tosana* (Steph.) S. Hatt.

丛生光萼苔原变种

Porella caespitans (Steph.) S. Hatt. var. **caespitans**, J. Hattori Bot. Lab. 33: 50. 1970.

植物体绿色至棕绿色，带叶宽 3.6—4.8 mm，羽状分枝，分枝耳叶苔型。茎横切面 16—19 个细胞厚，皮部细胞 2—3 层，厚壁，内部细胞薄壁。假根少，成束生于腹叶基部。侧叶覆瓦状排列，少倾斜或近水平伸展，背瓣椭圆状卵形，顶端具顶生的长齿，边缘全缘，腹侧边缘略直，下半部内弯，上半部外弯，基部截形，背侧边缘稍拱起，近基部稍膨大至膨大，基部心形或稍耳状；腹瓣舌形，边缘全缘，或近顶部偶尔具齿，与背瓣相连处具明显的短脊，近轴侧基部沿茎长下延，下延部分全缘或有时具波状钝齿。侧叶细胞薄壁至稍厚壁，角质层平滑，三角体小至大，简单三角形或肿胀形（基部），中部球状加厚缺。油体均一型。腹叶毗邻至远生，三角状卵形，宽为茎的 1.0—1.5 倍，顶端钝尖或近截形，全缘或具齿，两侧边缘全缘，基部两侧沿茎长下延，下延部分全缘。雄苞和雌苞未见。

生境：石生，树基。海拔 1125—1730 m。

雅长分布：草王山，大棚屯，黄猄洞天坑，拉洞天坑。

中国分布：全国大部分省区均有分布。

Plants green to brownish green when fresh, 3.6−4.8 mm wide, pinnately branched, branching of the *Frullania*-type. Stem in transverse section 16−19 cells thick, cortical cells of 2−3 layers, thick-walled, medullary cells thin-walled. Rhizoids few, in tufts from underleaf bases. Leaves imbricate, obliquely or horizontally spreading, lobe oblong-ovate, apex with a apical long tooth, margin entire, dorsal margin straight, incurved at lower part, recurved at upper part, base truncate, ventral margin arched, slightly ampliate to ampliate near base, base cordate to auriculate; lobule lingulate, margin entire, occasionally toothed near apex, attached to the lobe by a short but distinct keel, the proximal base long-decurrent, entire or sometomes crispate in decurrent area. Leaf cells thin-walled to slightly thick-walled, trigones small to large, simple-triangulate or nodulate at base, intermediate thickenings absent. Cuticle smooth. Oil bodies homogeneous. Underleaves contigous to distant, triangular-ovate, 1.0−1.5 times as wide as the stem, apex acute-obtuse to subtruncate, entire or toothed, lateral margins entire, base long-decurrent, entire in decurrent area. Androecia and gynoecia not seen.

Habitat: On rocks and tree bases. Elev. 1125−1730 m.

Distribution in Yachang: Caowangshan, Dapeng Tun, Huangjingdong Tiankeng, Ladong Tiankeng.

Distribution in China: Widely distributed in most provinces of China.

▶ A. 种群；B. 植物体；C. 植物体一段；D. 侧叶；E. 茎一段，示腹瓣和腹叶；F. 茎横切面；G. 腹瓣；H. 腹叶；I. 叶中部细胞，示油体；J. 腹瓣和背瓣连接处。（凭证标本：*韦玉梅等 191013-85*）

A. Population; B. Plant; C. Portion of plant; D. Leaves; E. Portion of stem showing leaf lobules and underleaves; F. Transverse section of stem; G. Leaf lobules; H. Underleaves; I. Median cells of leaf lobe showing oil bodies; J. Connection between the leaf lobe and lobule. (All from *Wei et al. 191013-85*)

图 79a 丛生光萼苔原变种

Fig. 79a *Porella caespitans* (Steph.) S. Hatt. var. *caespitans*

丛生光萼苔心叶变种（丛生光萼苔细柄变种）

Porella caespitans var. **cordifolia** (Steph.) S. Hatt. ex T. Katag. et T. Yamag., Bryol. Res. 10(5): 133. 2011.

Porella caespitans var. *setigera* (Steph.) S. Hatt., J. Hattori Bot. Lab. 33: 53. 1970.

植物体绿色至棕褐色，带叶宽 3.6—4.8 mm，羽状分枝，分枝耳叶苔型。茎横切面 15—19 个细胞厚，皮部细胞 2—4 层。假根少，成束生于腹叶基部。侧叶覆瓦状排列，稍倾斜至近水平伸展，背瓣斜三角状卵形，顶端具顶生的长齿，边缘全缘，稀具齿，腹侧边缘稍拱起，下半部内弯，上半部外折，基部心形，背侧边缘拱起，近基部明显膨大，基部耳状下延；腹瓣舌形，边缘全缘，或近顶部偶尔具齿，与背瓣相连处具明显的短脊，近轴侧基部沿茎长下延，下延部分具裂片状齿且常卷曲。侧叶细胞稍厚壁，角质层平滑，三角体大，肿胀形，有时简单三角形，中部球状加厚缺。油体均一型。腹叶毗邻至远生，三角状卵形，宽为茎的 1.0—1.5 倍，顶端具长的双齿，两侧边缘全缘，基部两侧沿茎长下延，下延部分具波状钝齿或裂片状齿。

生境：石生环境常见，树干环境也有分布。海拔 1073—1823 m。

雅长分布：白岩坨屯，草王山，大棚屯，大宴坪天坑漏斗，逻家田屯，盘古王，下棚屯，下岩洞屯，悬崖天坑。

中国分布：全国大部分省区均有分布。

Plants green to dark brown when fresh, 3.6–4.8 mm wide, pinnately branched, branching of the *Frullania*-type. Stem in transverse section 15–19 cells thick, cortical cells of 2–4 layers. Rhizoids few, in tufts from underleaf bases. Leaves imbricate, obliquely to horizontally spreading, lobe obliquely triangular-ovate, apex with a apical long tooth, margin entire, dorsal margin slightly arched, incurved at lower part, recurved at upper part, base cordate, ventral margin arched, strongly ampliate near base, base auriculate; lobule lingulate, margin entire, occasionally toothed near apex, attached to the lobe by a short but distinct keel, the proximal base long-decurrent, laciniate-toothed and revolute in decurrent area. Leaf cells slightly thick-walled, trigones large, nodulate, occasionally simple-triangulate, intermediate thickenings absent. Cuticle smooth. Oil bodies homogeneous. Underleaves contiguous to distant, triangular-ovate, 1.0–1.5 times as wide as the stem, apex long-bilobed, lateral margins entire, base long-decurrent, crispate or laciniate-toothed in decurrent area.

Habitat: Often on rocks, sometimes on tree trunks. Elev. 1073–1823 m.

Distribution in Yachang: Baiyantuo Tun, Caowangshan, Dapeng Tun, Dayanping Tiankeng, Luojiatian Tun, Panguwang, Xiapeng Tun, Xiayandong Tun, Xuanya Tiankeng.

Distribution in China: Widely distributed in most provinces of China.

▶ A. 种群；B. 植物体；C. 植物体一段；D. 侧叶；E. 茎一段，示腹瓣和腹叶；F. 茎横切面；G. 腹瓣；H. 腹叶；I. 叶中部细胞，示油体；J. 腹瓣和背瓣连接处。（凭证标本：*韦玉梅等 201105-15*）

A. Population; B. Plant; C. Portion of plant; D. Leaves; E. Portion of stem showing leaf lobules and underleaves; F. Transverse section of stem; G. Leaf lobules; H. Underleaves; I. Median cells of leaf lobe showing oil bodies; J. Connection between the leaf lobe and lobule. (All from *Wei et al. 201105-15*)

图 79b 丛生光萼苔心叶变种

Fig. 79b *Porella caespitans* var. *cordifolia* (Steph.) S. Hatt. ex T. Katag. et T. Yamag.

密叶光萼苔原亚种

Porella densifolia (Steph.) S. Hatt. subsp. **densifolia**, J. Jap. Bot. 20: 109. 1944.

植物体绿色至棕褐色，带叶宽3.4—4.5 mm，羽状分枝，分枝耳叶苔型。茎横切面20—24个细胞厚，皮部细胞3—4层，厚壁，内部细胞薄壁。假根少，成束生于腹叶基部。侧叶覆瓦状排列，倾斜伸展，背瓣阔卵形或阔三角状卵形，顶端有时具明显的顶生齿，顶生齿两侧常具1—3个锐齿，有时顶生齿不明显，腹侧边缘直或稍拱起，下半部外折，上半部狭的内卷，基部耳状，背侧边缘拱起，近基部膨大，基部强烈反卷；腹瓣椭圆形，顶部钝或圆钝，边缘全缘，与背瓣相连处不具脊或脊不明显，两侧基部长下延，下延部分具裂片状齿且强烈卷曲。侧叶细胞稍厚壁，角质层平滑，三角体大，简单三角形或肿胀形，中部球状加厚少见。油体均一型。腹叶覆瓦状排列，卵形至椭圆状卵形，宽为茎的1.5—2.5倍，顶端圆截形，边缘全缘，基部两侧沿茎长下延，下延部分具深裂片状齿且常卷曲。雄苞和雌苞未见。

生境：石生环境常见，土生、腐木环境也有分布。海拔1052—1751 m。

雅长分布：大棚屯，大宴坪天坑漏斗，黄猄洞天坑，盘古王，旁墙屯，下岩洞屯，悬崖天坑，中井屯。

中国分布：除了东北以外的全国大部分省区均有分布。

Plants green to dark brown when fresh, 3.4–4.5 mm wide, pinnately branched, branching of the *Frullania*-type. Stem in transverse section 20–24 cells thick, cortical cells of 3–4 layers, thick-walled, medullary cells thin-walled. Rhizoids few, in tufts from underleaf bases. Leaves imbricate, obliquely spreading, lobe broad ovate or broad triangular-ovate, apex with a apical long tooth and 1–3 additional acute teeth, dorsal margin straight or slightly arhed, recurved at lower part, narrowly incurved at upper part, base auriculate, ventral margin arched, ampliate near base, base strongly revolute; lobule oblong, margin entire, attached to the lobe by a indistinct keel or without keel, base long-decurrent, laciniate-toothed and strongly revolute in decurrent area. Leaf cells slightly thick-walled, trigones large, simple-triangulate or nodulate, intermediate thickenings scarce. Cuticle smooth. Oil bodies homogeneous. Underleaves imbricate, ovate to oblong-ovate, 1.5–2.5 times as wide as the stem, apex rounded-truncate, margin entire, base long-decurrent, deeply laciniate-toothed and revolute in decurrent area. Androecia and gynoecia not seen.

Habitat: Often on rocks, sometimes on soil and rotten logs. Elev. 1052–1751 m.

Distribution in Yachang: Dapeng Tun, Dayanping Tiankeng, Huangjingdong Tiankeng, Panguwang, Pangqiang Tun, Xiayandong Tun, Xuanya Tiankeng, Zhongjing Tun.

Distribution in China: Widely distributed in most provinces of China except Northeast Region.

▶ A. 种群；B. 植物体；C. 植物体一段；D. 侧叶；E. 茎一段，示腹瓣和腹叶；F. 叶中部细胞，示油体；G. 腹叶；H. 腹瓣；I. 茎横切面；J. 腹瓣和背瓣连接处。（凭证标本：*韦玉梅等 201105-12*）

A. Population; B. Plant; C. Portion of plant; D. Leaves; E. Portion of stem showing leaf lobules and underleaves; F. Median cells of leaf lobe showing oil bodies; G. Underleaves; H. Leaf lobules; I. Transverse section of stem; J. Connection between the leaf lobe and lobule. (All from *Wei et al. 201105-12*)

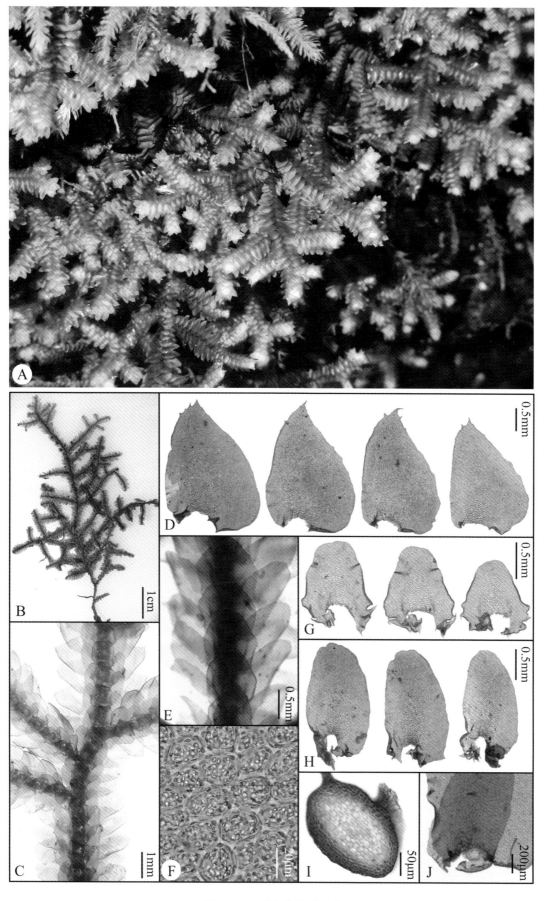

图 80a 密叶光萼苔原亚种

Fig. 80a *Porella densifolia* (Steph.) S. Hatt. subsp. *densifolia*

密叶光萼苔长叶亚种

Porella densifolia subsp. **appendiculata** (Steph.) S. Hatt., J. Hattori Bot. Lab. 32: 343. 1969.

植物体绿色至棕绿色，带叶宽 4.2—6.9 mm，羽状分枝，分枝耳叶苔型。茎横切面 20—23 个细胞厚，皮部细胞 2—3 层，厚壁，内部细胞薄壁。假根少，成束生于腹叶基部。侧叶覆瓦状排列，近水平伸展，背瓣椭圆状卵形至狭长椭圆状卵形，顶端具一个稍长的顶齿，具 2—6 个粗锐齿，腹侧边缘直，外折且常波曲状，基部耳状，背侧边缘拱起，近基部稍膨大，基部反卷；腹瓣椭圆形，边缘全缘，与背瓣相连处不具脊或脊不明显，两侧基部下延，下延部分有时具裂片状齿且强烈卷曲。侧叶细胞稍厚壁，角质层平滑，三角体大，肿胀形，有时简单三角形，中部球状加厚缺。油体均一型。腹叶覆瓦状排列，椭圆状卵形，宽为茎的 1—2 倍，顶端平截或圆钝，边缘全缘，或近顶部具齿，基部两侧沿茎长下延，下延部分全缘，强烈卷曲。雌雄异株。雄苞未见。雌苞生于侧短枝，雌苞叶 1—2 对，雌苞叶和雌苞腹叶边缘具齿。

生境：石生，树基。海拔 1079—1252 m。

雅长分布：黄猄洞天坑，下岩洞屯。

中国分布：重庆，福建，甘肃，广西，贵州，河南，湖北，陕西，四川，西藏，云南，浙江。

Plants green to brownish green when fresh, 4.2−6.9 mm wide, pinnately branched, branching of the *Frullania*-type. Stem in transverse section 20−23 cells thick, cortical cells of 2−3 layers, thick-walled, medullary cells thin-walled. Rhizoids few, in tufts from underleaf bases. Leaves imbricate, horizontally spreading, lobe oblong-ovate to narrow oblong-ovate, apex with a apical tooth and 2−6 additional acute teeth, sometimes apical tooth indistinct, dorsal margin straight, recurved and crispate, base auriculate, ventral margin arched, slightly ampliate near base, base strongly revolute; lobule oblong, margin entire, attached to the lobe by a indistinct keel or without keel, base decurrent, laciniate-toothed and strongly revolute in decurrent area. Leaf cells slightly thick-walled, trigones large, nodulate, sometimes simple-triangulate, intermediate thickenings absent. Cuticle smooth. Oil bodies homogeneous. Underleaves imbricate, oblong-ovate, 1−2 times as wide as the stem, apex truncate to rounded-obtuse, margin entire, or toothed near apex, base long-decurrent, entire and strongly revolute in decurrent area. Androecia and gynoecia not seen.

Habitat: On rocks and tree bases. Elev. 1079−1252 m.

Distribution in Yachang: Huangjingdong Tiankeng, Xiayandong Tun.

Distribution in China: Chongqing, Fujian, Gansu, Guangxi, Guizhou, Henan, Hubei, Shaanxi, Sichuan, Tibet, Yunnan, Zhejiang.

▶ A. 种群；B. 植物体；C. 植物体一段；D. 侧叶；E. 茎一段，示腹瓣和腹叶；F. 叶中部细胞，示油体；G. 腹叶；H. 茎横切面；I. 腹瓣和背瓣连接处；J. 腹瓣。（凭证标本：*唐启明 20180612-316*）

A. Population; B. Plant; C. Portion of plant; D. Leaves; E. Portion of stem showing leaf lobules and underleaves; F. Median cells of leaf lobe showing oil bodies; G. Underleaves; H. Transverse section of stem; I. Connection between the leaf lobe and lobule; J. Leaf lobules. (All from *Tang 20180612-316*)

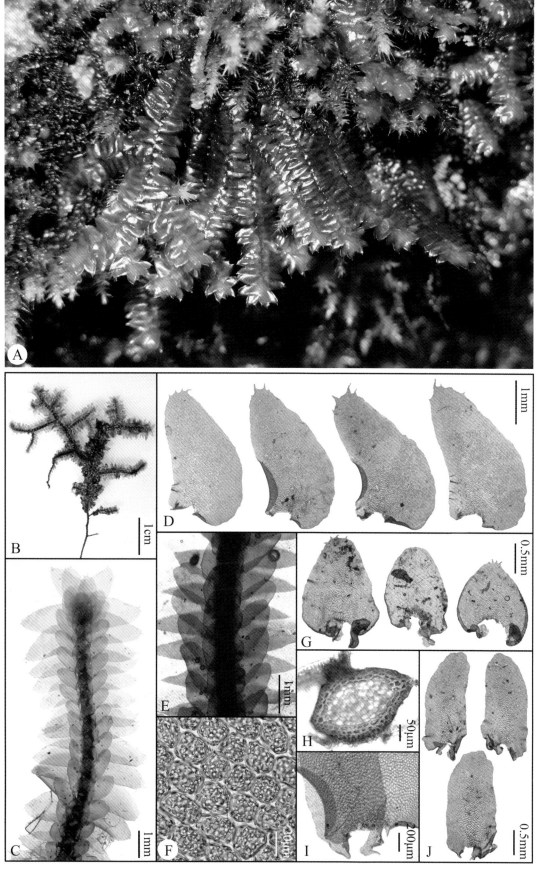

图 80b 密叶光萼苔长叶亚种

Fig. 80b *Porella densifolia* subsp. *appendiculata* (Steph.) S. Hatt.

大叶光萼苔

Porella grandifolia (Steph.) S. Hatt., J. Hattori Bot. Lab. 30: 136. 1967.

植物体深绿色，带叶宽 3.8—5.6 mm，不规则分枝，耳叶苔型。茎横切面 16—18 个细胞厚，皮部细胞 2—3 层，厚壁，内部细胞薄壁。假根多，成束生于腹叶基部。侧叶覆瓦状排列，略倾斜或近水平伸展，背瓣椭圆状长方形或椭圆状卵形，顶端圆，边缘全缘，常波曲，腹侧边缘直或稍拱起，内折，基部截形，背侧边缘拱起，近基部不膨大或稍膨大，基部稍呈耳状；腹瓣阔舌形，边缘全缘，与背瓣相连处具明显的短脊，基部不下延，但常具小片状附属物。侧叶细胞薄壁，但边缘常明显厚壁，角质层平滑，三角体小至大，简单三角形或肿胀形（基部），中部球状加厚缺。油体均一型。腹叶毗邻，舌形，与茎等宽或略大于茎，顶端圆或圆钝，边缘全缘，稀具齿，基部两侧沿茎略下延。雌雄异株。雄苞未见。雌苞生于侧短枝，雌苞叶 1—2 对。

生境：石生，树干。海拔 1240—1261 m。

雅长分布：大宴坪天坑漏斗，黄猄洞天坑，蓝家湾天坑，悬崖天坑，霄罗湾洞穴。

中国分布：贵州，台湾，云南。首次记录于广西。

Plants dark green when fresh, 3.8−5.6 mm wide, irregularly branched, branching of the *Frullania*-type. Stem in transverse section 16−18 cells thick, cortical cells of 2−3 layers, thick-walled, medullary cells thin-walled. Rhizoids numerous, in tufts from underleaf bases. Leaves imbricate, obliquely or horizontally spreading, lobe oblong-rectangular or oblong-ovate, apex rounded, margin entire, crispate, dorsal margin straight or slightly arched, incurved, base truncate, ventral margin arched, not amplicate or slightly ampliate near base, base slightly auriculate; lobule broad lingulate, margin entire, attached to the lobe by a short but distinct keel, the proximal base not decurrent, but appendiculate. Leaf cells thin-walled, except thick-walled along margin, trigones small to large, simple-triangulate or nodulate at base, intermediate thickenings absent. Cuticle smooth. Oil bodies homogeneous. Underleaves contigous, lingulate, as wide as the stem or wider than the stem, apex rounded-obtuse or rounded, entire, rarely toothed, base weak-decurrent, entire in decurrent area. Dioicous. Androecia not seen. Gynoecia lateral on short branches, bracts in 1−2 pairs, margins of bracts and bracteoles toothed.

Habitat: On rocks and tree trunks. Elev. 1240−1261 m.

Distribution in Yachang: Dayanping Tiankeng, Huangjingdong Tiankeng, Lanjiawan Tiankeng, Xiaoluowan Cave, Xuanya Tiankeng.

Distribution in China: Guizhou, Taiwan, Yunnan. New to Guangxi.

▶ A. 种群；B. 植物体；C. 植物体一段；D. 侧叶；E. 茎一段，示腹瓣和腹叶；F. 茎横切面；G. 腹瓣；H. 叶中部细胞，示油体；I. 腹叶；J. 腹瓣和背瓣连接处。（凭证标本：*唐启明等 20210721-42A*）

A. Population; B. Plant; C. Portion of plant; D. Leaves; E. Portion of stem showing leaf lobules and underleaves; F. Transverse section of stem; G. Leaf lobules; H. Median cells of leaf lobe showing oil bodies; I. Underleaves; J. Connection between the leaf lobe and lobule. (All from *Tang et al. 20210721-42A*)

图 81 大叶光萼苔
Fig. 81 *Porella grandifolia* (Steph.) S. Hatt.

尾尖光萼苔

Porella handelii S. Hatt., J. Hattori Bot. Lab. 33: 65. 1970.

植物体绿色至棕褐色，带叶宽 2.8—4.3 mm，羽状分枝，分枝耳叶苔型。茎横切面 16—18 个细胞厚，皮部细胞 1—2 层。假根少，成束生于腹叶基部。侧叶覆瓦状排列，倾斜伸展，背瓣卵形或三角状卵形，顶端具长尾尖状的顶生齿，边缘全缘，腹侧边缘稍拱起，反折，基部略心形，背侧边缘拱起，近基部明显膨大，基部稍呈耳状；腹瓣舌形，边缘全缘，与背瓣相连处具明显的短脊，近轴侧基部沿茎稍下延。侧叶细胞薄壁，但边缘常明显厚壁，角质层平滑，三角体小至大，简单三角形或肿胀形（基部），中部球状加厚缺。油体均一型。腹叶毗邻，三角状卵形，宽为茎的 1.0—1.5 倍，顶端具 2 个长齿，边缘全缘，基部两侧沿茎下延，下延部分全缘或具裂片状齿。

生境： 石生环境常见，腐木、树基环境也有分布。海拔 582—1389 m。

雅长分布： 草王山，吊井天坑，黄猄洞天坑，兰花园，老屋基天坑，里郎天坑，隆合朝屯，香当天坑，悬崖天坑，一沟，中井天坑。

中国分布： 安徽，广东，广西，贵州，海南，湖北，湖南，江西，四川，西藏，云南。

Plants green to dark brown when fresh, 2.8–4.3 mm wide, pinnately branched, branching of the *Frullania*-type. Stem in transverse section 16–18 cells thick, cortical cells of 1–2 layers. Rhizoids few, in tufts from underleaf bases. Leaves imbricate, obliquely spreading, lobe ovate to triangular-ovate, apex with a apical long tooth, margin entire, dorsal margin slightly arched, recurved, base slightly cordate, ventral margin arched, strongly ampliate near base, base slightly auriculate; lobule lingulate, margin entire, attached to the lobe by a short but distinct keel, the proximal base weak-decurrent. Leaf cells thin-walled, except thick-walled along margin, trigones small to large, simple-triangulate or nodulate at base, intermediate thickenings absent. Cuticle smooth. Oil bodies homogeneous. Underleaves contiguous, triangular-ovate, 1.0–1.5 times as wide as the stem, apex long-bilobed, lateral margin entire, base decurrent, entire or laciniate-toothed in decurrent area.

Habitat: Often on rocks, sometimes on rotten logs and tree bases. Elev. 582–1389 m.

Distribution in Yachang: Caowangshan, Diaojing Tiankeng, Huangjingdong Tiankeng, Laowuji Tiankeng, Lilang Tiankeng, Longhechao Tun, Orchid Garden, Xiangdang Tiankeng, Xuanya Tiankeng, Yigou, Zhongjing Tiankeng.

Distribution in China: Anhui, Guangdong, Guangxi, Guizhou, Hainan, Hubei, Hunan, Jiangxi, Sichuan, Tibet, Yunnan.

▶ A. 种群；B. 植物体；C. 植物体一段；D. 侧叶；E. 茎一段，示腹瓣和腹叶；F. 茎横切面；G. 腹瓣；H. 叶中部细胞，示油体；I. 腹叶；J. 腹瓣和背瓣连接处。（凭证标本：*唐启明等 20210723-48*）

A. Population; B. Plant; C. Portion of plant, D. Leaves; E. Portion of stem showing leaf lobules and underleaves; F. Transverse section of stem; G. Leaf lobules; H. Median cells of leaf lobe showing oil bodies; I. Underleaves; J. Connection between the leaf lobe and lobule. (All from *Tang et al. 20210723-48*)

图 82　尾尖光萼苔

Fig. 82　*Porella handelii* S. Hatt.

日本光萼苔

Porella japonica (Sande Lac.) Mitt., Trans. Linn. Soc. London, Bot. 3(3): 202. 1891.

植物体绿色至褐绿色，带叶宽 1.3—2.7 mm，羽状分枝，耳叶苔型。茎横切面 13—16 个细胞厚，皮部细胞 1—2 层。假根少，生于腹叶基部。侧叶覆瓦状排列，稍倾斜至近水平伸展，背瓣卵形至椭圆状卵形，顶端圆钝，常具 1—3 个短齿，稀全缘，腹侧边缘直，稍内折，基部截形，背侧边缘稍拱起，近基部轻微膨大，基部稍呈耳状；腹瓣舌形，顶部钝，全缘或具齿，两侧边缘全缘，偶尔近基部具齿，与背瓣相连处不具脊或脊不明显，两侧基部短下延，下延部分边缘具纤毛状齿。侧叶细胞薄壁，角质层平滑，三角体小至大，简单三角形，中部球状加厚缺。油体均一型。腹叶远生，三角状卵形，与茎等宽或略大于茎，顶端平截，具齿，有时全缘，两侧边缘全缘，或近基部具齿，基部两侧沿茎短下延，下延部分常具纤毛状齿。蒴萼钟形，表面平滑无脊，口部截形，边缘具长纤毛状齿。孢蒴椭圆形，成熟时 4 瓣开裂。孢子球形，表面具刺状纹饰。弹丝 2 螺旋加厚。

生境： 石生。海拔 1342—1509 m。

雅长分布： 盘古王，下棚屯。

中国分布： 全国大部分省区均有分布。

Plants green to brownish green when fresh, 1.3–2.7 mm wide, pinnately branched, branching of the *Frullania*-type. Stem in transverse section 13–16 cells thick, cortical cells of 1–2 layers. Rhizoids few, in tufts from underleaf bases. Leaves imbricate, obliquely or horizontally spreading, lobe oblong-ovate to ovate, apex rounded-obtuse with 1–3 short teeth, rarely entire, dorsal margin straight, slightly incurved, base truncate, ventral margin arched, slightly ampliate near base, base slightly auriculate; lobule lingulate, apex obtuse, toothed or entire, lateral margin entire, except toothed near base, attached to the lobe by a indistinct keel or without keel, base short-decurrent, ciliate-toothed in decurrent area. Leaf cells thin-walled, trigones small to large, simple-triangulate, intermediate thickenings absent. Cuticle smooth. Oil bodies homogeneous. Underleaves distant, triangular-ovate, as wide as the stem or slightly wider than the stem, apex truncate, toothed, occasionally entire, lateral margin entire, except toothed near base, base short-decurrent, ciliate-toothed in decurrent area. Perianths campanulate, surface smooth, mouth truncate, with ciliate teeth. Capsules ellipsoidal, longitudinally dehiscing by 4 regular valves at maturity. Spores globose, surface spinulate. Elaters with 2 spiral thickenings.

Habitat: On rocks. Elev. 1342–1509 m.

Distribution in Yachang: Panguwang, Xiapeng Tun.

Distribution in China: Widely distributed in most provinces of China.

▶ A. 种群；B. 植物体；C. 植物体一段；D. 侧叶；E. 弹丝；F. 茎一段，示腹瓣和腹叶；G. 茎横切面；H. 孢子；I. 腹瓣；J. 叶中部细胞，示油体；K. 腹叶；L. 腹瓣和背瓣连接处。（凭证标本：*韦玉梅等 201110-31*）

A. Population; B. Plant; C. Portion of plant; D. Leaves; E. Elaters; F. Portion of stem showing leaf lobules and underleaves; G. Transverse section of stem; H. Spores; I. Leaf lobules; J. Median cells of leaf lobe showing oil bodies; K. Underleaves; L. Connection between the leaf lobe and lobule. (All from *Wei et al. 201110-31*)

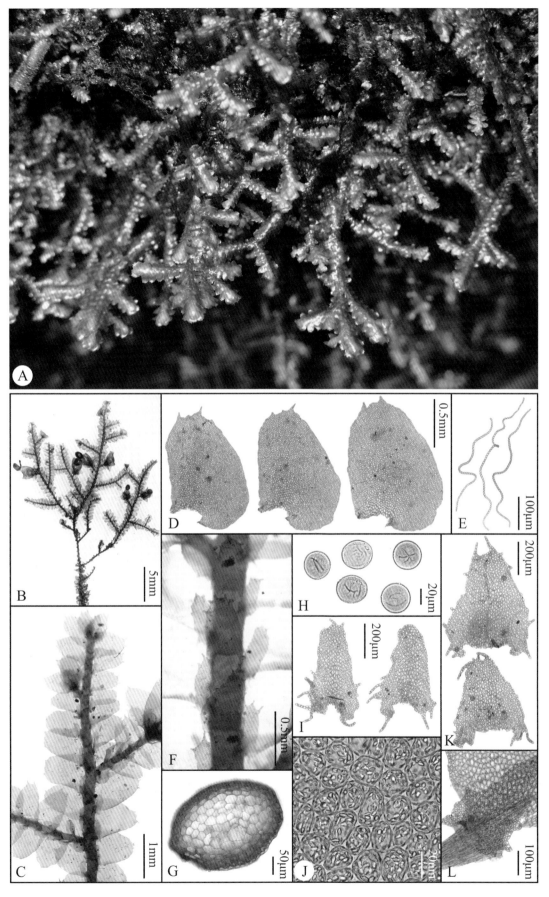

图 83 日本光萼苔

Fig. 83 *Porella japonica* (Sande Lac.) Mitt.

基齿光萼苔

Porella madagascariensis (Nees et Mont.) Trevis., Mem. Reale Ist. Lombardo Sci. (Ser. 3), C. Sci. Mat. 4(13): 407. 1877.

植物体绿色至深绿色，带叶宽 2.2—3.7 mm，羽状分枝，耳叶苔型。茎横切面 15—16 个细胞厚，皮部细胞 1—2 层，厚壁，内部细胞薄壁。假根多，生于腹叶基部。侧叶覆瓦状排列，稍倾斜至近水平伸展，背瓣椭圆形，顶端圆至圆截形，边缘全缘，略波曲，腹侧边缘直，基部楔形，背侧边缘稍拱起，近基部不膨大，基部心形或稍呈耳状；腹瓣舌形，顶部圆钝，全缘或具齿，下部边缘具长纤毛状齿，与背瓣相连处不具脊或脊不明显，两侧基部下延，下延部分边缘具长纤毛状齿。侧叶细胞薄壁，角质层平滑，三角体小至大，简单三角形，中部球状加厚缺。油体均一型。腹叶毗邻至远生，椭圆状卵形，与茎等宽或略大于茎，顶端近截形，全缘或具齿，中下部边缘具不规则长或短的齿，基部两侧沿茎下延，下延部分边缘具长纤毛状齿。雌雄异株。雄苞顶生在短枝上或间生在长枝上，雄苞叶 2—6 对，雄苞腹叶生于全枝。雌苞未见。

生境： 腐木，树干。海拔 1217—1470 m。

雅长分布： 蓝家湾天坑，盘古王。

中国分布： 广西，贵州，四川，西藏，云南。

Plants green to dark green when fresh, 2.2–3.7 mm wide, pinnately branched, branching of the *Frullania*-type. Stem in transverse section 15–16 cells thick, cortical cells of 1–2 layers, thick-walled, medullary cells thin-walled. Rhizoids numerous, in tufts from underleaf bases. Leaves imbricate, obliquely or horizontally spreading, lobe oblong, apex rounded or rounded-truncate, margin entire, slightly crispate, dorsal margin straight, base truncate, ventral margin slightly arched, not ampliate near base, base cordate or slightly auriculate; lobule lingulate, apex obtuse, toothed or entire, lateral margin entire, except ciliate-toothed near base, attached to the lobe by a indistinct keel or without keel, base short-decurrent, ciliate-toothed in decurrent area. Leaf cells thin-walled, trigones small to large, simple-triangulate, intermediate thickenings absent. Cuticle smooth. Oil bodies homogeneous. Underleaves contigous to distant, oblong-ovate, as wide as the stem or slightly wider than the stem, apex truncate, entire or toothed, lateral margin entire, except toothed near base, base decurrent, ciliate-toothed in decurrent area. Dioicous. Androecia terminal on short branches or intercalary on long branches, bracts in 2–6 pairs, bracteoles present throughout the androecium. Gynoecia not seen.

Habitat: On rotten logs and tree trunks. Elev. 1217–1470 m.

Distribution in Yachang: Lanjiawan Tiankeng, Panguwang.

Distribution in China: Guangxi, Guizhou, Sichuan, Tibet, Yunnan.

▶ A. 种群；B. 植物体；C. 植物体一段；D. 侧叶；E. 茎一段，示腹瓣和腹叶；F. 茎横切面；G. 腹瓣；H. 叶中部细胞，示油体；I. 腹叶；J. 腹瓣和背瓣连接处。（凭证标本：*唐启明 & 韦玉梅 20191019-363*）

A. Population; B. Plant; C. Portion of plant; D. Leaves, E. Portion of stem showing leaf lobules and underleaves; F. Transverse section of stem; G. Leaf lobules; H. Median cells of leaf lobe showing oil bodies; I. Underleaves; J. Connection between the leaf lobe and lobule. (All from *Tang & Wei 20191019-363*)

图 84　基齿光萼苔

Fig. 84　*Porella madagascariensis* (Nees & Mont.) Trevis.

亮叶光萼苔

Porella nitens (Steph.) S. Hatt., Fl. E. Himalaya 1: 525. 1966.

植物体绿色至褐绿色，带叶宽 3.0—4.5 mm，羽状分枝，耳叶苔型。茎横切面 15—17 个细胞厚，皮部细胞 2—3 层，厚壁，内部细胞薄壁。假根少，生于腹叶基部。侧叶覆瓦状排列，稍倾斜至近水平伸展，背瓣椭圆形，顶端圆至圆截形，边缘全缘，腹侧边缘直，反折，基部截形，背侧边缘稍拱起，近基部不膨大至轻微膨大，基部稍呈耳状；腹瓣舌形，顶部圆钝，边缘全缘，与背瓣相连处具明显的短脊，近轴侧基部下延，下延部分耳状或边缘具波状齿。侧叶细胞稍厚壁，角质层平滑，三角体大，肿胀形或简单三角形，中部球状加厚缺。油体均一型。腹叶毗邻至远生，舌形，与茎等宽或略大于茎，顶端近截形或圆，边缘全缘，基部两侧沿茎常下延，下延部分常具波状齿。雌雄异株。雄苞顶生在短枝上，雄苞叶 3—8 对，雄苞腹叶生于全枝。雌苞未见。

生境： 石生。海拔 1342 m。

雅长分布： 下棚屯。

中国分布： 全国大部分省区均有分布。

Plants green to brownish green when fresh, 3.0–4.5 mm wide, pinnately branched, branching of the *Frullania*-type. Stem in transverse section 15–17 cells thick, cortical cells of 2–3 layers, thick-walled, medullary cells thin-walled. Rhizoids few, in tufts from underleaf bases. Leaves imbricate, obliquely or horizontally spreading, lobe oblong, apex rounded or rounded-truncate, margin entire, dorsal margin straight, recurved, base truncate, ventral margin slightly arched, not ampliate to slightly ampliate near base, base slightly auriculate; lobule lingulate, apex rounded-obtuse, margin entire, attached to the lobe by a short but distinct keel, the proximal base short-decurrent, crispate-toothed in decurrent area. Leaf cells slightly thick-walled, trigones large, nodulate or simple-triangulate, intermediate thickenings absent. Cuticle smooth. Oil bodies homogeneous. Underleaves contiguous to distant, oblong-ovate, as wide as the stem or slightly wider than the stem, apex truncate or rounded, margin entire, base decurrent, crispate-toothed in decurrent area. Dioicous. Androecia terminal on short branches, bracts in 3–8 pairs, bracteoles present throughout the androecium. Gynoecia not seen.

Habitat: On rocks. Elev. 1342 m.

Distribution in Yachang: Xiapeng Tun.

Distribution in China: Widely distributed in most provinces of China.

▶ A. 种群；B. 植物体；C. 植物体一段；D. 侧叶；E. 茎一段，示腹瓣和腹叶；F. 茎横切面；G. 腹瓣；H. 叶中部细胞，示油体；I. 腹叶；J. 腹瓣和背瓣连接处。（凭证标本：*韦玉梅等 191015-169*）

A. Population; B. Plant; C. Portion of plant; D. Leaves; E. Portion of stem showing leaf lobules and underleaves; F. Transverse section of stem; G. Leaf lobules; H. Median cells of leaf lobe showing oil bodies; I. Underleaves; J. Connection between the leaf lobe and lobule. (All from *Wei et al. 191015-169*)

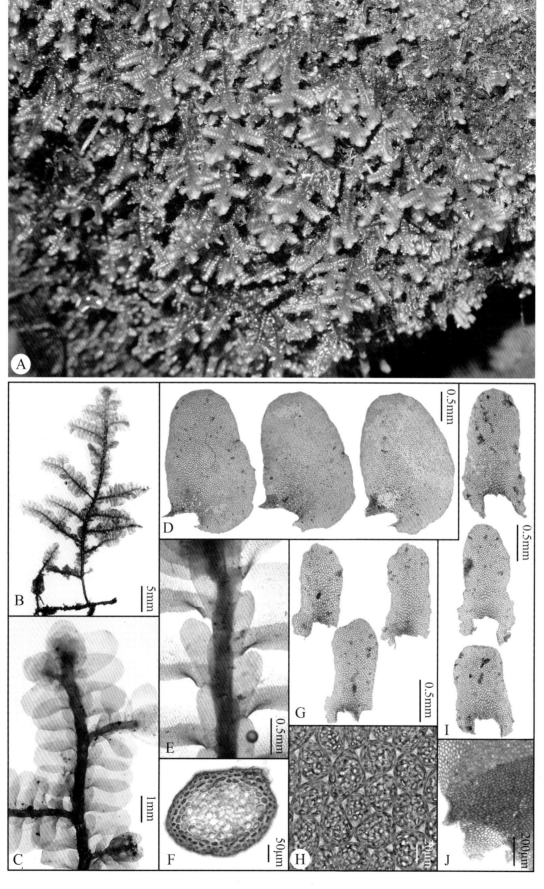

图 85 亮叶光萼苔

Fig. 85 *Porella nitens* (Steph.) S. Hatt.

钝叶光萼苔鳞叶变种

Porella obtusata var. **macroloba** (Steph.) S. Hatt. et M. X. Zhang, J. Jap. Bot. 60: 325. 1985.

植物体绿色至棕褐色，带叶宽 1.6—2.8 mm，羽状分枝，耳叶苔型。茎横切面 13—14 个细胞厚，皮部细胞 2—4 层。假根少，生于腹叶基部。侧叶覆瓦状排列，稍倾斜至近水平伸展，背瓣阔椭圆形，顶端圆，强烈内卷，边缘全缘，腹侧边缘略直，稍内折，基部稍耳状，背侧边缘稍拱起，近基部不膨大至轻微膨大，基部稍耳状；腹瓣阔舌形，顶部圆钝至钝，边缘全缘，与背瓣相连处不具脊或脊不明显，两侧基部短下延，下延部分全缘或有时具裂片状齿。侧叶细胞稍厚壁，角质层平滑，三角体小至大，简单三角形或肿胀形，中部球状加厚缺。油体均一型。腹叶毗邻至远生，阔卵形，宽为茎的 1.0—1.5 倍，顶端圆钝，常强烈下弯，边缘全缘，基部两侧沿茎长下延，下延部分常具裂片状齿。

生境：石生环境常见，潮湿石生、腐木、树基、藤茎环境也有分布。海拔 1051—1730 m。

雅长分布：草王山，达陇坪屯，大棚屯，大宴坪天坑漏斗，黄猄洞天坑，里郎天坑，逻家田屯，盘古王，下棚屯，香当天坑，悬崖天坑，中井屯。

中国分布：福建，甘肃，广西，贵州，河北，湖北，湖南，江西，陕西，山东，四川，台湾，西藏，新疆，云南，浙江。

Plants green to dark brown when fresh, 1.6–2.8 mm wide, pinnately branched, branching of the *Frullania*-type. Stem in transverse section 13–14 cells thick, cortical cells of 2–4 layers. Rhizoids few, in tufts from underleaf bases. Leaves imbricate, obliquely or horizontally spreading, lobe broad oblong, apex rounded, strongly involute, margin entire, dorsal margin straight, slightly incurved, base slightly auriculate, ventral margin slightly arched, not ampliate to slightly amplicate near base, base slightly auriculate; lobule broad lingulate, apex rounded-obtuse to obtuse, margin entire, attached to the lobe by a indistinct keel or without keel, base short-decurrent, entire or sometimes laciniate-toothed in decurrent area. Leaf cells slightly thick-walled, trigones small to large, simple-triangulate or nodulate, intermediate thickenings absent. Cuticle smooth. Oil bodies homogeneous. Underleaves contigous to distant, broad ovate, 1.0–1.5 times as wide as the stem, apex rounded-obtuse, margin entire, base long-decurrent, laciniate-toothed in decurrent area.

Habitat: Often on rocks, sometimes on wet rocks, rotten logs, tree bases and liana. Elev. 1051–1730 m.

Distribution in Yachang: Caowangshan, Dalongping Tun, Dapeng Tun, Dayanping Tiankeng, Huangjingdong Tiankeng, Lilang Tiankeng, Luojiatian Tun, Panguwang, Xiapeng Tun, Xiangdang Tiankeng, Xuanya Tiankeng, Zhongjing Tun.

Distribution in China: Fujian, Gansu, Guangxi, Guizhou, Hebei, Hubei, Hunan, Jiangxi, Shaanxi, Shandong, Sichuan, Taiwan, Tibet, Xinjiang, Yunnan, Zhejiang.

A. 种群；B. 植物体；C. 植物体一段；D. 侧叶；E. 茎一段，示腹瓣和腹叶；F. 茎横切面；G. 腹瓣；H. 腹瓣和背瓣连接处；I. 叶中部细胞，示油体；J. 腹叶。（凭证标本：韦玉梅等 201107-9）

A. Population; B. Plant; C. Portion of plant; D. Leaves; E. Portion of stem showing leaf lobules and underleaves; F. Transverse section of stem; G. Leaf lobules; H. Connection between the leaf lobe and lobule; I. Median cells of leaf lobe showing oil bodies; J. Underleaves. (All from *Wei et al. 201107-9*)

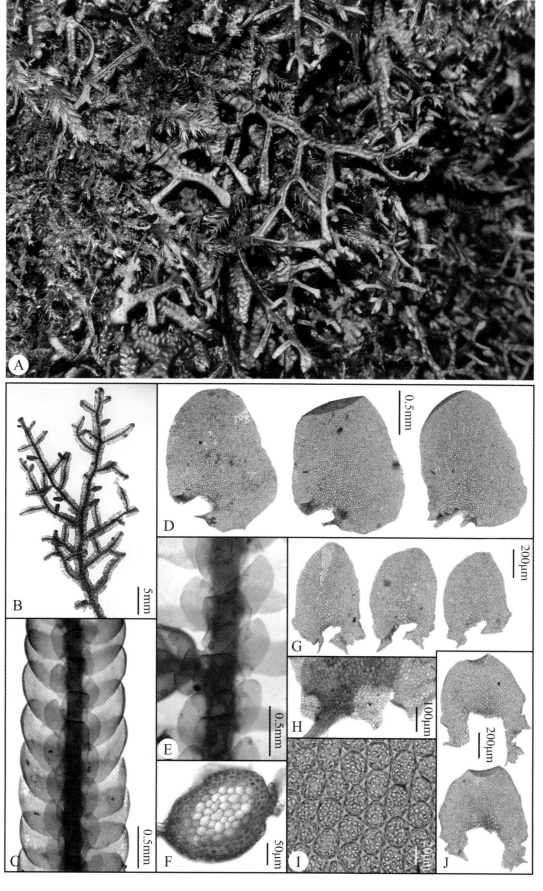

图 86 钝叶光萼苔鳞叶变种

Fig. 86 *Porella obtusata* var. *macroloba* (Steph.) S. Hatt. & M. X. Zhang

毛边光萼苔原变种

Porella perrottetiana (Mont.) Trevis. var. **perrottetiana**, Mem. Reale Ist. Lombardo Sci. (Ser. 3), C. Sci. Mat. 4(13): 408. 1877.

植物体绿色至棕褐色，带叶宽 2.7—3.9 mm，羽状分枝，分枝耳叶苔型。茎横切面 15—19 个细胞厚，皮部细胞 2—3 层，厚壁，内部细胞薄壁。假根少，成束生于腹叶基部。侧叶覆瓦状排列，倾斜伸展，背瓣卵形，顶端具长尾状尖，近顶部边缘具稀疏长齿，腹侧边缘稍拱起，外折，基部心形，背侧边缘拱起，近基部膨大至明显膨大，基部耳状；腹瓣舌形，顶部具 1—4 个长齿，两侧边缘全缘，与背瓣相连处具明显的短脊，近轴侧基部沿茎下延，下延部分具裂片状齿。侧叶细胞薄壁，角质层平滑，三角体小，简单三角形，中部球状加厚缺。油体均一型。腹叶覆瓦状排列至毗邻，三角状卵形至椭圆状卵形，宽为茎的 1.0—1.5 倍，顶端近截形，具不规则长齿，两侧边缘全缘或具稀疏长齿，基部两侧沿茎下延，下延部分具裂片状齿。雄苞和雌苞未见。

生境： 石生环境常见，腐木环境也有分布。海拔 1048—1324 m。

雅长分布： 大宴坪天坑漏斗，吊井天坑，黄猄洞天坑，拉洞天坑，里郎天坑，香当天坑。

中国分布： 除了东北以外的全国大部分省区均有分布。

Plants green to dark brown when fresh, 2.7–3.9 mm wide, pinnately branched, branching of the *Frullania*-type. Stem in transverse section 15–19 cells thick, cortical cells of 2–3 layers, thick-walled, medullary cells thin-walled. Rhizoids few, in tufts from underleaf bases. Leaves imbricate, obliquely spreading, lobe ovate, apex with a apical long tooth and few long additional spinose teeth, dorsal margin slightly arched, recurved, base cordate, ventral margin arched, ampliate to strongly ampliate near base, base auriculate; lobule lingulate, apex with 1–4 long teeth, lateral margin entire, attached to the lobe by a short but distinct keel, the proximal base decurrent, laciniate-toothed in decurrent area. Leaf cells thin-walled, trigones small, simple-triangulate, intermediate thickenings absent. Cuticle smooth. Oil bodies homogeneous. Underleaves imbricate to contigous, triangular-ovate to oblong-ovate, 1.0–1.5 times as wide as the stem, apex truncate, with irregular long teeth, lateral margin entire or toothed, base decurrent, laciniate-toothed in decurrent area. Androecia and gynoecia not seen.

Habitat: Often on rocks, sometimes on rotten logs. Elev. 1048–1324 m.

Distribution in Yachang: Dayanping Tiankeng, Diaojing Tiankeng, Huangjingdong Tiankeng, Ladong Tiankeng, Lilang Tiankeng, Xiangdang Tiankeng.

Distribution in China: Widely distributed in most provinces of China except Northeast Region.

▶ A. 种群；B. 植物体；C. 植物体一段；D. 侧叶；E. 茎一段，示腹瓣和腹叶；F. 茎横切面；G. 腹瓣；H. 腹叶；I. 叶中部细胞，示油体；J. 腹瓣和背瓣连接处。（凭证标本：*韦玉梅等 201107-8*）

A. Population; B. Plant; C. Portion of plant; D. Leaves; E. Portion of stem showing leaf lobules and underleaves; F. Transverse section of stem; G. Leaf lobules; H. Underleaves; I. Median cells of leaf lobe showing oil bodies; J. Connection between the leaf lobe and lobule. (All from *Wei et al. 201107-8*)

图 87a 毛边光萼苔原变种

Fig. 87a *Porella perrottetiana* (Mont.) Trevis. var. *perrottetiana*

毛边光萼苔狭叶变种（新拟中文名）

Porella perrottetiana var. **angustifolia** Pócs, J. Hattori Bot. Lab. 31: 75. 1968.

植物体浅绿色至黄绿色，带叶宽 4.2—7.4 mm，不规则分枝，耳叶苔型。茎横切面 15—17 个细胞厚，皮部细胞 1—2 层，厚壁，内部细胞薄壁。假根多，成束生于腹叶基部。侧叶覆瓦状排列，倾斜或近水平伸展，背瓣长椭圆形、椭圆状长方形或长椭圆状卵形，顶端斜截形，具 3—5 不规则长粗齿，腹侧边缘向内弯曲或略直，稍背弯且常波曲，基部心形，背侧边缘拱起，近顶端有时具 1—2 个锐齿，近基部稍膨大，基部耳状；腹瓣椭圆状长方形或长方形，边缘密生纤毛状长齿，与背瓣相连处不具脊或脊不明显，两侧基部短下延，下延部分边缘具纤毛状长齿。侧叶细胞薄壁至稍厚壁，角质层平滑，三角体小至大，简单三角形或肿胀形，中部球状加厚偶尔可见。油体均一型。腹叶覆瓦状排列，长椭圆状卵形，宽为茎的 1.0—1.5 倍，边缘密生纤毛状长齿，基部两侧沿茎下延，下延部分边缘具纤毛状长齿。雄苞和雌苞未见。

生境：石生。海拔 1315—1240 m。

雅长分布：大宴坪天坑漏斗，蓝家湾天坑。

中国分布：贵州。首次记录于广西。

Plants light green to yellowish green when fresh, 4.2–7.4 mm wide, irregularly branched, branching of the *Frullania*-type. Stem in transverse section 15–17 cells thick, cortical cells of 1–2 layers, thick-walled, medullary cells thin-walled. Rhizoids numerous, in tufts from underleaf bases. Leaves imbricate, obliquely or horizontally spreading, lobe oblong, oblong-retangular or oblong-ovate, apex obliquely truncate, with 3–5 irregularly long teeth, dorsal margin slightly sinuate or straight, slightly recurved, crispate, base cordate, ventral margin arched, with 1–2 teeth near apex, slightly ampliate near base, base auriculate; lobule oblong-retangular or retangular, margin densely toothed, attached to the lobe by a indistinct keel or without keel, base short-decurrent, ciliate-toothed in decurrent area. Leaf cells thin-walled to slightly thick-walled, trigones small to large, simple-triangulate or nodulate, intermediate thickenings absent. Cuticle smooth. Oil bodies homogeneous. Underleaves imbricate, oblong-ovate, 1.0–1.5 times as wide as the stem, margin densely toothed, base decurrent, ciliate-toothed in decurrent area. Androecia and gynoecia not seen.

Habitat: On rocks. Elev. 1315–1240 m.

Distribution in Yachang: Dayanping Tiankeng, Lanjiawan Tiankeng.

Distribution in China: Guizhou. New to Guangxi.

A. 种群；B. 植物体；C. 植物体一段；D. 侧叶；E. 茎一段，示腹瓣和腹叶；F. 茎横切面；G. 腹瓣；H. 叶中部细胞，示油体；I. 腹叶；J. 腹瓣和背瓣连接处。（凭证标本：*唐启明等 20210718-79*）

A. Population; B. Plant; C. Portion of plant; D. Leaves; E. Portion of stem showing leaf lobules and underleaves; F. Transverse section of stem; G. Leaf lobules; H. Median cells of leaf lobe showing oil bodies; I. Underleaves; J. Connection between the leaf lobe and lobule. (All from *Tang et al. 20210718-79*)

图 87b 毛边光萼苔狭叶变种

Fig. 87b *Porella perrottetiana* var. *angustifolia* Pócs

毛边光萼苔齿叶变种

Porella perrottetiana var. **ciliatodentata** (P. C. Chen et P. C. Wu) S. Hatt., J. Hattori Bot. Lab. 30: 144. 1967.

植物体绿色至棕褐色，带叶宽 3.2—4.5 mm，羽状分枝，耳叶苔型。茎横切面 23—25 个细胞厚，皮部细胞 3—4 层，厚壁，内部细胞薄壁。假根少，生于腹叶基部。侧叶覆瓦状排列，倾斜伸展，背瓣宽三角状卵形，顶端具顶生的长齿，顶生齿两侧具 4—10 个长纤毛状齿，腹侧边缘拱起，反折，基部心形，背侧边缘拱起，近基部明显膨大，基部耳状；腹瓣圆方形，边缘密生纤毛状长齿，与背瓣相连处具明显的短脊，近轴侧基部下延，下延部分全缘有时强烈卷曲。侧叶细胞稍厚壁，角质层平滑，三角体大，肿胀形，中部球状加厚缺。油体均一型。腹叶毗邻，圆卵形，宽为茎的 1—2 倍，边缘密生纤毛状长齿，基部两侧沿茎长下延，下延部分全缘。雄苞和雌苞未见。

生境： 石生。海拔 1185—1612 m。

雅长分布： 九十九堡，老屋基天坑，逻家田屯，悬崖天坑。

中国分布： 安徽，重庆，福建，甘肃，广东，贵州，河南，湖北，湖南，江西，陕西，四川，云南，浙江。首次记录于广西。

Plants green to dark brown when fresh, 3.2–4.5 mm wide, pinnately branched, branching of the *Frullania*-type. Stem in transverse section 23–25 cells thick, cortical cells of 3–4 layers, thick-walled, medullary cells thin-walled. Rhizoids few, in tufts from underleaf bases. Leaves imbricate, obliquely spreading, lobe broad triangular-ovate, apex with a apical long tooth and 4–10 additional ciliate teeth, dorsal margin arched, recurved, base cordate, ventral margin arched, strongly ampliate near base, base auriculate; lobule rounded-quadrate, margin densely ciliate-toothed, attached to the lobe by a short but distinct keel, the proximal base long-decurrent, entire and sometimes strongly revolute in decurrent area. Leaf cells slightly thick-walled, trigones large, nodulate, intermediate thickenings absent. Cuticle smooth. Oil bodies homogeneous. Underleaves contiguous, rounded-ovate, 1–2 times as wide as the stem, margin densely ciliate-toothed, base long-decurrent, laciniate-toothed in decurrent area, entire in decurrent area. Androecia and gynoecia not seen.

Habitat: On rocks. Elev. 1185–1612 m.

Distribution in Yachang: Jiushijiubao, Laowuji Tiankeng, Luojiatian Tun, Xuanya Tiankeng.

Distribution in China: Anhui, Chongqing, Fujian, Gansu, Guangdong, Guizhou, Henan, Hubei, Hunan, Jiangxi, Shaanxi, Sichuan, Yunnan, Zhejiang. New to Guangxi.

▶ A. 种群；B. 植物体；C. 植物体一段；D. 侧叶；E. 腹瓣和背瓣连接处；F. 茎一段，示腹瓣和腹叶；G. 茎横切面；H. 腹瓣；I. 叶中部细胞，示油体；J. 腹叶。（凭证标本：*韦玉梅等 201107-27*）

A. Population; B. Plant; C. Portion of plant; D. Leaves; E. Connection between the leaf lobe and lobule; F. Portion of stem showing leaf lobules and underleaves; G. Transverse section of stem; H. Leaf lobules; I. Median cells of leaf lobe showing oil bodies; J. Underleaves. (All from *Wei et al. 201107-27*)

图 87c 毛边光萼苔齿叶变种

Fig. 87c *Porella perrottetiana* var. *ciliatodentata* (P. C. Chen & P. C. Wu) S. Hatt.

小瓣光萼苔

Porella plumosa (Mitt.) Parihar, Univ. Allahabad Stud., Bot. Sect. 1961-62(Bot. sect.): 17. 1962.

植物体绿色至棕绿色，带叶宽 1.9—3.0 mm，羽状分枝，耳叶苔型。茎横切面 14—18 个细胞厚，皮部细胞 2—3 层，厚壁，内部细胞薄壁。假根少，生于腹叶基部。侧叶疏松覆瓦状排列，倾斜伸展，背瓣椭圆状卵形，顶端钝或钝尖，具不规则锐齿，齿的数量变化大，有时仅 1—2 个，有时 3—7 个，腹侧边缘稍拱起，下半部内弯，上半部背弯，基部截形，背侧边缘拱起，近基部不膨大或略微膨大，基部稍耳状；腹瓣舌形，边缘全缘，与背瓣相连处具短脊，近轴侧基部不下延或略下延，但常小片状突出。侧叶细胞薄壁，仅边缘稍厚壁，角质层平滑，三角体小，简单三角形，中部球状加厚缺。油体均一型。腹叶远生，椭圆状长方形，与茎等宽或略大于茎，顶端截形，边缘全缘，基部两侧略微下延。雄苞和雌苞未见。

生境： 石生环境常见，潮湿石生、潮湿土生、树干、树基环境也有分布。海拔 1240—1860 m。

雅长分布： 草王山，大宴坪天坑漏斗，黄猄洞天坑，拉洞屯，蓝家湾天坑，盘古王，下棚屯，霄罗湾洞穴。

中国分布： 甘肃，广西，贵州，江西，陕西，四川，台湾，云南，浙江。

Plants green to brownish green when fresh, 1.9−3.0 mm wide, pinnately branched, branching of the *Frullania*-type. Stem in transverse section 14−18 cells thick, cortical cells of 2−3 layers, thick-walled, medullary cells thin-walled. Rhizoids few, in tufts from underleaf bases. Leaves imbricate, obliquely spreading, lobe oblong-ovate, apex obtuse or obtuse-acute, with irregular teeth, sometimes entire, number of teeth vary, sometimes with 1−2 teeth, sometimes 3−7, dorsal margin slightly arched, incurved at lower part, recurved at upper part, base truncate, ventral margin arched, not ampliate to slightly ampliate near base, base slightly auriculate; lobule lingulate, margin entire, attached to the lobe by a short but distinct keel, the proximal base not decurrent or weak-decurrent, but appendiculate. Leaf cells thin-walled, except thick-walled along margin, trigones small, simple-triangulate, intermediate thickenings absent. Cuticle smooth. Oil bodies homogeneous. Underleaves distant, oblong-rectangular, as wide as the stem or slightly wider than the stem, apex truncate, margins entire, base slightly decurrent, entire in decurrent area. Androecia and gynoecia not seen.

Habitat: Often on rocks, sometimes on wet rocks, wet soil, tree trunks and tree bases. Elev. 1240−1860 m.

Distribution in Yachang: Caowangshan, Dayanping Tiankeng, Huangjingdong Tiankeng, Ladong Tun, Lanjiawan Tiankeng, Panguwang, Xiapeng Tun, Xiaoluowan Cave.

Distribution in China: Gansu, Guangxi, Guizhou, Jiangxi, Shaanxi, Sichuan, Taiwan, Yunnan, Zhejiang.

A. 种群；B. 植物体；C. 植物体一段；D. 侧叶；E. 茎一段，示腹瓣和腹叶；F. 茎横切面；G. 腹瓣；H. 叶中部细胞，示油体；I. 腹叶；J. 腹瓣和背瓣连接处。（凭证标本：*韦玉梅等 191013-96*）

A. Population; B. Plant; C. Portion of plant; D. Leaves; E. Portion of stem showing leaf lobules and underleaves; F. Transverse section of stem; G. Leaf lobules; H. Median cells of leaf lobe showing oil bodies; I. Underleaves; J. Connection between the leaf lobe and lobule. (All from *Wei et al. 191013-96*)

图 88 小瓣光萼苔

Fig. 88 *Porella plumosa* (Mitt.) Parihar

齿边光萼苔

Porella stephaniana (C. Massal.) S. Hatt., J. Hattori Bot. Lab. 5: 81. 1951.

植物体绿色至暗绿色，带叶宽 2.3—3.7 mm，不规则分枝，耳叶苔型。茎横切面 14—18 个细胞厚，皮部细胞 2—3 层，厚壁，内部细胞薄壁。假根多，生于匍匐茎腹叶基部。侧叶覆瓦状排列，倾斜伸展，背瓣卵形，顶端急尖，腹侧边缘直或稍拱起，强烈内弯且波曲，边缘具粗齿，基部截形，背侧边缘拱起，近基部略微膨大，近顶部边缘具齿，基部稍耳状；腹瓣舌形，全边缘具齿，与背瓣相连处具明显的短脊，近轴侧基部短下延，下延部分具裂片状齿。侧叶细胞稍厚壁，角质层平滑，三角体小至大，简单三角形，有时基部呈肿胀形，中部球状加厚缺。油体均一型。腹叶远生，舌形，与茎等宽或略小于茎，顶端常 2 裂，通常还有 1—2 个小齿在中间，两侧边缘具齿，基部两侧短下延，下延部分具裂片状齿。雄苞和雌苞未见。

生境： 石生。海拔 1240 m。

雅长分布： 蓝家湾天坑。

中国分布： 甘肃，广西，贵州，湖北，陕西，四川，云南，浙江。

Plants green to dark green when fresh, 2.3–3.7 mm wide, irregularly branched, branching of the *Frullania*-type. Stem in transverse section 14–18 cells thick, cortical cells of 2–3 layers, thick-walled, medullary cells thin-walled. Rhizoids numerous, in tufts from underleaf bases. Leaves imbricate, obliquely spreading, lobe ovate, apex acute, dorsal margin straight or slightly arched, strongly incurved, crispate, toothed along margin, base truncate, ventral margin arched, toothed near apex, slightly ampliate near base, base slightly auriculate; lobule lingulate, margin toothed, attached to the lobe by a short but distinct keel, the proximal base short-decurrent, laciniate-toothed in decurrent area. Leaf cells slightly thick-walled, trigones small to large, simple-triangulate, or sometimes nodulate at base, intermediate thickenings absent. Cuticle smooth. Oil bodies homogeneous. Underleaves distant, lingulate, as wide as the stem or slightly smaller than the stem, apex bilobed, usually with 1–2 teeth in between, lateral margin toothed, base short-decurrent, laciniate-toothed in decurrent area. Androecia and gynoecia not seen.

Habitat: On rocks. Elev. 1240 m.

Distribution in Yachang: Lanjiawan Tiankeng.

Distribution in China: Gansu, Guangxi, Guizhou, Hubei, Shaanxi, Sichuan, Yunnan, Zhejiang.

▶ A. 种群；B. 植物体；C. 植物体一段；D. 侧叶；E. 茎一段，示腹瓣和腹叶；F. 茎横切面；G. 腹瓣；H. 叶中部细胞，示油体；I. 腹叶；J. 腹瓣和背瓣连接处。（凭证标本：*韦玉梅等 191018-331*）

A. Population; B. Plant; C. Portion of plant; D. Leaves; E. Portion of stem showing leaf lobules and underleaves; F. Transverse section of stem; G. Leaf lobules; H. Median cells of leaf lobe showing oil bodies; I. Underleaves; J. Connection between the leaf lobe and lobule. (All from *Wei et al. 191018-331*)

图 89 齿边光萼苔

Fig. 89 *Porella stephaniana* (C. Massal.) S. Hatt.

多瓣光萼苔（多瓣苔）

Porella ulophylla (Steph.) S. Hatt., Bull. Tokyo Sci. Mus. 11: 92. 1944.

Macvicaria ulophylla (Steph.) S. Hatt., J. Hattori Bot. Lab. 5: 81. 1951.

植物体绿色至棕褐色，带叶宽 3.0—4.6 mm，羽状分枝，耳叶苔型。茎横切面 14—16 个细胞厚，皮部细胞 2—3 层，厚壁，内部细胞薄壁。假根多，生于腹叶基部。侧叶覆瓦状排列，近水平或稍倾斜伸展，背瓣卵形至卵圆形，顶端圆至圆钝，边缘全缘，但强烈波状卷曲，腹侧边缘稍拱起，基部截形，背侧边缘拱起，近基部不膨大，基部耳状；腹瓣舌形或三角状舌形，边缘全缘，或偶尔具少数齿，波状卷曲，与背瓣相连处具明显的短脊，近轴侧基部长下延，下延部分具裂片状齿。侧叶细胞薄壁，角质层平滑，三角体小，简单三角形，中部球状加厚缺。油体均一型。腹叶毗邻，卵形至三角状卵形，宽为茎的 1—2 倍，边缘全缘，波状卷曲，基部两侧长下延，下延部分具裂片状齿。雄苞和雌苞未见。

生境： 树干。海拔 1215 m。

雅长分布： 香当天坑。

中国分布： 全国大部分省区均有分布。首次记录于广西。

Plants green to brownish green when fresh, 3.0–4.6 mm wide, pinnately branched, branching of the *Frullania*-type. Stem in transverse section 14–16 cells thick, cortical cells of 2–3 layers, thick-walled, medullary cells thin-walled. Rhizoids numerous, in tufts from underleaf bases. Leaves imbricate, obliquely or horizontally spreading, lobe ovate to rounded-ovate, apex obtuse to rounded-obtuse, margin entire, but strongly crispate, dorsal margin slightly arched, base truncate, ventral margin arched, not ampliate near base, base auriculate; lobule lingulate to triangular-lingulate, margin entire, or occasionally with few teeth, crispate, attached to the lobe by a short but distinct keel, the proximal base long-decurrent, laciniate-toothed in decurrent area. Leaf cells thin-walled, trigones small, simple-triangulate, intermediate thickenings absent. Cuticle smooth. Oil bodies homogeneous. Underleaves contiguous, ovate to triangular-ovate, 1–2 times as wide as the stem, margin entire, or occasionally with few teeth, crispate, base long-decurrent, laciniate-toothed in decurrent area. Androecia and gynoecia not seen.

Habitat: On tree trunk. Elev. 1215 m.

Distribution in Yachang: Xiangdang Tiankeng.

Distribution in China: Widely distributed in most provinces of China. New to Guangxi.

▶ A. 种群；B. 植物体；C. 植物体一段；D. 侧叶；E. 叶边缘细胞；F. 腹瓣和背瓣连接处；G. 茎横切面；H. 腹瓣；I. 叶中部细胞，示油体；J. 腹叶。（凭证标本：*唐启明等 20201107-190*）

A. Population; B. Plant; C. Portion of plant; D. Leaves; E. Marginal cells of leaf lobe; F. Connection between the leaf lobe and lobule; G. Transverse section of stem; H. Leaf lobules; I. Median cells of leaf lobe showing oil bodies; J. Underleaves. (All from *Tang et al. 20201107-190*)

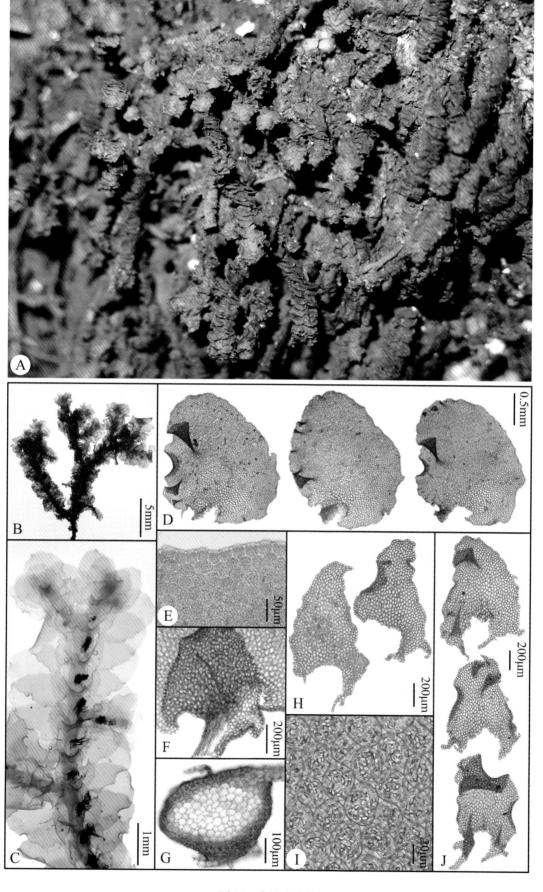

图 90 多瓣光萼苔

Fig. 90 *Porella ulophylla* (Steph.) S. Hatt.

尖舌扁萼苔

Radula acuminata Steph., Sp. Hepat. (Stephani) 4: 230. 1910.

植物体绿色，带叶宽 1.03—1.60 mm，不规则羽状分枝，分枝扁萼苔型。茎横切面 4—5 个细胞厚，表皮细胞和内部细胞几乎等大，薄壁，具小三角体；侧叶覆瓦状排列，背瓣卵形，顶端圆至圆钝，边缘全缘，背侧基部稍覆盖茎；腹瓣近方形，偶近舌形，为背瓣长的 1/3—1/2，顶端常突出呈钝尖状，近轴边缘略向内弯曲，基部略拱起，不盖茎，龙骨区略膨起；假根着生区稍凸起，假根束状，丛生；背脊与茎呈 40°—50° 角，向外拱起，不下延，弯缺狭，呈锐角。叶细胞薄壁，角质层平滑，三角体小，不明显，中部球状加厚缺。油体聚合型，每个细胞 1—2 个。腹叶缺。雄苞和雌苞未见。芽胞圆盘形，常垂直着生在背瓣腹面。

生境：树干，叶附生。海拔 620—931 m。

雅长分布：二沟，深洞。

中国分布：安徽，福建，广东，广西，贵州，海南，湖北，湖南，江西，四川，台湾，西藏，云南，浙江。

Plants green when fresh, 1.03–1.60 mm wide, irregularly pinnately branched, branching of the *Radula*-type. Stem in transverse section 4–5 cells thick, epidermal cells equal to medullary cells in size, both cells thin-walled with small trigones. Leaves imbricate, lobe ovate, apex rounded to rounded-obtuse, margin entire, dorsal base slightly covering the stem; lobule subquadrate, rarely sublingulate, 1/3–1/2 as long as the lobe, apex usually elongating to a blunt tip, adaxial margin sinuate toward the arched base, not covering the stem; carinal region weakly inflated; rhizoid-initial area convex, with bundle of rhizoids; keel extending at angles of 40°–50° with the stem, arched, not decurrent, the sinus subacute, narrow. Lobe cells thin-walled, trigones small, indistinct, intermediate thickenings absent. Cuticle smooth. Oil bodies segmented, 1–2 per median cell. Underleaves absent. Androecia and gynoecia not seen. Gemmae discoid, occurring vertically on ventral surface of leaf lobes.

Habitat: On tree trunks and leaves. Elev. 620–931 m.

Distribution in Yachang: Ergou, Shendong.

Distribution in China: Anhui, Fujian, Guangdong, Guangxi, Guizhou, Hainan, Hubei, Hunan, Jiangxi, Sichuan, Taiwan, Tibet, Yunnan, Zhejiang.

▶ A. 种群；B. 植物体；C. 植物体一段；D. 侧叶带芽胞；E. 背瓣中部细胞，示油体；F. 茎一段，示腹瓣；G. 芽胞；H. 油体。（凭证标本：*韦玉梅等 201109-29*）

A. Population; B. Plant; C. Portion of plant; D. Leaves with gemmae; E. Median cells of leaf lobe showing oil bodies; F. Portion of stem showing leaf lobules; G. Gemmae; H. Transverse section of stem. (All from *Wei et al. 201109-29*)

图 91　尖舌扁萼苔

Fig. 91　*Radula acuminata* Steph.

大瓣扁萼苔

Radula cavifolia Hampe ex Gottsche, Lindenb. et Nees, Syn. Hepat. 2: 259. 1845.

植物体绿色至黄绿色，带叶宽 0.65—1.05 mm，不规则羽状分枝，分枝扁萼苔型。茎横切面 5—6 个细胞厚，表皮细胞和内部细胞几乎等大，薄壁，具大三角体。侧叶疏松覆瓦状排列至毗邻，背瓣近圆形，内凹，顶端圆钝，内弯，边缘全缘，背侧基部完全盖茎；腹瓣大，为背瓣长的 5/6，近轴边缘略向内弯曲，基部直，不盖茎，龙骨区略膨起；假根着生区稍凸起，假根束状，丛生；背脊与茎呈 60°—70°角，向外拱起，不下延，弯缺无。叶细胞薄壁，角质层具细疣，三角体小，不明显或简单三角形，中部球状加厚缺。油体聚合型，每个细胞 1 个。腹叶缺。雄苞和雌苞未见。

生境：树干，树基以及倒木生。海拔 1861—1878 m。

雅长分布：草王山。

中国分布：安徽，重庆，福建，广东，广西，贵州，海南，江西，四川，台湾，香港，云南，浙江。

Plants green to yellowish green when fresh, 0.65–1.05 mm wide, irregularly pinnately branched, branching of the *Radula*-type. Stem in transverse section 5–6 cells thick, epidermal cells equal to medullary cells in size, both cells thin-walled with large trigones. Leaves loosely imbricate, lobe subrounded, apex rounded-obtuse, concave, margin entire, dorsal base covering the stem; lobule large, 5/6 as long as the lobe, adaxial margin sinuate toward the arched base, not covering the stem; carinal region weakly inflated; rhizoid-initial area convex, with bundle of rhizoids; keel extending at angles of 60°–70° with the stem, arched, not decurrent, the sinus absent. Lobe cells thin-walled, trigones small, indistinct or simple-triangulate, intermediate thickenings absent. Cuticle minutely verrucose. Oil bodies segmented, 1 per median cell. Underleaves absent. Androecia and gynoecia not seen.

Habitat: On tree trunks, tree bases, and fallen trees. Elev. 1861–1878 m.

Distribution in Yachang: Caowangshan.

Distribution in China: Anhui, Chongqing, Fujian, Guangdong, Guangxi, Guizhou, Hainan, Hongkong, Jiangxi, Sichuan, Taiwan, Yunnan, Zhejiang.

▶ A. 种群；B–C. 植物体一段；D. 侧叶；E. 茎横切面；F. 腹瓣；G. 背瓣中部细胞，示角质层细疣；H. 背瓣中部细胞，示油体；I. 植物体一段（背面观）。（凭证标本：*韦玉梅等 211112-43*）

A. Population; B–C. Portions of plants; D. Leaves; E. Transverse section of stem; F. Leaf lobule; G. Median cells of leaf lobe showing dorsal cuticle; H. Median cells of leaf lobe showing oil bodies; I. Portion of plant (dorsal view). (All from *Wei et al. 211112-43*)

图 92 大瓣扁萼苔

Fig. 92 *Radula cavifolia* Hampe ex Gottsche, Lindenb. et Nees

扁萼苔

Radula complanata (L.) Dumort., Syll. Jungerm. Europ.: 38. 1831.

植物体浅绿色至绿色，带叶宽 1.10—1.73 mm，不规则羽状分枝，分枝扁萼苔型。茎横切面约 6 个细胞厚，表皮细胞和内部细胞几乎等大，薄壁，具小三角体；侧叶覆瓦状排列至毗邻，背瓣卵圆形，顶端圆，略内弯，边缘全缘，背侧基部完全盖茎；腹瓣方形至近方形，约为背瓣长的 1/2，顶端钝，近轴边缘直，基部覆盖茎宽的 1/3—1/2，龙骨区扁平或略膨起。假根着生区不明显或稍凸起，假根少。背脊与茎呈 70°—80° 角，直，不下延，弯缺钝或无。叶细胞薄壁，角质层平滑，三角体小，不明显，中部球状加厚缺。油体聚合型，每个细胞 1 个。腹叶缺。雌雄同株。雄苞常间生，雄苞叶 2—4 对。雌苞具 1—2 个新生枝。蒴萼和芽胞未见。

生境：潮湿石生，石生。海拔 1136—1730 m。

雅长分布：草王山，黄猄洞天坑，蓝家湾天坑。

中国分布：全国范围广布。

备注：该种有时背瓣的背侧边缘长有盘状芽胞，但是雅长的种群均未发现有长芽胞的情况。

Plants light green to green when fresh, 1.10–1.73 mm wide, irregularly pinnately branched, branching of the *Radula*-type. Stem in transverse section ca. 6 cells thick, epidermal cells equal to medullary cells in size, both cells thin-walled with small trigones. Leaves imbricate to contiguous, lobe rounded-ovate, apex rounded, slightly incurved, margin entire, dorsal base fully covering the stem; lobule quadrate to subquadrate, ca. 1/2 as long as the lobe, apex obtuse, adaxial margin straight, base covering the stem 1/3–1/2 the stem-width; carinal region not or weakly inflated; rhizoid-initial area indistinct or slightly convex, rhizoids few; keel extending at angles of 70°–80° with the stem, straight, not decurrent, the sinus very wide or almost none. Lobe cells thin-walled, trigones small, indistinct, intermediate thickenings absent. Cuticle smooth. Oil bodies segmented, 1 per median cell. Underleaves absent. Monoicous. Androecia usually intercalary on branches, bracts in 2–4 pairs. Gynoecia with 1–2 innovations. Perianth and gemmae not seen.

Habitat: On (wet) rocks. Elev. 1136–1730 m.

Distribution in Yachang: Caowangshan, Huangjingdong Tiankeng, Lanjiawan Tiankeng.

Distribution in China: Widely distributed in China.

Note: Discoid gemmae of R. complanata sometimes occur on dorsal margin of leaf lobes, but have not been seen in all specimens from Yachang.

▶ A. 种群；B. 植物体；C. 植物体一段；D. 侧叶；E. 雌苞叶；F. 茎一段；G. 茎横切面；H. 背瓣中部细胞，示油体。（凭证标本：*韦玉梅等 191018-315*）

A. Population; B. Plant; C. Portion of plant; D. Leaves; E. Female bracts; F. Portion of stem; G. Transverse section of stem; H. Median cells of leaf lobe showing oil bodies. (All from *Wei et al. 191018-315*)

图 93 扁萼苔

Fig. 93 *Radula complanata* (L.) Dumort.

爪哇扁萼苔

Radula javanica Gottsche, Syn. Hepat. 2: 257. 1845.

植物体绿色，带叶宽1.59—2.20 mm，不规则羽状分枝，分枝扁萼苔型。茎横切面9—10个细胞厚，表皮细胞比内部细胞略小，薄壁，具三角形状的三角体；侧叶覆瓦状排列，背瓣有时易脱落，椭圆形至椭圆状卵形，顶端圆，边缘全缘，背侧基部完全盖茎；腹瓣近方形或近菱形，约为背瓣长的1/3，顶端圆钝，近轴边缘稍拱起呈弧形，基部稍突出，覆盖茎宽的1/3—3/4，龙骨区略膨起，假根着生区稍凸起，假根少。背脊与茎呈45°—60°角，直或略向内弯曲，不下延，弯缺钝角至锐角。叶细胞薄壁，角质层平滑，三角体小，不明显，中部球状加厚缺。油体聚合型，每个细胞1个。腹叶缺。雌雄异株。雄苞常顶生，雄苞叶3—9对。雌苞未见。

生境： 石生。海拔1444 m。

雅长分布： 盘古王。

中国分布： 福建，广东，广西，贵州，海南，湖北，江西，台湾，西藏，香港，云南，浙江。

Plants green when fresh, 1.59−2.20 mm wide, irregularly pinnately branched, branching of the *Radula*-type. Stem in transverse section 9−10 cells thick, epidermal cells equal to medullary cells in size, both cells thin-walled with triangular trigones. Leaves imbricate, lobe sometimes caducous, oblong to oblong-ovate, apex rounded, margin entire, dorsal base fully covering the stem; lobule subquadrate to subrhombic, ca. 1/3 as long as the lobe, apex rounded-obtuse, adaxial margin slightly arched, base covering the stem 1/3−3/4 the stem-width; carinal region weakly inflated; rhizoid-initial area slightly convex, rhizoids few; keel extending at angles of 45°−60° with the stem, straight or slightly sinuate, not decurrent, the sinus obtuse to subacute. Lobe cells thin-walled, trigones small, indistinct, intermediate thickenings absent. Cuticle smooth. Oil bodies segmented, 1 per median cell. Underleaves absent. Dioicous. Androecia usually terminal on branches, bracts in 3−9 pairs. Gynoecia not seen.

Habitat: On rocks. Elev. 1444 m.

Distribution in Yachang: Panguwang.

Distribution in China: Fujian, Guangdong, Guangxi, Guizhou, Hainan, Hongkong, Hubei, Jiangxi, Taiwan, Tibet, Yunnan, Zhejiang.

▶ A. 种群；B. 植物体；C. 植物体一段；D. 侧叶；E. 茎横切面；F. 背瓣中部细胞，示油体；G. 茎一段。（凭证标本：*唐启明 & 张仕艳 20181004-428B*）

A. Population; B. Plant; C. Portion of plant; D. Leaves; E. Transverse section of stem; F. Median cells of leaf lobe showing oil bodies; G. Portion of stem. (All from *Tang & Zhang 20181004-428B*)

图 94 爪哇扁萼苔

Fig. 94 *Radula javanica* Gottsche

尖叶扁萼苔

Radula kojana Steph., Bull. Herb. Boissier 5(2): 105. 1897.

植物体浅绿色至绿色，带叶宽 1.10—1.80 mm，不规则羽状分枝，分枝扁萼苔型。茎横切面约 8 个细胞厚，表皮细胞比内部细胞几乎等大，薄壁，无三角体；侧叶覆瓦状排列至毗邻，背瓣卵形，顶端细尖，边缘全缘，背侧基部覆盖茎宽的 1/2—3/4；腹瓣近方形，约为背瓣长的 1/3，顶端圆钝，近轴边缘拱起呈弧形，基部稍突出，覆盖约茎宽的 1/5，龙骨区膨起，假根着生区扁平或稍凸起，假根少。背脊与茎呈 55°—60° 角，向外拱起，不下延，弯缺呈锐角。叶细胞薄壁，角质层平滑，三角体小，不明显，中部球状加厚缺。油体眼球型，每个细胞 2—3 个。腹叶缺。雌雄异株。雄苞未见。雌苞具 1—2 个新生枝，雌苞叶 1—2 对。蒴萼未见。芽胞盘状，着生在背瓣边缘（*韦玉梅等 191018-289*）。

生境：潮湿石生，石生，潮湿土生，腐殖质。海拔 931—1450 m。

雅长分布：蓝家湾天坑，里郎天坑，盘古王，深洞。

中国分布：安徽，重庆，福建，广东，广西，贵州，海南，湖北，湖南，江西，四川，台湾，香港，新疆，云南，浙江。

Plants light green to green when fresh, 1.10−1.80 mm wide, irregularly pinnately branched, branching of the *Radula*-type. Stem in transverse section ca 8 cells thick, epidermal cells equal to medullary cells in size, both cells thin-walled without trigones. Leaves imbricate to contiguous, lobe ovate, apex apiculate, margin entire, dorsal base covering the stem 1/2−3/4 the stem-width; lobule subquadrate, ca. 1/3 as long as the lobe, apex rounded-obtuse, adaxial margin arched, base covering the stem ca. 1/5 the stem-width; carinal region inflated; rhizoid-initial area not or slightly convex, rhizoids few; keel extending at angles of 55°−60° with the stem, arched, not decurrent, the sinus subacute. Lobe cells thin-walled, trigones small, indistinct, intermediate thickenings absent. Cuticle smooth. Oil bodies eyeball-like, 2−3 per median cell. Underleaves absent. Dioicous. Androecia not seen. Gynoecia with 1−2 innovations. Perianth not seen. Gemmae discoid, occurring on the margins of leaf lobes (*Wei et al. 191018-289*).

Habitat: On (wet) rocks, wet soil and humus. Elev. 931−1450 m.

Distribution in Yachang: Lanjiawan Tiankeng, Lilang Tiankeng, Panguwang, Shendong.

Distribution in China: Anhui, Chongqing, Fujian, Guangdong, Guangxi, Guizhou, Hainan, Hongkong, Hubei, Hunan, Jiangxi, Sichuan, Taiwan, Xinjiang, Yunnan, Zhejiang.

A. 种群；B. 植物体带雌苞；C. 植物体一段；D. 侧叶；E. 雌苞叶；F. 茎横切面；G. 茎一段；H. 背瓣中部细胞，示油体。（凭证标本：*韦玉梅等 201110-21*）

A. Population; B. Plant with gynoecia; C. Portion of plant; D. Leaves; E. Female bracts; F. Transverse section of stem; G. Portion of stem; H. Median cells of leaf lobe showing oil bodies. (All from *Wei et al. 201110-21*)

图 95　尖叶扁萼苔

Fig. 95　*Radula kojana* Steph.

刺边扁萼苔

Radula lacerata Steph., Rev. Bryol. 35(2): 33. 1908.

植物体绿色，带叶宽 0.78—1.16 mm，规则羽状分枝，分枝扁萼苔型。茎横切面 5—6 个细胞厚，表皮细胞与内部细胞几乎等大，稍厚壁，具小三角体；侧叶覆瓦状排列，背瓣卵形或三角形，顶端长渐尖，边缘具不规则的稀疏的齿，背侧基部覆盖茎宽的 3/4—4/5；腹瓣近方形至卵形，为背瓣长的 1/3—1/2，顶端圆钝，近轴边缘直或稍拱起呈弧形，基部覆盖茎宽的 1/5—1/4，龙骨区膨起，假根着生区稍凸起，假根少。背脊与茎呈 60°—70° 角，明显向外拱起，不下延，弯缺钝角至锐角。叶细胞稍厚壁，角质层平滑，三角体小至大，简单三角形，中部球状加厚偶尔可见。油体眼球型，每个细胞 2—3 个。腹叶缺。雄苞和雌苞未见。

生境：叶附生。海拔 1240 m。

雅长分布：蓝家湾天坑。

中国分布：广西，海南。

Plants green when fresh, 0.78–1.16 mm wide, regularly pinnately branched, branching of the *Radula*-type. Stem in transverse section 5–6 cells thick, epidermal cells equal to medullary cells in size, both cells slightly thick-walled with minute trigones. Leaves imbricate, lobe ovate or triangular, apex acuminate, margin irregularly and sparsely toothed, dorsal base covering the stem 3/4–4/5 the stem-width; lobule subquadrate to ovate, 1/3–1/2 as long as the lobe, apex rounded-obtuse, adaxial margin straight to slightly arched, base covering the stem 1/5–1/4 the stem-width; carinal region inflated; rhizoid-initial area slightly convex, rhizoids few; keel extending at angles of 60°–70° with the stem, strongly arched, not decurrent, the sinus obtuse to subacute. Lobe cells slightly thick-walled, trigones small to large, simple-triangulate, intermediate thickenings occasionally present. Cuticle smooth. Oil bodies eyeball-like, 2–3 per median cell. Underleaves absent. Androecia and gynoecia not seen.

Habitat: Epiphyllous. Elev. 1240 m.

Distribution in Yachang: Lanjiawan Tiankeng.

Distribution in China: Guangxi, Hainan.

A. 种群；B. 植物体；C. 植物体一段；D. 侧叶；E. 背瓣中部细胞，示油体；F. 茎横切面；G. 茎一段（背面观）。（凭证标本：*韦玉梅等 191018-280A*）

A. Population; B. Plant; C. Portion of plant; D. Leaves; E. Median cells of leaf lobe showing oil bodies; F. Transverse section of stem; G. Portion of stem (doesal view). (All from *Wei et al. 191018-280A*)

图 96 刺边扁萼苔

Fig. 96 *Radula lacerata* Steph.

芽胞扁萼苔

Radula lindenbergiana Gottsche ex C. Hartm., Handb. Skand. Fl. (ed.9): 98 1864.

植物体绿色，带叶宽 1.43—2.40 mm，不规则羽状分枝，分枝扁萼苔型。茎横切面 7—8 个细胞厚，表皮细胞和内部细胞几乎等大，薄壁，具大的三角体；侧叶覆瓦状排列，背瓣阔圆形，顶端圆，边缘全缘，背侧基部完全盖茎；腹瓣方形或近方形，约为背瓣长的 1/2，顶端突出呈短尖状或不突出，近轴边缘直，常反折，基部覆盖茎宽的 1/2—3/4，龙骨区膨起。假根着生区常凸起，假根束状。背脊与茎呈 45°—60° 角，直或略向外拱起，不下延，弯缺钝或无。叶细胞薄壁，角质层平滑，三角体小，不明显，中部球状加厚缺。油体聚合型，每个细胞 1 个。腹叶缺。雄苞和雌苞未见。芽胞盘状，大量着生于背瓣边缘。

生境： 树干环境常见，石生、倒木、树基环境也有分布。海拔 1073—1368 m。

雅长分布： 白岩坨屯，吊井天坑，黄猄洞天坑，蓝家湾天坑，逻家田屯，下岩洞屯，菁罗湾洞穴，悬崖天坑，中井屯竖井。

中国分布： 全国大部分省区均有分布。

Plants green when fresh, 1.43–2.40 mm wide, irregularly pinnately branched, branching of the *Radula*-type. Stem in transverse section ca. 7–8 cells thick, epidermal cells equal to medullary cells in size, both cells thin-walled with large trigones. Leaves imbricate, lobe broad ovate, apex rounded, margin entire, dorsal base fully covering the stem; lobule quadrate to subquadrate, ca. 1/2 as long as the lobe, apex slightly or not elongate, subacute, adaxial margin straight, usually recurved, base covering the stem 1/2–3/4 the stem-width; carinal region inflated; rhizoid-initial area usually convex, rhizoids in a bundle; keel extending at angles of 45°–60° with the stem, straight or arched, not decurrent, the sinus obtuse or none. Lobe cells thin-walled, trigones small, indistinct, intermediate thickenings absent. Cuticle smooth. Oil bodies segmented, 1 per median cell. Underleaves absent. Androecia and gynoecia not seen. Gemmae discoid, abundantly occurring on the margins of leaf lobes.

Habitat: Often on tree trunks, sometimes on rocks, fallen trees and tree bases. Elev. 1073–1368 m.

Distribution in Yachang: Baiyantuo Tun, Diaojing Tiankeng, Huangjingdong Tiankeng, Lanjiawan Tiankeng, Luojiatian Tun, Xiayandong Tun, Xiaoluowan Cave, Xuanya Tiankeng, Zhongjing Tun.

Distribution in China: Widely distributed in most provinces of China.

▶ A. 种群；B. 植物体；C. 植物体一段；D. 植物体一段（背面观）；E. 背瓣中部细胞，示油体；F. 侧叶；G. 叶边缘芽胞；H. 茎横切面。（凭证标本：*韦玉梅等 201107-5*）

A. Population; B. Plant; C. Portion of plant; D. Portion of plant (dorsal view); E. Median cells of leaf lobe showing oil bodies; F. Leaf; G. Gemmae on the margin of leaf lobe; H. Transverse section of stem. (All from *Wei et al. 201107-5*)

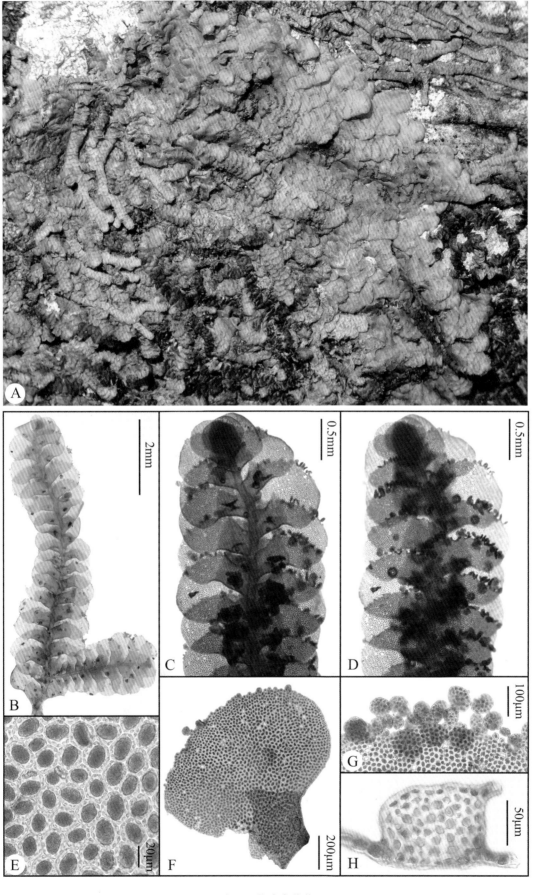

图 97　芽胞扁萼苔

Fig. 97　*Radula lindenbergiana* Gottsche ex C. Hartm.

星苞扁萼苔

Radula stellatogemmipara C. Gao et Y. H. Wu., Nova Hedwigia 80(1/2): 239. 2005.

植物体绿色，带叶宽 0.82—1.58 mm，不规则羽状分枝，分枝扁萼苔型。茎横切面约 6 个细胞厚，表皮细胞与内部细胞几乎等大，薄壁，具大的三角体；侧叶覆瓦状排列，背瓣易脱落，椭圆形至椭圆状卵形，顶端圆，常内弯，边缘全缘，背侧基部完全盖茎；腹瓣长方形，约为背瓣长的 1/2，顶端常突出呈钝尖状，近轴边缘直或略向内弯曲，基部几乎不盖茎，龙骨区膨起。假根着生区稍凸起，假根少。背脊与茎呈 55°—75° 角，略向外拱起，不下延，弯缺钝。叶细胞薄壁，角质层具细疣，三角体小，不明显，中部球状加厚缺。油体聚合型，每个细胞 1 个。腹叶缺。雄苞未见。雌苞常具 2 个新生枝。脱落的背瓣边缘常产生无性芽。

生境：腐木，树干。海拔 1073—1083 m。

雅长分布：白岩坨屯，黄猄洞天坑。

中国分布：福建，广西，贵州。

Plants green when fresh, 0.82–1.58 mm wide, irregularly pinnately branched, branching of the *Radula*-type. Stem in transverse section ca. 6 cells thick, epidermal cells equal to medullary cells in size, both cells thin-walled with large trigones. Leaves usually caducous, imbricate, lobe oblong to oblong-ovate, apex rounded, usually incurved, margin entire, dorsal base fully covering the stem; lobule rectangular, ca. 1/2 as long as the lobe, apex usually elongating to a blunt tip, adaxial margin straight or slightly sinuate, base slightly covering the stem; carinal region inflated; rhizoid-initial area weakly convex, rhizoids few; keel extending at angles of 55°–75° with the stem, slightly arched, not decurrent, the sinus obtuse. Lobe cells thin-walled, trigones small, indistinct, intermediate thickenings absent. Cuticle minutely verrucose. Oil bodies segmented, 1 per median cell. Underleaves absent. Dioicous. Androecia not seen. Gynoecia usually with 2 innovations. Propagules usually occurring on the margins of caducous leaf lobes.

Habitat: On rotten logs and tree trunks. Elev. 1073–1083 m.

Distribution in Yachang: Baiyantuo Tun, Huangjingdong Tiankeng.

Distribution in China: Fujian, Guangxi, Guizhou.

▶ A. 种群；B. 植物体；C. 植物体一段；D. 侧叶；E–K. 脱落的背瓣带无性芽；F. 茎一段；G. 雌苞叶；H. 茎横切面；I. 背瓣中部细胞，示角质层细疣；J. 背瓣中部细胞，示油体。（凭证标本：*韦玉梅等 191012-18*）

A. Population; B. Plant; C. Portion of plant; D. Leaf; E–K. Caducous leaf lobes with propagules; F. Portion of stem; G. Female bracts; H. Transverse section of stem; I. Median cells of leaf lobe showing dorsal cuticle; J. Median cells of leaf lobe showing oil bodies. (All from *Wei et al. 191012-18*)

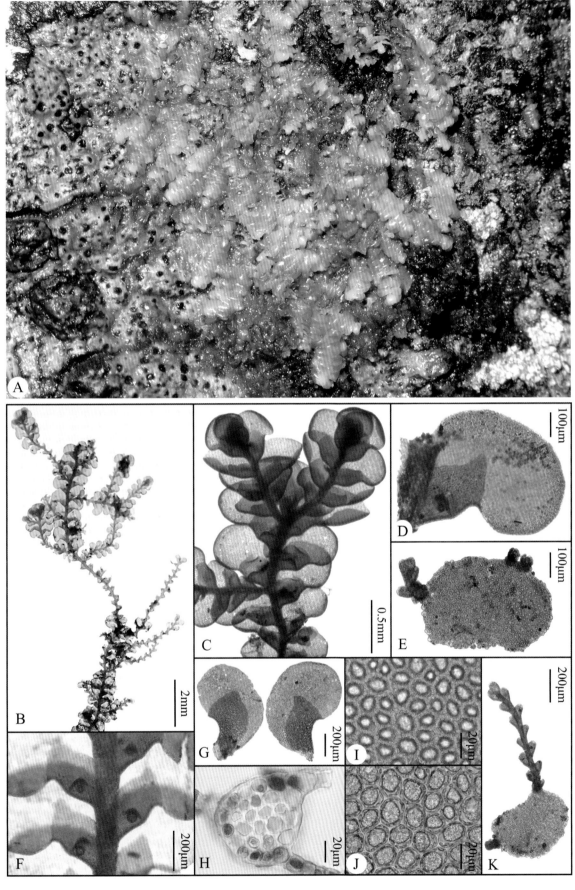

图 98　星苞扁萼苔

Fig. 98　*Radula stellatogemmipara* C. Gao & Y. H. Wu.

黑耳叶苔

Frullania amplicrania Steph., Sp. Hepat. 4: 404. 1910.

植物体绿色、棕绿色至黑棕色，带叶宽0.70—1.25 mm，不规则羽状分枝，分枝耳叶苔型。侧叶疏松覆瓦状排列至毗邻，背瓣近圆形，顶端圆，边缘全缘，轻微反折，腹侧基部心形，背侧基部耳形；腹瓣头骨状，紧贴茎着生并与茎平行，口部收缩变小，斜截形，不具喙；副体丝状，2—3个细胞长。侧叶细胞波曲状，角质层平滑，三角体小至大，简单三角形或肿胀形，中部球状加厚常可见。油体聚合型，每个细胞2—3个。腹叶远生，宽为茎的1—2倍，顶端2裂达1/3—2/5，两侧边缘常具1—2个齿，基部楔形。雌雄异株。雄苞未见。雌苞常侧生于分枝上，雌苞叶2—3对，边缘全缘。蒴萼未见。

生境：树干，树基。海拔1133—1238 m。

雅长分布：黄猄洞天坑，悬崖天坑，下岩洞屯。

中国分布：广西，台湾，浙江。

Plants green, browish green to brackish brown when fresh, 0.70−1.25 mm wide, irregularly pinnately branched, branching of the *Frullania*-type. Leaves loosely imbricate to contiguous, lobe suborbicular, apex rounded, margin entire, slightly recurved, ventral margin cordate, dorsal base auriculate; lobule skull-shaped, contiguous to and parallel to stem, mouth narrowed, obliquely truncate, without rostrum; stylus filiform, in a row of 2−3 cells. Lobe cells flexuose, trigones small to large, simple-triangulate or nodulate, intermediate thickenings frequent. Cuticle smooth. Oil bodies segmented, 2−3 per median cell. Underleaves distant, 1−2 times as wide as the stem, bilobed to 1/3−2/5 underleaf length, usually with 1−2 teeth at outer margins, base cuneate. Dioicous. Androecia not seen. Gynoecia usually lateral on branches, bracts in 2−3 pairs, margins entire. Perianths not seen.

Habitat: On tree trunks and tree bases. Elev. 1133−1238 m.

Distribution in Yachang: Huangjingdong Tiankeng, Xiayandong Tun, Xuanya Tiankeng.

Distribution in China: Guangxi, Taiwan, Zhejiang.

▶ A. 种群；B. 植物体；C. 植物体一段带雌苞；D. 背瓣；E. 腹瓣和副体；F. 植物体一段；G. 茎一段，示腹瓣；H. 腹叶；I. 背瓣中部细胞，示油体。（凭证标本：*韦玉梅等 201107-37*）

A. Population; B. Plant; C. Portion of plant with a gynoecium; D. Leaf lobes; E. Leaf lobule and stylus; F. Portion of plant; G. Portion of stem showing leaf lobules; H. Underleaves; I. Median cells of leaf lobe showing oil bodies. (All from *Wei et al. 201107-37*)

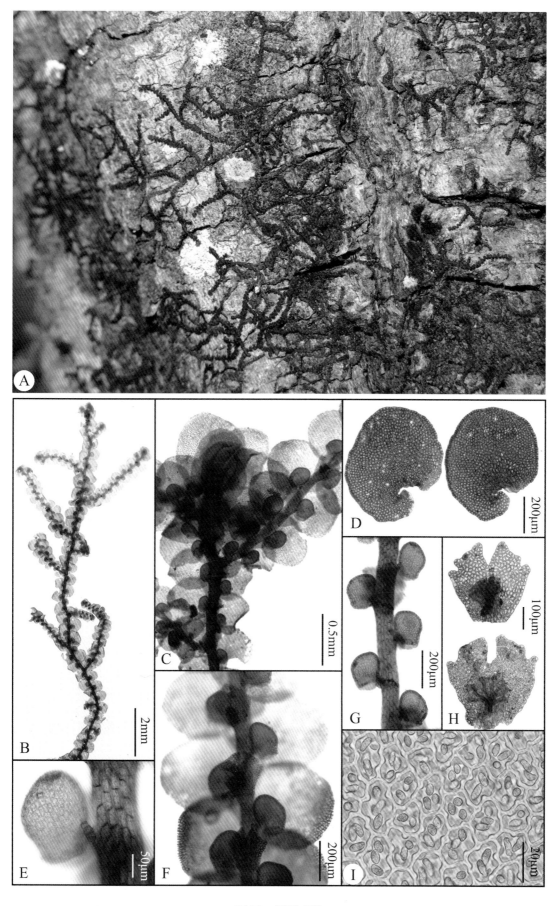

图 99 黑耳叶苔

Fig. 99 *Frullania amplicrania* Steph.

细茎耳叶苔

Frullania bolanderi Austin, Proc. Acad. Nat. Sci. Philadelphia 21(4): 226. 1869.

植物体绿色至红棕色，带叶宽 0.55—0.85 mm，不规则羽状分枝，分枝耳叶苔型。侧叶疏松覆瓦状排列至毗邻，背瓣易脱落，近圆形，顶端圆，边缘全缘（易脱落的侧叶边缘常生假根），腹侧基部截形，背侧基部心形；腹瓣兜状，紧贴茎着生，口部略下倾，一侧内折，具短钝的喙；副体丝状，2—3 个细胞长。叶细胞稍波曲状，角质层平滑，三角体大，简单三角形或肿胀形，中部球状加厚常可见。油体聚合型，每个细胞 3—6 个。腹叶远生，宽为茎的 1.0—1.5 倍，顶端 2 裂达 1/3—1/2 深，侧缘常具 1—2 个钝齿，基部楔形。雄苞和雌苞未见。

生境： 石生，树基。海拔 837—1342 m。

雅长分布： 二沟，下棚屯，中井天坑。

中国分布： 福建，甘肃，广西，贵州，湖南，吉林，内蒙古，陕西，四川，云南。

Plants green to redish brown when fresh, 0.55–0.85 mm wide, irregularly pinnately branched, branching of the *Frullania*-type. Leaves loosely imbricate to contiguous, lobe usually caducous, suborbicular, apex rounded, margin entire (rhizoids borne usually at margin of caducous leaf), ventral margin truncate, dorsal base cordate; lobule cucullate, contiguous to stem, mouth slightly inclined toward the stem, hemifold, with short obtuse rostrum; stylus filiform, in a row of 2–3 cells. Lobe cells slightly flexuose, trigones large, simple-triangulate or nodulate, intermediate thickenings frequent. Cuticle smooth. Oil bodies segmented, 3–6 per median cell. Underleaves distant, 1.0–1.5 times as wide as the stem, bilobed to 1/3–1/2 underleaf length, usually with 1–2 blunt teeth at outer margins, base cuneate. Androecia and gynoecia not seen.

Habitat: On rocks and tree bases. Elev. 837–1342 m.

Distribution in Yachang: Ergou, Xiapeng Tun, Zhongjing Tiankeng.

Distribution in China: Fujian, Gansu, Guangxi, Guizhou, Hunan, Inner Mongolia, Jilin, Shaanxi, Sichuan, Yunnan.

▶ A. 种群；B. 植物体；C. 植物体一段；D. 易脱落的侧叶；E. 背瓣中部细胞，示油体；F. 茎一段，示腹叶；G. 茎一段，示腹瓣；H. 腹瓣和副体。（凭证标本：*韦玉梅等 191015-188*）

A. Population; B. Plant; C. Portion of plant; D. Caducous leaves; E. Median cells of leaf lobe showing oil bodies; F. Portion of stem showing underleaves; G. Portion of stem showing leaf lobules; H. Leaf lobule and styllus. (All from *Wei et al. 191015-188*)

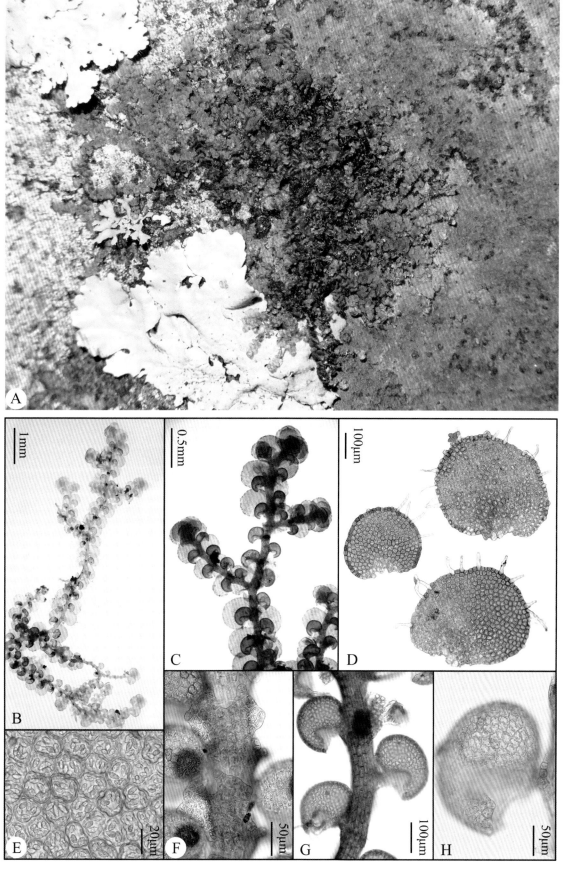

图 100　细茎耳叶苔

Fig. 100　*Frullania bolanderi* Austin

达乌里耳叶苔

Frullania davurica Hampe ex Gottsche, Lindenb. et Nees, Syn. Hepat.: 422. 1845.

植物体深绿色至红棕色，带叶宽 1.2—2.0 mm，不规则羽状分枝，分枝耳叶苔型。侧叶覆瓦状排列，背瓣阔卵形，顶端圆，边缘全缘，腹侧基部心形，背侧基部耳形；腹瓣兜状，紧贴茎着生，口部明显下倾，一侧内折，具短钝的喙；副体丝状或三角状，3—11 个细胞长，基部常 2 个细胞宽。叶细胞稍波曲状，角质层平滑，三角体大，肿胀形，中部球状加厚常可见。油体聚合型，每个细胞 5—8 个。腹叶疏松覆瓦状排列至毗邻，宽为茎的 3.5—5.0 倍，顶端不裂，边缘全缘，基部稍心形。雌雄异株。雄苞未见。雌苞顶生或侧生于分枝上，雌苞叶 3 对，边缘具齿，最内层雌苞叶腹叶顶端 2 裂达 1/3—1/2，边缘具齿。蒴萼长梨形，表面密被纤毛状的瘤，具 3 个脊，喙 5—12 个细胞长。

生境：石生。海拔 1185—1505 m。

雅长分布：盘古王，悬崖天坑。

中国分布：全国大部分省区均有分布。

Plants dark green to redish brown when fresh, 1.2–2.0 mm wide, irregularly pinnately branched, branching of the *Frullania*-type. Leaves imbricate, lobe broad ovate, apex rounded, margin entire, ventral base cordate, dorsal base auriculate; lobule cucullate, contiguous to stem, mouth distinctly inclined toward the stem, hemifold, with short obtuse rostrum; stylus filiform or triangular, consisting of 3–11 cells, 2 cells wide at base.. Lobe cell slightly flexuose, trigones large, nodulate, intermediate thickenings frequent. Cuticle smooth. Oil bodies segmented, 5–8 per median cell. Underleaves loosely imbricate to contiguous, 3.5–5.0 times as wide as the stem, apex undivided, margins entire, base slightly cordate. Dioicous. Androecia not seen. Gynoecia terminal or lateral on branches, bracts in 3 pairs, margins toothed, innermost bracteole bilobed to 1/3–1/2 its length, margin toothed. Perianths long pyriform, with densely ciliate protuberances on the surface, with 3 keels, beak 5–12 cells long.

Habitat: On rocks. Elev. 1185–1505 m.

Distribution in Yachang: Panguwang, Xuanya Tiankeng.

Distribution in China: Widely distributed in most provinces of China.

▶ A. 种群；B. 植物体；C. 植物体一段；D. 茎一段，示腹瓣；E. 腹瓣；F. 腹叶；G. 蒴萼横切面；H. 腹瓣和副体；I. 背瓣中部细胞，示油体；J. 蒴萼；K. 蒴萼表面；L. 最内层雌苞叶和雌苞腹叶。（凭证标本：G, J–K 拍摄自 *唐启明 & 张仕艳 20181004-441A*，其余拍摄自 *韦玉梅等 201107-40*）

A. Population; B. Plant; C. Portion of plant; D. Portion of stem showing leaf lobules; E. Leaf lobes; F. Underleaf; G. Transverse section of perianth; H. Leaf lobules and Styli; I. Median cells of leaf lobe showing oil bodies; J. Perianth; K. Surface of perianth; L. Innermost female bracts and innermost female bracteole. (G, J–K from *Tang & Zhang 20181004-441A*, the others from *Wei et al. 201107-40*)

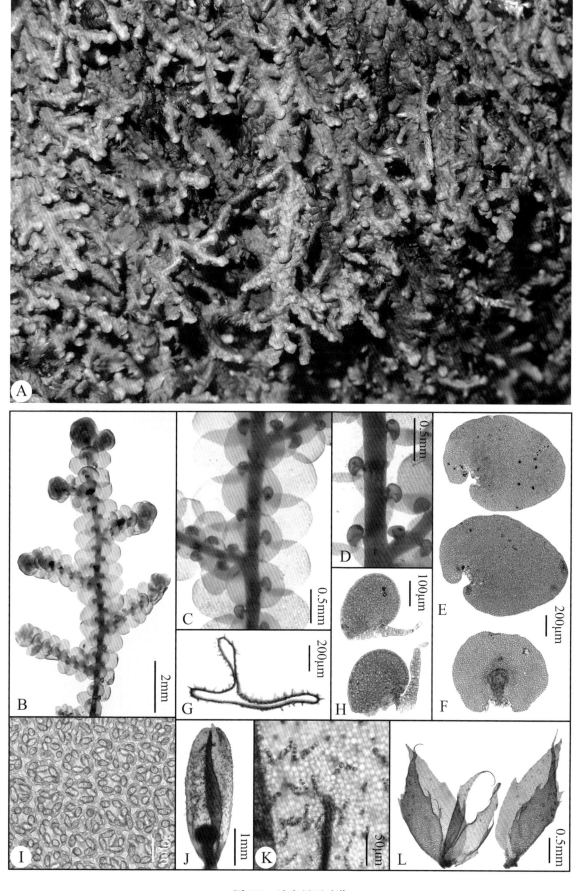

图 101 达乌里耳叶苔

Fig. 101 *Frullania davurica* Hampe

皱叶耳叶苔

Frullania ericoides (Nees) Mont., Ann. Sci. Nat. Bot. (sér. 2)12: 51. 1839.

植物体绿色至红棕色，带叶宽 1.0—1.6 mm，不规则羽状分枝，分枝耳叶苔型。侧叶覆瓦状排列，干时包裹茎，湿时常背仰，背瓣易碎，卵圆形，顶端圆，边缘全缘，腹侧和背侧基部耳形；腹瓣兜状或片状，紧贴茎着生，兜状时口部与茎近乎垂直，一侧内折，具钝尖的喙；副体三角状，4—7 个细胞长，基部 2—3 个细胞宽。侧叶细胞波曲状，角质层平滑，三角体大，肿胀形，中部球状加厚常可见。油体聚合型，每个细胞 2—5 个。腹叶疏松覆瓦状排列至毗邻，宽为茎的 2—4 倍，顶端 2 裂达 1/4—1/3，侧缘常具 1—2 个钝齿，基部楔形。雌雄异株。雄苞未见。雌苞顶生于分枝上，雌苞叶 2 对，边缘全缘，雌苞叶和雌苞腹叶易碎。蒴萼倒卵形，表面密被裂片状的瘤，具 3 个脊，喙 5—10 个细胞长。孢蒴球形，成熟时 4 瓣开裂。孢子球形至椭球形，表面具莲座状纹饰。弹丝单螺旋加厚，附着在每个孢蒴裂瓣的上部。

生境： 树干和石生环境常见，倒木、腐木、树基、树枝、藤茎上也有分布。海拔 754—1724 m。

雅长分布： 广泛分布于保护区各处。

中国分布： 除了东北以外的全国大部分省区均有分布。

Plants green, redish brown to dark brown when fresh, 1.0–1.6 mm wide, irregularly pinnately branched, branching of the *Frullania*-type. Leaves imbricate, convolute the stem when dry, widely spreading and strongly squarrose when moist, lobe highly fragmenting, broad ovate, apex rounded, margin entire, ventral and dorsal base auriculate; lobule cucullate or explanate, contiguous to stem, mouth nearly perpendicular to the stem, hemifold, with obtuse-acute rostrum; stylus triangular, consisting of 4–7 cells, 2–3 cells wide at base. Lobe cell flexuose, trigones large, nodulate, intermediate thickenings frequent. Cuticle smooth. Oil bodies segmented, 2–5 per median cell. Underleaves loosely imbricate to contiguous, 2–4 times as wide as the stem, bilobed to 1/4–1/3 underleaf length, with 1–2 blunt teeth at outer margins, base cuneate. Dioicous. Androecia not seen. Gynoecia terminal on branches, bracts in 2 pairs, margins entire, bracts and bracteoles highly fragmenting. Perianths obovate, with densly laciniate protuberances on the surface, with 3 keels, beak 5–10 cells long. Capsules spherical, longitudinally dehiscing by 4 regular valves at maturity. Spores globose to ellipsoidal, surface with rosettes. Elaters with 1 spiral thickening, stick to the upper 1/3 of each capsule-valve.

Habitat: Often on tree trunks and rocks, sometimes on fallen trees, rotten logs, tree bases, tree branches and lianas. Elev. 754–1724 m.

Distribution in Yachang: Widely distributed in Yachang.

Distribution in China: Widely distributed in most provinces of China except Northeast Region.

▶ A. 种群；B. 植物体（背面观）；C. 植物体一段；D. 腹叶；E. 背瓣；F. 裂开的孢蒴和弹丝；G. 背瓣中部细胞，示油体；H. 茎一段，示腹瓣；I. 孢子；J. 蒴萼；K. 蒴萼横切面；L. 蒴萼表面。（凭证标本：*韦玉梅等 201105-17*）

A. Population; B. Plant (dorsal view); C. Portion of plant; D. Underleaves; E. Leaf lobes; F. Dehisced capsule and elaters; G. Median cells of leaf lobe showing oil bodies; H. Portion of stem showing leaf lobules; I. Spores; J. Perianth; K. Transverse section of perianth; L. Surface of perianth. (All from *Wei et al. 201105-17*)

图 102　皱叶耳叶苔

Fig. 102　*Frullania ericoides* (Nees) Mont.

细瓣耳叶苔

Frullania hypoleuca Nees, Nov. Actorum Acad. Caes. Leop.-Carol. Nat. Cur. 19(Suppl. 1): 470. 1843.

植物体干时暗棕色，带叶宽 0.80—1.65 mm，规则羽状分枝，分枝耳叶苔型。侧叶覆瓦状排列，背瓣卵圆形，顶端圆钝，常内弯，边缘全缘，腹侧和背侧基部截形；腹瓣圆柱状，远离茎着生，与茎平行或略向外倾斜，口部圆弧形，不具喙；副体叶状，平贴腹瓣基部的柄着生。侧叶细胞不呈波曲状，角质层平滑，三角体大，简单三角形或肿胀形（近基部），中部球状加厚偶可见。油体未见。腹叶覆瓦状排列至毗邻，宽为茎的 2—3 倍，顶端 2 裂达 1/4—1/3，边缘全缘，基部楔形。雌雄同株。雄苞顶生于短分枝上，雄苞叶 1—3 对，雄苞腹叶仅生于雄苞基部。雌苞顶生或侧生于分枝上，雌苞叶 2—3 对，边缘全缘，最内层雌苞叶腹叶顶端 2 裂约达 1/2，边缘全缘。蒴萼长倒卵形，表面平滑，具 3 个脊，喙 4—6 个细胞长。

生境： 石生。海拔 1643 m。

雅长分布： 大棚屯。

中国分布： 福建，台湾，云南。首次记录于广西。

Plants dark brown in herbarium specimens, 0.80–1.65 mm wide, pinnately branched, branching of the *Frullania*-type. Leaves imbricate, lobe rounded-ovate, apex rounded-obtuse, usually incurved, margin entire, ventral and dorsal base truncate; lobule cylindrical, remote to stem, parallel or oblique to stem, mouth rounded, arched, without rostrum; stylus foliaceous, appressed to lobules. Lobe cell not flexuose, trigones large, simple-triangulate or nodulate (at base), intermediate thickenings occasionally present. Cuticle smooth. Oil bodies not seen. Underleaves imbricate to contiguous, 2–3 times as wide as the stem, bilobed to 1/4–1/3 underleaf length, margins entire, base cuneate. Monoicous. Androecia terminal on short branches, bracts in 1–3 pairs, bracteoles present only at the base of the androecium. Gynoecia terminal on branches, bracts in 2–3 pairs, margins entire, innermost bracteole bilobed to ca. 1/2 its length, margin entire. Perianths long obovate, surface smooth, with 3 keels, beak 4–6 cells long.

Habitat: On rocks. Elev. 1643 m.

Distribution in Yachang: Dapeng Tun.

Distribution in China: Fujian, Taiwan, Yunnan. New to Guangxi.

A. 种群；B. 植物体；C. 植物体一段带雌苞和雄苞；D. 背瓣（左）和侧叶（右）；E. 最内层雌苞叶；F. 最内层雌苞腹叶；G. 背瓣中部细胞；H. 植物体一段；I. 蒴萼横切面；J. 蒴萼；K. 腹叶；L. 腹瓣。（凭证标本：唐启明等 20190520-392）

A. Population; B. Plant; C. Portion of plant with a gynoecium and an androecium; D. Leaf lobe (left) and leaf (right); E. Innermost female bracts; F. Innermost female bracteole; G. Median cells of leaf lobe; H. Portion of plant; I. Transverse section of perianth; J. Perianth; K. Underleaves; L. Lobules. (All from *Tang et al.* 20190520-392)

图 103 细瓣耳叶苔

Fig. 103 *Frullania hypoleuca* Nees

石生耳叶苔

Frullania inflata Gottsche, Syn. Hepat. 3: 424. 1845.

植物体浅绿色至棕绿色，带叶宽 1.05—1.78 mm，不规则羽状分枝，分枝耳叶苔型。侧叶覆瓦状排列至毗邻，背瓣宽卵形，顶端圆，边缘全缘，腹侧和背侧基部心形；腹瓣片状，稍远离茎着生并与茎平行或略向外倾斜；副体丝状，2—3 个细胞长。侧叶细胞不呈波曲状，角质层平滑，三角体小，简单三角形，中部球状加厚缺。油体聚合型，每个细胞 1—4 个。腹叶远生，宽为茎的 1.0—1.5 倍，顶端 2 裂达 1/3—1/2，侧缘常具 1—3 个钝或裂片状齿，有时全缘，基部楔形。雌雄同株。雄苞顶生于侧短枝上，雄苞叶 2—3 对，雄苞腹叶仅生于雄苞基部。雌苞顶生于分枝上，雌苞叶 1—2 对，边缘全缘，最内层雌苞叶腹叶顶端 2 裂约达 1/2，两侧边缘各具 1 个长刺状齿。蒴萼倒卵形，表面平滑，具 5 个脊，有时还存在少量的短的脊，喙 2—3 个细胞长。

生境： 石生。海拔 1142—1324 m。

雅长分布： 大宴坪天坑漏斗，吊井天坑，黄猄洞天坑，拉洞天坑，蓝家湾天坑，里郎天坑，悬崖天坑，中井天坑，中井屯。

中国分布： 全国大部分省区均有分布。首次记录于广西。

Plants light green to brownish green when fresh, 1.05–1.78 mm wide, irregularly pinnately branched, branching of the *Frullania*-type. Leaves imbricate to contiguous, lobe broad ovate, apex rounded, margin entire, ventral and dorsal base cordate; lobule explanate, slightly remote to stem, parallel or oblique to stem; stylus filiform, in a row of 2–3 cells. Lobe cell not flexuose, trigones small, simple-triangulate, intermediate thickenings absent. Cuticle smooth. Oil bodies segmented, 1–4 per median cell. Underleaves distant, 1.0–1.5 times as wide as the stem, bilobed to 1/3–1/2 underleaf length, usually with 1–3 blunt or laciniate teeth at outer margins, base cuneate. Monoicous. Androecia terminal on short branches, bracts in 2–3 pairs, bracteoles present only at the base of the androecium. Gynoecia terminal on branches, bracts in 1–2 pairs, margins entire, innermost bracteole bilobed to ca. 1/2 its length, usually with 1 lateral long-spinose tooth on each margin. Perianths obovate, surface smooth, with 5 keels, sometimes also with a few, much smaller additional keels, beak 2–3 cells long.

Habitat: On rocks. Elev. 1142–1324 m.

Distribution in Yachang: Dayanping Tiankeng, Diaojing Tiankeng, Huangjingdong Tiankeng, Ladong Tiankeng, Lanjiawan Tiankeng, Lilang Tiankeng, Xuanya Tiankeng, Zhongjing Tiankeng, Zhongjing Tun.

Distribution in China: Widely distributed in most provinces of China. New to Guangxi.

▶ A. 种群；B. 植物体；C. 植物体一段带雌苞和雄苞；D. 植物体一段；E. 背瓣；F. 副体；G. 蒴萼；H. 蒴萼横切面；I. 最内层雌苞腹叶；J. 最内层雌苞叶；K. 腹叶；L. 背瓣中部细胞，示油体。（凭证标本：黄萍等 210723-6）

A. Population; B. Plant; C. Portion of plant with an androecium and a gynoecium; D. Portion of plant; E. Leaf lobes; F. Styllus; G. Perianth; H. Transverse section of perianth; I. Innermost female bracteole; J. Innermost female bracts; K. Underleaves; L. Median cells of leaf lobe showing oil bodies. (All from *Huang et al. 210723-6*)

图 104　石生耳叶苔

Fig. 104　*Frullania inflata* Gottsche

列胞耳叶苔

Frullania moniliata (Reinw., Blume et Nees) Mont., Ann. Sci. Nat. Bot. (sér. 2) 18: 13. 1842.

植物体绿色、红棕色至暗棕色，带叶宽 0.95—1.45 mm，规则羽状分枝，分枝耳叶苔型。侧叶覆瓦状排列，背瓣阔卵形，顶端细尖，常内弯，边缘全缘，腹侧基部心形，背侧基部耳形；腹瓣圆柱状或片状，远离茎着生并略向外倾斜，口部圆，拱形，不具喙；副体丝状，3—5 个细胞长，基部具半圆形的附属物。侧叶细胞不呈波曲状，角质层平滑，三角体小，简单三角形，中部球状加厚偶可见。油体聚合型，每个细胞 2—4 个。油胞 15—28 个排成一列，有时存在散生油胞。腹叶常远生，宽为茎的 3—4 倍，顶端 2 裂达 1/5—1/3，常下弯，边缘全缘或偶具钝齿，基部略心形，常具披针形或耳形附属物。雌雄异株。雄苞未见。雌苞顶生于分枝上，雌苞叶 3 对，边缘具长刺状齿，最内层雌苞叶腹叶顶端 2 裂约达 2/3，边缘具长刺状齿。蒴萼长梨形，表面平滑，具 3 个脊，喙 15—23 个细胞长。孢蒴球形，成熟时 4 瓣开裂。孢子球形至椭球形，表面具莲座状纹饰。弹丝单螺旋加厚。

生境： 石生，树干。海拔 1336—1747 m。

雅长分布： 草王山，大棚屯，逻家田屯，下棚屯。

中国分布： 全国范围广布。

Plants green, redish brown to dark brown when fresh, 0.95–1.45 mm wide, regularly pinnately branched, branching of the *Frullania*-type. Leaves imbricate, lobe rounded-ovate, apex apiculate, usually incurved, margin entire, ventral base cordate, dorsal base auriculate; lobule cylindrical or explanate, remote to stem, parallel or oblique to stem, mouth rounded, arched, without rostrum; stylus filiform, in a row of 3–5 cells, with semicircular disc-like appendages. Lobe cell not flexuose, trigones small, simple-triangulate, intermediate thickenings occasionally present. Cuticle smooth. Oil bodies segmented, 2–4 per median cell. Ocelli usually in 1 row 15–28 cells long, and few scattered or almost none. Underleaves distant, 3–4 times as wide as the stem, bilobed to ca. 1/5–1/3 underleaf length, apex recurved, entire or occasionally with 1–2 teeth at outer margins, base slightly cordate, sometimes with lanceolate or auriculate appendages. Dioicous. Androecia not seen. Gynoecia terminal on branches, bracts in 3 pairs, margins spinose-dentate, innermost bracteole bilobed to ca. 2/3 its length, margin spinose-dentate. Perianths pyriform, surface smooth, with 3 keels, beak 15–23 cells long. Capsules spherical, longitudinally dehiscing by 4 regular valves at maturity. Spores globose to ellipsoidal, surface with rosettes. Elaters with 1 spiral thickening.

Habitat: On rocks and tree trunks. Elev. 1336–1747 m.

Distribution in Yachang: Caowangshan, Dapeng Tun, Luojiatian Tun, Xiapeng Tun.

Distribution in China: Widely distributed in China.

▶ A. 种群；B. 植物体；C. 植物体一段；D. 腹叶；E. 背瓣；F. 副体；G. 裂开的孢蒴和弹丝；H. 孢子；I. 茎一段，示腹瓣；J. 蒴萼横切面；K. 蒴萼；L. 最内层雌苞腹叶；M. 最内层雌苞叶；N. 背瓣中部细胞，示油体和油胞。（凭证标本：韦玉梅等 191013-109A）

A. Population; B. Plant; C. Portion of plant; D. Underleaves; E. Leaf lobes; F. Styllus; G. Dehisced capsule and elaters; H. Spores; I. Portion of stem showing leaf lobules; J. Transverse section of perianth; K. Perianth; L. Innermost female bracteole; M. Innermost female bracts; N. Median cells of leaf lobe showing oil bodies and ocelli. (All from *Wei et al. 191013-109A*)

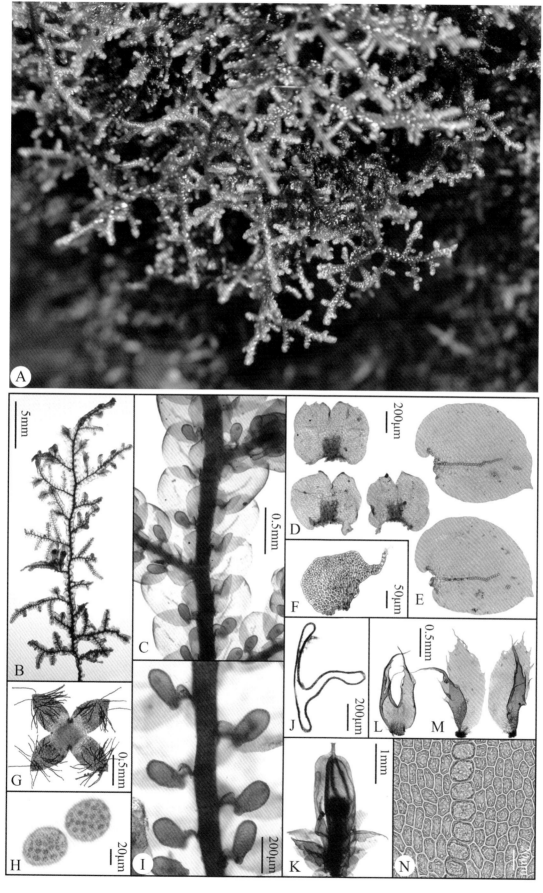

图 105 列胞耳叶苔

Fig. 105 *Frullania moniliata* (Reinw., Blume et Nees) Mont.

羊角耳叶苔喙尖变种（喙尖耳叶苔）

Frullania monocera var. **acutiloba** (Mitt.) Hentschel et von Konrat, Phytotaxa 220(2): 136. 2015.

Frullania acutiloba Mitt., J. Proc. Linn. Soc., Bot. 5(18): 120. 1860[1861].

植物体绿色至深绿色，带叶宽 1.30—2.26 mm，不规则羽状分枝，分枝耳叶苔型。侧叶覆瓦状排列，背瓣椭圆形至椭圆状卵形，顶端圆，边缘全缘，腹缘有时反折，腹侧基部截形，背侧基部耳形；腹瓣兜状或片状，稍远离茎着生，口部略下倾，一侧内折，具长尾尖的喙；副体丝状或三角状，4—5 个细胞长，基部 2—3 个细胞宽。叶细胞波曲状，角质层平滑，三角体大，肿胀形，中部球状加厚常可见。油体聚合型。腹叶覆瓦状排列至毗邻，宽为茎的 2.0—3.5 倍，顶端 2 裂达 1/4—2/5，边缘全缘或偶具钝齿，基部楔形。雌雄异株。雄苞未见。雌苞顶生或侧生于分枝上，雌苞叶 3 对，边缘全缘，常反折，最内层雌苞叶腹叶顶端 2 裂达 1/2—2/3，两侧边缘各具 1 个粗齿，常外折。蒴萼长梨形，表面具长刺状或裂片状的瘤，具 3 个脊，有时还存在少量短的脊，喙 3—5 个细胞长。

生境：树干。海拔 1331—1747 m。

雅长分布：草王山，中井屯。

中国分布：重庆，福建，甘肃，广西，贵州，江西，台湾，西藏，云南。

Plants green to dark green when fresh, 1.30–2.26 mm wide, irregularly pinnately branched, branching of the *Frullania*-type. Leaves imbricate, lobe oblong to oblong-ovate, apex rounded, margin entire, ventral margin usually recurved, ventral base truncate, dorsal base auriculate; lobule cucullate or explanate, slightly remote to stem, mouth slightly inclined toward the stem, hemifold, with long caudate rostrum; stylus filiform or triangular, consisting of 4–5 cells, 2–3 cells wide at base. Lobe cell flexuose, trigones large, nodulate, intermediate thickenings frequent. Cuticle smooth. Oil bodies segmented. Underleaves imbricate to contiguous, 2.0–3.5 times as wide as the stem, bilobed to 1/4–2/5 underleaf length, margins entire or occasionally with 1–2 teeth at outer margins, base cuneate. Dioicous. Androecia not seen. Gynoecia terminal or lateral on branches, bracts in 3 pairs, margins entire, often recurved, innermost bracteole bilobed to 1/2–2/3 its length, usually with 1 lateral lancelate tooth on each margin, recurved. Perianths long pyriform, with spinate or laciniate protuberances on the surface, with 3 keels, sometimes also with a few, much smaller additional keels, beak 3–5 cells long.

Habitat: On tree trunk. Elev. 1331–1747 m.

Distribution in Yachang: Caowangshan, Zhongjing Tun.

Distribution in China: Chongqing, Fujian, Gansu, Guangxi, Guizhou, Jiangxi, Taiwan, Tibet, Yunnan.

▶ A. 种群；B. 植物体；C. 植物体一段；D. 腹叶；E. 背瓣；F. 最内层雌苞腹叶；G. 最内层雌苞叶；H. 茎一段，示腹瓣；I. 副体；J. 蒴萼；K. 蒴萼横切面；L. 蒴萼表面；M. 背瓣中部细胞，示油体。（凭证标本：*唐启明 & 张仕艳 20181007-596*）

A. Population; B. Plant; C. Portion of plant; D. Underleaves; E. Leaf lobes; F. Innermost female bracteole; G. Innermost female bracts; H. Portion of stem showing leaf lobules; I. Styli; J. Perianth; K. Transverse section of perianth; L. Surface of perianth; M. Median cells of leaf lobe showing oil bodies. (All from *Tang & Zhang 20181007-596*)

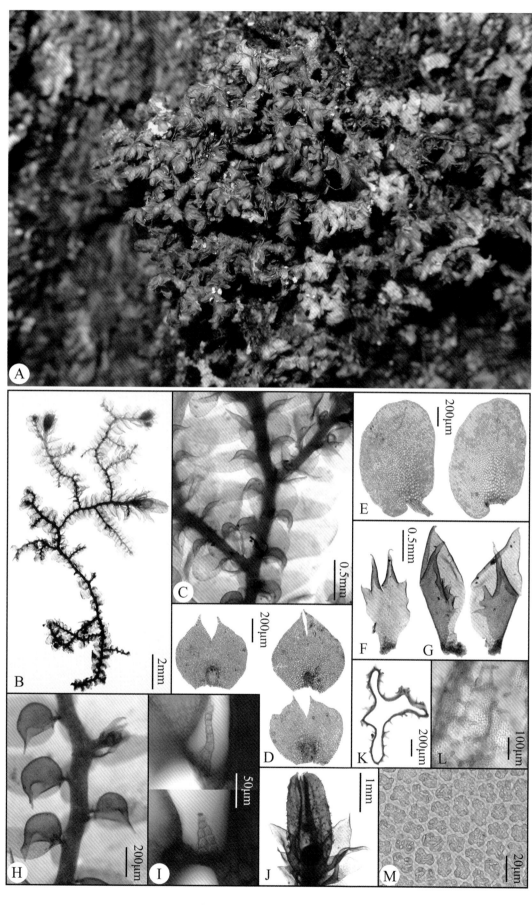

图 106 羊角耳叶苔喙尖变种

Fig. 106 *Frullania monocera* var. *acutiloba* (Mitt.) Hentschel et von Konrat

盔瓣耳叶苔

Frullania muscicola Steph., Hedwigia 33(3): 146. 1894.

植物体深绿色至红棕色，带叶宽 0.75—1.35 mm，不规则羽状分枝，分枝耳叶苔型。侧叶覆瓦状排列，背瓣近圆形至卵圆形，顶端圆，内弯，边缘全缘，腹侧基部心形，背侧基部耳形；腹瓣兜状或片状，紧贴茎着生，口部与茎近乎垂直，一侧内折，具短钝的喙，有时喙不明显；副体丝状或三角状，3—6 个细胞长，基部 2—3 个细胞宽。侧叶细胞波曲状，角质层平滑，三角体大，肿胀形，中部球状加厚常可见。油体聚合型，每个细胞 2—5 个。腹叶毗邻至远生，宽为茎的 2.0—3.5 倍，顶端 2 裂达 1/4—1/3，边缘具 1—2 个钝齿，基部楔形。雌雄异株。雄苞未见。雌苞顶生或侧生于分枝上，雌苞叶 3 对，边缘全缘，最内层雌苞叶腹叶顶端 2 裂达 1/2—2/3，边缘全缘或两侧边缘各具 1 个长齿。蒴萼倒卵形，表面平滑，具 5 个脊，有时还存在少量短的脊，喙 4—6 个细胞长。

生境： 树干环境常见，腐木和倒木上也有分布。海拔 769—1747 m。
雅长分布： 草王山，拉雅沟，逻家田屯，中井屯，香当天坑。
中国分布： 全国范围广布。

Plants dark green to redish brown when fresh, 0.75−1.35 mm wide, irregularly pinnately branched, branching of the *Frullania*-type. Leaves imbricate, lobe suborbicular to rounded-ovate, apex rounded, incurved, margin entire, ventral base cordate, dorsal base auriculate; lobule cucullate or explanate, contiguous to stem, mouth nearly perpendicular to the stem, hemifold, with short obtuse rostrum or indistinct rostrum; stylus filiform or triangular, consisting of 3−6 cells, 2−3 cells wide at base. Lobe cell flexuose, trigones large, nodulate, intermediate thickenings frequent. Cuticle smooth. Oil bodies segmented, 2−5 per median cell. Underleaves contiguous to distant, 2.0−3.5 times as wide as the stem, bilobed to 1/4−1/3 underleaf length, with 1−2 teeth at outer margins, base cuneate. Dioicous. Androecia not seen. Gynoecia terminal or lateral on branches, bracts in 3 pairs, margins entire, innermost bracteole bilobed to 1/2−2/3 its length, margin entire or with 1 lateral long tooth on each margin. Perianths obovate, surface smooth, with 5 keels, sometimes with a few, much smaller additional keels, beak 4−6 cells long.

Habitat: Often on tree trunks, sometimes on fallen tree and rotten logs. Elev. 769−1747 m.
Distribution in Yachang: Caowangshan, Layagou, Luojiatian Tun, Xiangdang Tiankeng, Zhongjing Tun.
Distribution in China: Widely distributed in China.

▶ A. 种群；B. 植物体；C&E. 植物体一段；D. 侧叶（左）和背瓣（右）；F. 腹叶；G. 最内层雌苞腹叶；H. 最内层雌苞叶；I. 副体；J. 蒴萼；K. 蒴萼横切面；L. 背瓣中部细胞，示油体。（凭证标本：*韦玉梅等 191016-206*）

A. Population; B. Plant; C&E. Portions of plants, D. Leaf (left) and leaf lobe (right); F. Underleaves; G. Innermost female bracteole; H. Innermost female bracts; I. Styli; J. Perianth; K. Transverse section of perianth; L. Median cells of leaf lobe showing oil bodies. (All from *Wei et al. 191016-206*)

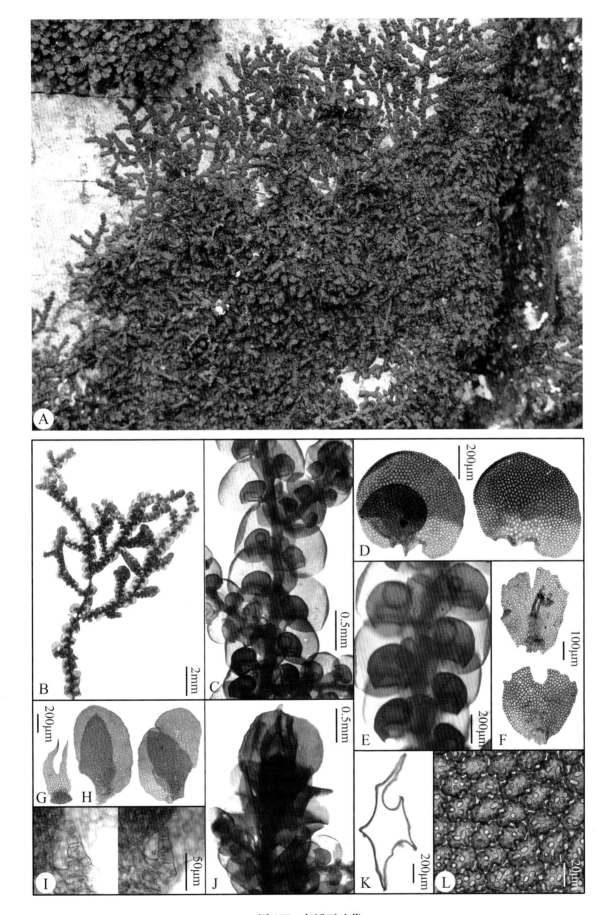

图 107　盔瓣耳叶苔

Fig. 107　*Frullania muscicola* Steph.

尼泊尔耳叶苔

Frullania nepalensis (Spreng.) Lehm. et Lindenb., Syn. Hepat. 3: 422. 1845.

植物体棕绿色至红棕色，带叶宽 0.70—1.20 mm，规则羽状分枝，分枝耳叶苔型。侧叶覆瓦状排列，背瓣卵形至椭圆状卵形，顶端圆，强烈内弯，边缘全缘，常稍反折，腹侧基部心形，背侧基部大的耳形；腹瓣兜状，紧贴茎着生，口部略下倾，一侧内折，但不具喙；副体丝状或三角状，3—5 个细胞长，基部 2（—3）宽。侧叶细胞波曲状，角质层平滑，三角体大，肿胀形，中部球状加厚常可见。油体聚合型，每个细胞 2—5 个。腹叶毗邻至远生，宽为茎的 2.5—3.5 倍，顶端 2 裂达 1/6—1/5，边缘全缘，基部耳形。雌雄异株。雄苞未见。雌苞顶生或侧生于分枝上，雌苞叶 3 对，边缘全缘。蒴萼未见。

生境：石生，腐木，树基。海拔 837—1348 m。

雅长分布：二沟，黄猄洞天坑，李家坨屯，盘古王，旁墙屯。

中国分布：除了东北以外的全国大部分省区均有分布。

Plants browish green to redish brown when fresh, 0.70−1.20 mm wide, regularly pinnately branched, branching of the *Frullania*-type. Leaves imbricate, lobe ovate to oblong-ovate, apex rounded, strongly incurved, margin entire, slightly revurved, ventral base cordate, dorsal base large auriculate; lobule cucullate, contiguous to stem, mouth slightly inclined toward the stem, hemifold, without rostrum; stylus filiform or triangular, consisting of 3−5 cells, 2(−3) cells wide at base. Lobe cell flexuose, trigones large, nodulate, intermediate thickenings frequent. Cuticle smooth. Oil bodies segmented, 2−5 per median cell. Underleaves contiguous to distant, 2.5−3.5 times as wide as the stem, bilobed to 1/6−1/5 underleaf length, margins entire, base auriculate. Dioicous. Androecia not seen. Gynoecia terminal or lateral on branches, bracts in 3 pairs, margins entire. Perianths not seen.

Habitat: On rocks, rotten logs and tree bases. Elev. 837−1348 m.

Distribution in Yachang: Ergou, Huangjingdong Tiankeng, Lijiatuo Tun, Panguwang, Pangqiang Tun.

Distribution in China: Widely distributed in most provinces of China except Northeast Region.

▶ A. 种群；B. 植物体；C. 植物体一段；D. 茎一段，示腹瓣；E. 腹叶；F. 背瓣；G. 腹瓣和副体；H. 雌苞；I. 背瓣中部细胞，示油体。（凭证标本：韦玉梅等 201110-3）

A. Population; B. Plant; C. Portion of plant; D. Portion of stem showing leaf lobules; E. Underleaves; F. Leaf lobes; G. Leaf lobules and styli; H. Gynoecium; I. Median cells of leaf lobe showing oil bodies. (All from *Wei et al. 201110-3*)

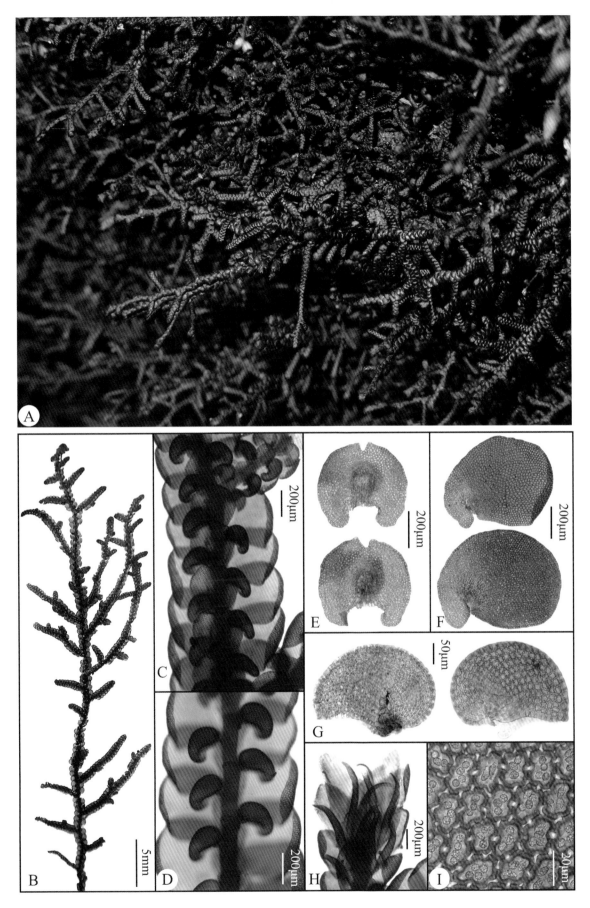

图 108 尼泊尔耳叶苔

Fig. 108 *Frullania nepalensis* (Spreng.) Lehm. & Lindenb.

大隅耳叶苔

Frullania osumiensis (S. Hatt.) S. Hatt., J. Hattori Bot. Lab. 16: 87. 1956.

植物体浅绿色至红棕色，带叶宽 1.2—2.4 mm，不规则羽状分枝，分枝耳叶苔型。侧叶覆瓦状排列，背瓣卵圆形或椭圆状卵形，顶端圆，常内弯，边缘全缘，腹侧基部截形，背侧基部耳形；腹瓣兜状，紧贴茎着生，口部与茎近乎垂直，一侧内折，具长钝尖的喙；副体丝状或三角形状，4—7 个细胞长，基部 2（—3）个细胞宽。侧叶细胞稍波曲状，角质层平滑，三角体大，简单三角形或肿胀形，中部球状加厚常可见。油体聚合型，每个细胞 4—7 个。腹叶覆瓦状排列至毗邻，宽为茎的 4—6 倍，顶端 2 裂约达 1/4，边缘全缘或具不明显钝齿（雄株），上部边缘具齿（雌株），基部楔形或微心形。雌雄异株。雄苞顶生或间生于分枝上，雄苞叶 2—8 对，雄苞腹叶仅生于雄苞基部。雌苞顶生或侧生于分枝上，雌苞叶 2—3 对，边缘全缘，最内层雌苞叶腹叶顶端 2 裂达 1/2—2/5，边缘具全缘，稀具齿。蒴萼倒卵形，表面平滑，具 3 个脊，脊平滑、皱波状或具瘤，喙 7—13 个细胞长。

生境： 倒木，腐木，树干。海拔 1383—1823 m。

雅长分布： 草王山，九十九堡。

中国分布： 安徽，福建，台湾。首次记录于广西。

Plants light green to redish brown when fresh, 1.2–2.4 mm wide, irregularly pinnately branched, branching of the *Frullania*-type. Leaves imbricate, lobe rounded-ovate to oblong-ovate, apex rounded, margin entire, ventral base truncate, dorsal base auriculate; lobule cucullate, contiguous to stem, mouth nearly perpendicular to the stem, hemifold, with long acute rostrum; stylus filiform or triangular, consisting of 4–7 cells, 2(–3) cells wide at base. Lobe cell slightly flexuose, trigones large, simple-triangulate or nodulate, intermediate thickenings frequent. Cuticle smooth. Oil bodies segmented, 4–7 per median cell. Underleaves distant, 4–6 times as wide as the stem, bilobed to ca. 1/4 underleaf length, margins entire or indistinctly toothed (male plants), upper margins usually with 1–3 teeth (female plants), base cuneate or slightly cordate. Dioicous. Androecia terminal or intercalary on branches, bracts in 2–8 pairs, bracteoles present only at the base of the androecium. Gynoecia terminal or lateral on branches, bracts in 2–3 pairs, margins entire, innermost bracteole bilobed to 1/2–2/5 its length, margin entire, rarely toothed. Perianths obovate, surface smooth, with 5 keels, keels smooth, crispate or tuberculate, beak 7–13 cells long.

Habitat: On fallen trees, rotten logs and tree trunks. Elev. 1383–1823 m.

Distribution in Yachang: Caowangshan, Jiushijiubao.

Distribution in China: Anhui, Fujian, Taiwan. New to Guangxi.

▶ A. 种群；B. 植物体；C. 植物体一段；D. 茎一段，示腹瓣；E. 雄苞；F. 背瓣中部细胞，示油体；G. 侧叶（左）和背瓣（右）；H. 腹叶（雌株）；I. 腹叶（雄株）；J. 最内层雌苞叶；K. 最内层雌苞腹叶；L. 蒴萼；M. 蒴萼横切面。（凭证标本：韦玉梅等 201108-18）

A. Population; B. Plant; C. Portion of plant; D. Portion of stem showing leaf lobules; E. Androecium; F. Median cells of leaf lobe showing oil bodies; G. Leaf (left) and leaf lobe (right); H. Underleaves (female plant); I. Underleaves (male plant); J. Innermost female bracts; K. Innermost female bracteole; L. Perianth; M. Transverse section of perianth. (All from *Wei et al. 201108-18*)

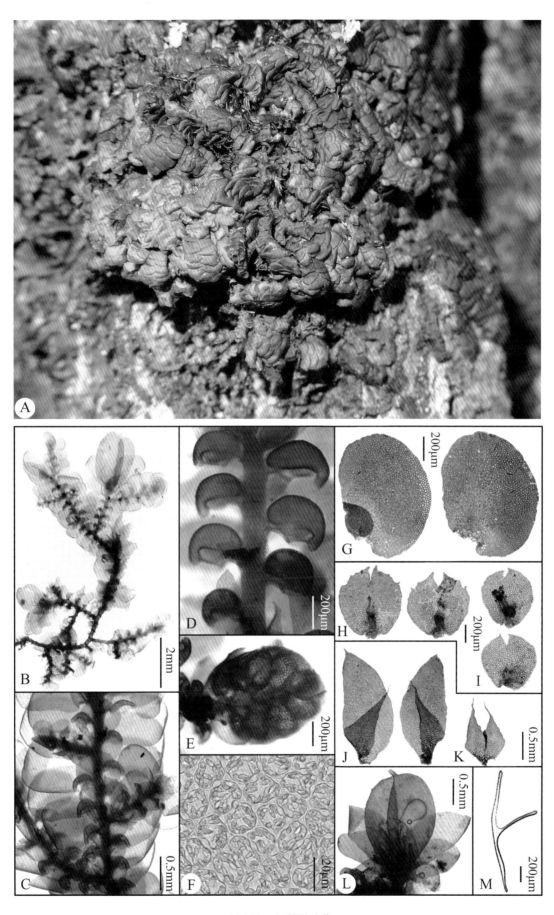

图 109 大隅耳叶苔

Fig. 109 *Frullania osumiensis* (S. Hatt.) S. Hatt.

钟瓣耳叶苔

Frullania parvistipula Steph., Sp. Hepat. (Stephani) 4: 397. 1910.

植物体浅绿色至红棕色，带叶宽 0.60—0.85 mm，不规则羽状分枝，分枝耳叶苔型。侧叶毗邻至远生，背瓣常易脱落，近圆形至卵圆形，顶端圆，边缘全缘，腹侧基部截形，背侧基部稍耳形；腹瓣钟状，紧贴茎着生并与茎平行，口部截形，不具喙；副体丝状，3—5 个细胞长。侧叶细胞波曲状，角质层平滑，三角体小至大，简单三角形或肿胀形，中部球状加厚常可见。油体聚合型，每个细胞 2—4 个。腹叶远生，宽为茎的 1.0—1.5 倍，顶端 2 裂达 1/2—2/3，侧缘常具 1 个齿，基部楔形。雌雄异株。雄苞顶生于分枝上，雄苞叶 2—6 对，雄苞腹叶仅生于雄苞基部。雌苞顶生或侧生于分枝上，雌苞叶 2—3 对，边缘全缘，最内层雌苞叶腹叶顶端 2 裂约达 1/2，侧缘常具 1 个裂片状齿。蒴萼倒卵形，表面平滑，具 3（—4）个脊，喙 4—6 个细胞长。

生境： 树干环境常见，钙华基质、土生、倒木、树基环境也有分布。海拔 1048—1643 m。

雅长分布： 广泛分布于保护区各处。

中国分布： 安徽，贵州，黑龙江，湖北，湖南，吉林，江西，陕西，山东，四川，台湾，西藏，云南，浙江。首次记录于广西。

Plants light green to redish brown when fresh, 0.60−0.85 mm wide, irregularly pinnately branched, branching of the *Frullania*-type. Leaves contiguous to distant, lobe usually caducous, suborbicular to rounded-ovate, apex rounded, margin entire, ventral margin truncate, dorsal base slightly auriculate; lobule campanulate, contiguous to and parallel to stem, mouth truncate, without rostrum; stylus filiform, in a row of 3−5 cells. Lobe cells flexuose, trigones small to large, simple-triangulate or nodulate, intermediate thickenings frequent. Cuticle smooth. Oil bodies segmented, 2−4 per median cell. Underleaves distant, 1.0−1.5 times as wide as the stem, bilobed to 1/2−2/3 underleaf length, usually with 1 blunt tooth at outer margins, base cuneate. Dioicous. Androecia terminal on branches, bracts in 2−6 pairs, bracteoles present only at the base of the androecium. Gynoecia terminal or lateral on branches, bracts in 2−3 pairs, margins entire, innermost bracteole bilobed to ca. 1/2 its length, usually with 1 lateral long tooth on each margin. Perianths obovate, surface smooth, with 3(−4) keels, beak 4−6 cells long.

Habitat: Often on tree trunks, sometimes on calcareous substrates, soil, fallen trees and tree bases. Elev. 1048−1643 m.

Distribution in Yachang: Widely distributed in Yachang.

Distribution in China: Anhui, Guizhou, Heilongjiang, Hubei, Hunan, Jilin, Jiangxi, Shaanxi, Shandong, Sichuan, Taiwan, Tibet, Yunnan, Zhejiang. New to Guangxi.

A. 种群；B. 植物体；C. 植物体一段；D. 背瓣；E. 最内层雌苞腹叶；F. 最内层雌苞叶；G. 背瓣中部细胞，示油体；H. 雄苞；I. 蒴萼横切面；J. 蒴萼；K. 腹叶；L. 副体。（凭证标本：*韦玉梅等 201105-1*）

A. Population; B. Plant; C. Portion of plant; D. Leaf lobe; E. Innermost female bracteole; F. Innermost female bracts; G. Median cells of leaf lobe showing oil bodies; H. Androecium; I. Transverse section of perianth; J. Perianth; K. Underleaf; L. Stylus. (All from *Wei et al. 201105-1*)

图 110　钟瓣耳叶苔

Fig. 110　*Frullania parvistipula* Steph.

喙瓣耳叶苔

Frullania pedicellata Steph., Bull. Herb. Boissier 5(2): 90, 1897.

植物体浅绿色至绿色，带叶宽 1.00—1.71 mm，不规则羽状分枝，分枝耳叶苔型。侧叶覆瓦状排列，背瓣椭圆形至椭圆状卵形，顶端圆，边缘全缘，侧缘稍反折，腹侧基部心形，背侧基部耳形；腹瓣兜状，有时片状，稍远离茎着生，口部明显下倾，一侧内折，具长钝尖的喙；副体丝状，3—5 个细胞长。侧叶细胞波曲状，角质层平滑，三角体大，肿胀形，中部球状加厚偶可见。油体聚合型，每个细胞 4—7 个。腹叶疏松覆瓦状排列至毗邻，宽为茎的 2.5—3.5 倍，顶端 2 裂达 1/3—1/2，边缘具 1—2 齿，基部楔形。雌雄异株。雄苞未见。雌苞顶生于分枝上，雌苞叶 3 对，边缘全缘。蒴萼未见。

生境： 树干。海拔 1073—1240 m。

雅长分布： 白岩坨屯，蓝家湾天坑。

中国分布： 福建，甘肃，黑龙江，湖北，吉林，江西，台湾，浙江。首次记录于广西。

Plants light green to green when fresh, 1.00−1.71 mm wide, irregularly pinnately branched, branching of the *Frullania*-type. Leaves imbricate, lobe oblong to oblong-ovate, apex rounded, margin entire, ventral base cordate, dorsal base auriculate; lobule cucullate, sometimes explanate, slightly remote to stem, mouth inclined toward the stem, hemifold, with obtuse-acute rostrum; stylus filiform, in a row of 3−5 cells. Lobe cell flexuose, trigones large, nodulate, intermediate thickenings occasionally present. Cuticle smooth. Oil bodies segmented, 4−7 per median cell. Underleaves loosely imbricate to contiguous, 2.5−3.5 times as wide as the stem, bilobed to 1/3−1/2 underleaf length, with 1−2 teeth at outer margins, base cuneate. Dioicous. Androecia not seen. Gynoecia terminal on branches, bracts in 3 pairs, margins entire. Perianths not seen.

Habitat: On tree trunk. Elev. 1073−1240 m.

Distribution in Yachang: Baiyantuo Tun, Lanjiawan Tiankeng.

Distribution in China: Fujian, Gansu, Heilongjiang, Hubei, Jilin, Jiangxi, Taiwan, Zhejiang. New to Guangxi.

▶ A. 种群；B. 植物体；C. 植物体一段带雌苞；D. 背瓣；E. 背瓣中部细胞，示油体；F. 茎一段，示腹瓣；G. 腹叶；H. 副体。（凭证标本：*韦玉梅等 191020-433*）

A. Population; B. Plant; C. Portion of plant with a gynoecium; D. Leaf lobes; E. Median cells of leaf lobe showing oil bodies; F. Portion of stem showing leaf lobules; G. Underleaf; H. Stylus. (All from *Wei et al. 191020-433*)

图 111 喙瓣耳叶苔

Fig. 111 *Frullania pedicellata* Steph.

大萼耳叶苔

Frullania physantha Mitt., J. Proc. Linn. Soc., Bot. 5(18): 121. 1860.

植物体绿色、红棕色至棕褐色，带叶宽 1.20—1.80 mm，不规则羽状分枝，分枝耳叶苔型。侧叶覆瓦状排列，背瓣阔椭圆形至椭圆状卵形，顶端圆，边缘全缘，常狭内弯，腹侧基部截形，背侧基部耳形；腹瓣兜状，有时片状，紧贴茎着生，口部与茎近乎垂直，一侧内折，具长钝尖的喙；副体三角状，4—13 个细胞长，基部 2—3 个细胞宽。侧叶细胞波曲状，角质层平滑，三角体大，肿胀形，中部球状加厚偶可见。油体聚合型，每个细胞 3—6 个。腹叶疏松覆瓦状排列至毗邻，宽为茎的 3—5 倍，顶端不裂，有时微凹，边缘全缘，常反折，基部心形。雌雄异株。雄苞未见。雌苞顶生或侧生于分枝上，雌苞叶 3 对，边缘全缘，最内层雌苞叶腹叶顶端浅 2 裂，边缘全缘。萼苞阔椭圆形，表面平滑，顶部具 5 个短脊，喙 12—18 个细胞长。

生境： 树干环境常见，石生、倒木、腐木、树基、藤茎环境也有分布。海拔 1048—1380 m。

雅长分布： 大宴坪天坑漏斗，黄猄洞天坑，蓝家湾天坑，旁墙屯，悬崖天坑，中井天坑，中井屯。

中国分布： 湖南，四川，西藏，云南。首次记录于广西。

Plants green, redish brown to dark brown when fresh, 1.20–1.80 mm wide, irregularly pinnately branched, branching of the *Frullania*-type. Leaves imbricate, lobe broad oblong to oblong-ovate, apex rounded, margin entire, narrowly incurved, ventral base truncate, dorsal base auriculate; lobule cucullate, sometimes explanate, contiguous to stem, mouth nearly perpendicular to the stem, hemifold, with obtuse-acute rostrum; stylus triangular, consisting of 4–13 cells, 2–3 cells wide at base. Lobe cell flexuose, trigones large, nodulate, intermediate thickenings occasionally present. Cuticle smooth. Oil bodies segmented, 3–6 per median cell. Underleaves loosely imbricate to contiguous, 3–5 times as wide as the stem, apex undivided or retuse, margins entire, base cordate. Dioicous. Androecia not seen. Gynoecia terminal or lateral on branches, bracts in 3 pairs, margins entire, innermost bracteole shallowly bilobed, margin entire. Perianths broad elliptic, surface smooth, apex with 5 short keels, beak 12–18 cells long.

Habitat: Often on tree trunks, sometimes on rocks, fallen trees, rotten logs, tree bases and lianas. Elev. 1048–1380 m.

Distribution in Yachang: Dayanping Tiankeng, Huangjingdong Tiankeng, Lanjiawan Tiankeng, Pangqiang Tun, Xuanya Tiankeng, Zhongjing Tiankeng, Zhongjing Tun.

Distribution in China: Hunan, Sichuan, Tibet, Yunnan. New to Guangxi.

▶ A. 种群；B. 植物体；C. 植物体一段；D. 侧叶；E. 腹叶；F. 茎一段，示腹瓣；G. 副体；H. 最内层雌苞叶和雌苞腹叶；I. 背瓣中部细胞，示油体；J. 萼苞；K. 萼苞上部横切面；L. 萼苞中部横切面。（凭证标本：*韦玉梅等 201107-39*）

A. Population; B. Plant; C. Portion of plant; D. Leaves; E. Underleaves; F. Portion of stem showing leaf lobules; G. Stylus; H. Innermost female bracts and bracteole; I. Median cells of leaf lobe showing oil bodies; J. Perianth; K. Transverse section of upper part of perianth; L. Transverse section of median part of perianth. (All from *Wei et al. 201107-39*)

图 112　大蒴耳叶苔

Fig. 112　*Frullania physantha* Mitt.

微齿耳叶苔

Frullania rhytidantha S. Hatt., J. Hattori Bot. Lab. 47: 97. 1980.

植物体浅绿色至红棕色，带叶宽 1.15—1.68 mm，不规则羽状分枝，分枝耳叶苔型。侧叶覆瓦状排列，背瓣卵形至卵圆形，顶端圆，稍内弯，边缘全缘，腹侧基部心形，背侧基部稍耳形；腹瓣兜状或片状，远离茎着生，口部与茎近乎垂直，一侧内折或不内折，具短钝的喙，或无喙；副体丝状，3—6 个细胞长。侧叶细胞不呈波曲状，角质层平滑，三角体小至大，简单三角形或肿胀形，中部球状加厚常可见。油体聚合型，每个细胞 4—7 个。腹叶毗邻至远生，宽为茎的 2.0—3.5 倍，顶端 2 裂达 1/4—1/3，边缘具 1—3 个钝齿，基部楔形。雌雄异株。雌苞常侧生于分枝上，雌苞叶 2—3 对，边缘全缘，最内层雌苞叶腹叶顶端 2 裂达 2/5—1/2，侧缘常具 1 个裂片状齿。蒴萼倒卵形，表面被疏松的瘤，具 5 个脊，有时还存在少量短的脊，脊波曲状，具细圆齿，喙 8—13 个细胞长。

生境：树干环境常见，石生、树基环境也有分布。海拔 1185—1759 m。

雅长分布：白岩垱漏斗，九十九堡，老屋基天坑，龙坪村，盘古王，下棚屯，香当天坑，中井屯。

中国分布：广西，贵州，湖北，陕西，云南。

Plants light green to redish brown when fresh, 1.15–1.68 mm wide, irregularly pinnately branched, branching of the *Frullania*-type. Leaves imbricate, lobe ovate to rounded-ovate, apex rounded, slightly incurved, margin entire, ventral base cordate, dorsal base slightly auriculate; lobule cucullate or explanate, remote to stem, mouth nearly perpendicular to the stem, hemifold, with short obtuse rostrum, or not hemifold, without rostrum; stylus filiform, in a row of 3–6 cells. Lobe cell not flexuose, trigones small to large, simple-triangulate or nodulate, intermediate thickenings frequent. Cuticle smooth. Oil bodies segmented, 4–7 per median cell. Underleaves contiguous to remote, 2.0–3.5 times as wide as the stem, bilobed to 1/4–1/3 underleaf length, with 1–3 teeth at outer margins, base cuneate. Dioicous. Gynoecia usually lateral on branches, bracts in 2–3 pairs, margins entire, innermost bracteole bilobed to 2/5–1/2 its length, usually with 1 lateral tooth on each margin. Perianths obovate, with single-celled or laciniate protuberances on the surface, with 5 crispate and crenulate keels, sometimes also with a few, much smaller additional keels, beak 8–13 cells long.

Habitat: Often on tree trunks, sometimes on rocks and tree bases. Elev. 1185–1759 m.

Distribution in Yachang: Baiyandangloudou, Jiushijiubao, Laowuji Tiankeng, Longping Village, Panguwang, Xiapeng Tun, Xiangdang Tiankeng, Zhongjing Tun.

Distribution in China: Guangxi, Guizhou, Hubei, Shaanxi, Yunnan.

▶ A. 种群；B. 植物体；C. 植物体一段；D. 背瓣（左）和侧叶（右）；E. 最内层雌苞腹叶；F. 最内层雌苞叶；G. 茎一段，示腹瓣；H. 蒴萼横切面；I. 蒴萼；J. 蒴萼的脊；K. 蒴萼表面；L. 腹叶；M. 背瓣中部细胞，示油体。（凭证标本：韦玉梅等 191015-190）

A. Population; B. Plant; C. Portion of plant; D. Leaf lobe (left) and leaf (right); E. Innermost female bracteole; F. Innermost female bracts; G. Portion of stem showing leaf lobules; H. Transverse section of perianth; I. Perianth; J. Part of perianth keel; K. Surface of perianth; L. Underleaves; M. Median cells of leaf lobe showing oil bodies. (All from *Wei et al. 191015-190*)

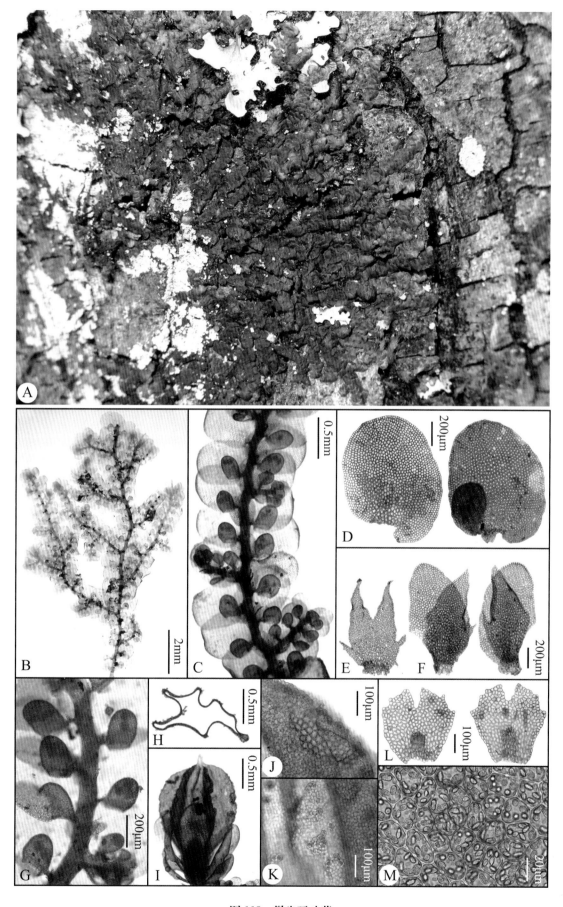

图 113 微齿耳叶苔

Fig. 113 *Frullania rhytidantha* S. Hatt.

陕西耳叶苔

Frullania schensiana C. Massal., Mem. Accad. Agric. Verona 73(2): 40. 1897.

植物体浅绿色至红棕色，带叶宽 1.26—1.90 mm，不规则羽状分枝，分枝耳叶苔型。侧叶疏松覆瓦状排列，背瓣卵形至卵圆形，顶端圆，稍内弯，边缘全缘，腹侧基部心形，背侧基部耳形；腹瓣兜状，紧贴茎着生，口部明显下倾，一侧内折，无喙或喙不明显；副体三角状，3—7 个细胞长，基部 2—4 个细胞宽。侧叶细胞稍波曲状，角质层平滑，三角体大，肿胀形，中部球状加厚偶可见。油体聚合型，每个细胞 5—8 个。腹叶毗邻至远生，宽为茎的 3—5 倍，顶端 2 裂达 1/5—1/4，边缘全缘，基部楔形。雌雄异株。雄苞未见。雌苞常侧生于分枝上，雌苞叶 2 对，边缘全缘，最内层雌苞叶腹叶顶端 2 裂约达 1/2，边缘强烈反折。蒴萼梨形，表面平滑，具 3 个脊，喙 8—12 个细胞长。

生境：石生，树干，树基。海拔 1215—1813 m。

雅长分布：草王山，香当天坑。

中国分布：安徽，重庆，甘肃，广西，贵州，河北，湖北，湖南，江西，内蒙古，山东，陕西，四川，台湾，西藏，云南。

Plants light green to redish brown when fresh, 1.26−1.90 mm wide, irregularly pinnately branched, branching of the *t*-type. Leaves loosely imbricate, lobe ovate to rounded-ovate, apex rounded, slightly incurved, margin entire, ventral base cordate, dorsal base auriculate; lobule cucullate, contiguous to stem, mouth inclined toward the stem, hemifold, without rostrum; stylus triangular, consisting of 3−7 cells, 2−4 cells wide at base. Lobe cell slightly flexuose, trigones large, simple-triangulate or nodulate, intermediate thickenings occasionally present. Cuticle smooth. Oil bodies segmented, 5−8 per median cell. Underleaves contiguous to remote, 3−5 times as wide as the stem, bilobed to 1/5−1/4 underleaf length, margins entire, base cuneate. Dioicous. Androecia not seen. Gynoecia usually lateral on branches, bracts in 2 pairs, margins entire, innermost bracteole bilobed to ca. 1/2 its length, margin entire, strongly recurved. Perianths pyriform, surface smooth, with 3 keels, beak 8−12 cells long.

Habitat: On rocks, tree trunks and tree bases. Elev. 1215−1813 m.

Distribution in Yachang: Caowangshan, Xiangdang Tiankeng.

Distribution in China: Anhui, Chongqing, Gansu, Guangxi, Guizhou, Hebei, Hubei, Hunan, Inner Mongolia, Jiangxi, Shaanxi, Shandong, Sichuan, Taiwan, Tibet, Yunnan.

▶ A. 种群；B. 植物体；C. 植物体一段；D. 茎一段，示腹瓣；E. 最内层雌苞叶和雌苞腹叶；F. 背瓣；G. 蒴萼横切面；H. 副体；I. 腹叶；J. 蒴萼；K. 背瓣中部细胞，示油体。（凭证标本：A 拍摄自唐启明 & 韦玉梅 20191013-115，其余拍摄自韦玉梅等 191013-109B）

A. Population; B. Plant; C. Portion of plant; D. Portion of stem showing leaf lobules; E. Innermost female bracts and bracteole; F. Leaf lobules; G. Transverse section of Perianth; H. Stylus; I. Underleaves; J. Perianth; K. Median cells of leaf lobe showing oil bodies. (A from *Tang & Wei 20191013-115*, the others from *Wei et al. 191013-109B*)

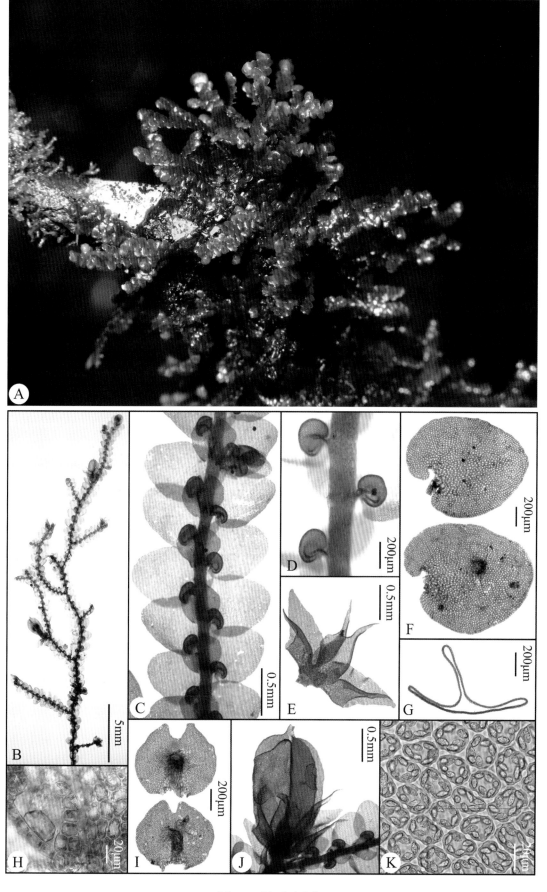

图 114 陕西耳叶苔

Fig. 114 *Frullania schensiana* C. Massal.

欧耳叶苔长叶变种

Frullania tamarisci var. **elongatistipula** (Vard.) S. Hatt., J. Hattori Bot. Lab. 59: 162. 1985.

植物体绿色、红棕色至暗棕色，带叶宽 1.5—2.7 mm，规则羽状分枝，分枝耳叶苔型。侧叶覆瓦状排列，背瓣长椭圆形，顶端渐尖或急长尖，常内弯，边缘全缘，腹侧基部略心形，背侧基部耳形；腹瓣圆柱状或片状，远离茎着生并略向外倾斜，口部圆，拱形，不具喙；副体丝状或三角状，可达 15 个细胞长，基部 2—7 宽，基部具半圆形的附属物。侧叶细胞不呈波曲状，角质层平滑，三角体小，简单三角形，中部球状加厚偶见。油体聚合型，每个细胞 2—5 个。油胞 13—22 个排成一列。腹叶覆瓦状排列，长明显大于宽，宽为茎的 2.0—3.5 倍，顶端 2 裂达 1/5—1/3，常下弯，边缘全缘或偶具钝齿，基部略心形，常具披针形附属物。雌雄异株。雄苞未见。雌苞顶生于分枝上，雌苞叶 2—3 对，边缘全缘，最内层雌苞叶腹叶顶端 2 裂约达 1/3，近基部边缘具裂片状齿。蒴萼长梨形，表面平滑，具 3 个脊，喙 18—25 个细胞长。

生境：倒木，石生。海拔 1625—1902 m。

雅长分布：草王山。

中国分布：福建，贵州，西藏，云南。首次记录于广西。

Plants green, redish brown to dark brown when fresh, 1.5–2.7 mm wide, regularly pinnately branched, branching of the *Frullania*-type. Leaves imbricate, lobe oblong-ovate, apex acuminate to acute, usually incurved, margin entire, ventral base slightly cordate, dorsal base auriculate; lobule cylindrical or explanate, remote and oblique to stem, mouth rounded, arched, without rostrum; stylus filiform or triangular, up to 15 cells long, 2–7 cells wide at base, with semicircular disc-like appendages. Lobe cell not flexuose, trigones small, simple-triangulate, intermediate thickenings occasionally present. Cuticle smooth. Oil bodies segmented, 2–5 per median cell. Ocelli usually in 1 row 13–22 cells long. Underleaves imbricate, 2.0–3.5 times as wide as the stem, bilobed to 1/5–1/3 underleaf length, apex recurved, entire or occasionally with 1–2 teeth at outer margins, base slightly cordate, with lanceolate appendages. Dioicous. Androecia not seen. Gynoecia terminal on branches, bracts in 2–3 pairs, margins entire, innermost bracteole bilobed to ca.1/3 its length, margin with small lobes at base. Perianths long pyriform, surface smooth, with 3 keels, beak 18–25 cells long.

Habitat: On fallen tree and rocks. Elev. 1625–1902 m.

Distribution in Yachang: Caowangshan.

Distribution in China: Fujian, Guizhou, Tibet, Yunnan. New to Guangxi.

▶ A. 种群；B. 植物体；C. 植物体一段；D. 侧叶；E. 腹叶；F. 背瓣中部细胞，示油体和油胞；G. 蒴萼横切面；H. 茎一段，示腹瓣；I. 腹瓣和副体；J. 蒴萼；K. 蒴萼横切面；L. 最内层雌苞叶；M. 最内层雌苞腹叶。（凭证标本：韦玉梅等 *221112-67*）

A. Population; B. Plant; C. Portion of plant; D. Leaves; E. Underleaves; F. Median cells of leaf lobe showing oil bodies and ocelli; G. Transverse section of Perianth; H. Portion of stem showing leaf lobules; I. Leaf lobule and stylus; J. Perianth; K. Transverse section of perianth; L. Innermost female bracts; M. Innermost female bracteole. (All from *Wei et al. 221112-67*)

图 115　欧耳叶苔长叶变种

Fig. 115　*Frullania tamarisci* var. *elongatistipula* (Vard.) S. Hatt.

云南耳叶苔密叶变种

Frullania yuennanensis var. **siamensis** (N. Kitag., Thaithong et S. Hatt.) S. Hatt. et P. J. Lin, J. Hattori Bot. Lab. 59: 133. 1985.

植物体干时黑棕色至黑色，带叶宽 1.15—2.03 mm，不规则羽状分枝，分枝耳叶苔型。侧叶覆瓦状排列，背瓣椭圆形，顶端圆，内弯，边缘全缘，侧缘轻微反折，腹侧基部心形，背侧基部大的耳形；腹瓣兜状，紧贴茎着生，口部明显下倾，一侧内折，但不具喙；副体三角状，5—10 个细胞长，基部 3—6 个细胞宽。侧叶细胞波曲状，角质层平滑，三角体大，肿胀形，中部球状加厚常可见。油体未见。腹叶覆瓦状排列，宽为茎的 2.0—3.5 倍，顶端不裂，边缘全缘，稍反折，基部耳形。雌雄异株。雄苞未见。雌苞顶生于侧短枝上，雌苞叶 3 对，边缘全缘。蒴萼未见。

生境：石生。海拔 1643 m。

雅长分布：大棚屯。

中国分布：广西，贵州，四川，西藏，云南。

Plants blackish brown to black in herbarium specimens, 1.15–2.03 mm wide, irregularly pinnately branched, branching of the *Frullania*-type. Leaves imbricate, lobe oblong, apex rounded, incurved, margin entire, ventral and dorsal margin slightly recurved, ventral base cordate, dorsal base large auriculate; lobule cucullate, contiguous to stem, mouth inclined toward the stem, hemifold, without rostrum; stylus triangular, consisting of 5–10 cells, 3–6 cells wide at base. Lobe cell flexuose, trigones large, nodulate, intermediate thickenings frequent. Cuticle smooth. Oil bodies not seen. Underleaves imbricate, 2.0–3.5 times as wide as the stem, apex undivided, margins entire, slightly recurved, base auriculate. Dioicous. Androecia not seen. Gynoecia usually terminal on short branches, bracts in 3 pairs, margins entire. Perianths not seen.

Habitat: On rocks. Elev. 1643 m.

Distribution in Yachang: Dapeng Tun.

Distribution in China: Guangxi, Guizhou, Sichuan, Tibet, Yunnan.

▶ A. 植物体；B. 腹叶；C. 植物体一段带雌苞；D. 植物体一段；E. 茎一段，示腹瓣；F. 背瓣；G. 副体；H. 背瓣中部细胞。（凭证标本：*唐启明等 20190520-396A*）

A. Plant; B. Underleaves; C. Portion of plant with gynoecium; D. Portion of plant; E. Portion of stem showing leaf lobules; F. Leaf lobes; G. Styli; H. Median cells of leaf lobe. (All from *Tang et al. 20190520-396A*)

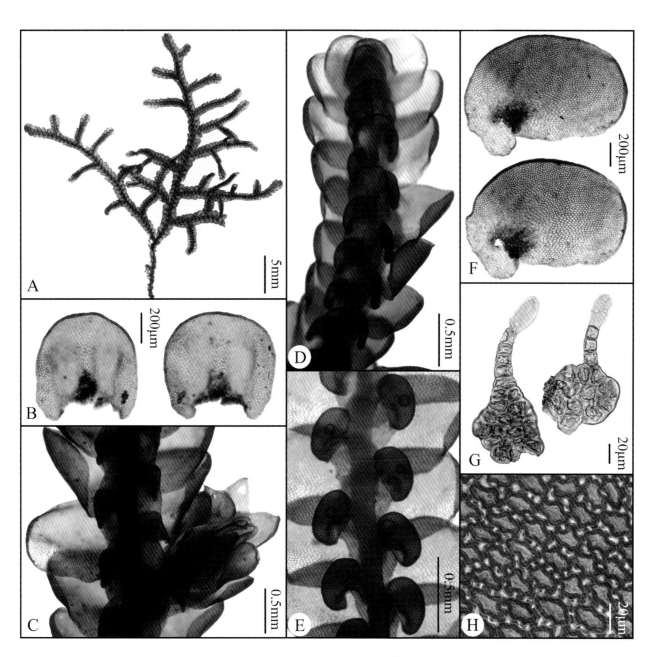

图 116　云南耳叶苔密叶变种

Fig. 116　*Frullania yuennanensis* var. *siamensis* (N. Kitag., Thaithong & S. Hatt.) S. Hatt. & P. J. Lin

汤泽耳叶苔

Frullania yuzawana S. Hatt., J. Hattori Bot. Lab. 49: 157. 1981.

植物体浅绿色至红棕色，带叶宽 1.3—2.1 mm，不规则羽状分枝，分枝耳叶苔型。侧叶覆瓦状排列，背瓣椭圆形或椭圆状卵形，顶端圆至圆钝，边缘全缘，腹缘轻微反折，腹侧基部微心形，背侧基部大的舌形；腹瓣兜状，紧贴茎着生，口部明显下倾，一侧内折，具短钝尖的喙；副体丝状，5—7 个细胞长。侧叶细胞波曲状，角质层平滑，三角体大，肿胀形，中部球状加厚常可见。油体聚合型，每个细胞 5—9 个。腹叶覆瓦状排列至毗邻，宽为茎的 3—5 倍，顶端 2 裂达 1/5—1/4，边缘波曲，近基部常反折，基部楔形。雌雄异株。雄苞未见。雌苞常顶生于分枝上，雌苞叶 3 对，边缘全缘。蒴萼未见。

生境：石生，岩面薄土生。海拔 1240 m。
雅长分布：蓝家湾天坑。
中国分布：广西，台湾。

Plants light green to redish brown when fresh, 1.3–2.1 mm wide, irregularly pinnately branched, branching of the *Frullania*-type. Leaves imbricate, lobe oblong to oblong-ovate, apex rounded to rounded-obtuse, margin entire, ventral margin slightly recurved, ventral base slightly cordate, dorsal base large ligulate; lobule cucullate, contiguous to stem, mouth distinctly inclined toward the stem, hemifold, with short obtuse-acute rostrum; stylus filiform, in a row of 5–7 cells. Lobe cell slightly flexuose, trigones large, nodulate, intermediate thickenings frequent. Cuticle smooth. Oil bodies segmented, 5–9 per median cell. Underleaves imbricate to contiguous, 3–5 times as wide as the stem, bilobed to 1/5–1/4 underleaf length, margins entire, usually recurved near base, base cuneate. Dioicous. Androecia not seen. Gynoecia usually terminal on branches, bracts in 3 pairs, margins entire. Perianths not seen.

Habitat: On rocks and rocks with a thin layer of soil. Elev. 1240 m.
Distribution in Yachang: Lanjiawan Tiankeng.
Distribution in China: Guangxi, Taiwan.

▶ A. 种群；B. 植物体；C. 植物体一段；D. 背瓣；E. 腹瓣和副体；F. 茎一段，示腹瓣；G. 雌苞；H. 背瓣中部细胞，示油体；I. 腹叶。（凭证标本：韦玉梅等 *191018-285*）

A. Population; B. Plant; C. Portion of plant with a gynoecium; D. Leaf lobes; E. Leaf lobule and stylus; F. Portion of stem showing leaf lobules; G. Gynoecium; H. Median cells of leaf lobe showing oil bodies; I. Underleaves. (All from *Wei et al. 191018-285*)

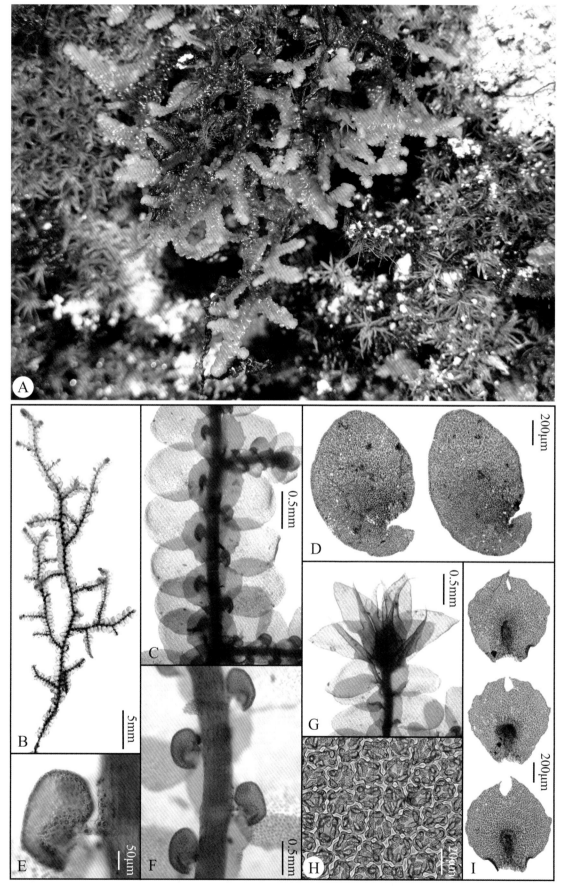

图 117　汤泽耳叶苔

Fig. 117　*Frullania yuzawana* S. Hatt.

南亚顶鳞苔（南亚瓦鳞苔）

Acrolejeunea sandvicensis (Gottsche) Steph., Bot. Jahrb. Syst. 23(1–2): 312. 1986.

Trocholejeunea sandvicensis (Gottsche) Mizut., Misc. Bryol. Lichenol. 2(12): 169. 1961.

植物体绿色至黄绿色，带叶宽 1.14—1.70 mm，不规则分枝，分枝多为耳叶苔型，偶尔为细鳞苔型。茎横切面约 13 个表皮细胞和 26 个内部细胞；腹面局部植物体 4—6 个细胞宽。侧叶密集覆瓦状排列，干时包裹茎，湿时常背仰，呈鱼鳃状，背瓣卵形，顶端阔圆形，边缘全缘；腹瓣卵形，长约为背瓣的 2/5，具 3—5 个齿，透明疣生于腹瓣顶端第一个齿基部的内表面。侧叶细胞薄壁至稍厚壁，角质层平滑，三角体大，心形，中部球状加厚常见。油体均一型，每个细胞含 15 个以上。腹叶密集覆瓦状排列，宽为茎的 3—5 倍，顶端不裂，稍下弯，边缘全缘。雌雄同株。雄苞顶生或间生，雄苞叶和雄苞腹叶大小、形状与侧叶、茎腹叶相似。雌苞具 1—2 个耳叶苔型新生枝或缺。蒴萼倒卵形，具 6—10 个平滑的脊，脊略弯曲，喙 2 个细胞长。孢子球形至椭球形，表面具密集乳突状纹饰及少数莲座状纹饰。弹丝单螺旋加厚。

生境：树干和石生环境常见，腐殖质、腐木、叶附生、树基环境也有分布。海拔 778—1829 m。

雅长分布：广泛分布于保护区各处。

中国分布：全国范围广布。

Plants green to yellowish green when fresh, 1.14−1.70 mm wide, irregularly branched, branches of the *Frullania*-type, rarely of the *Lejeunea*-type. Stem in transverse section with ca. 13 epidermal cells and ca. 26 medullary cells; ventral merophytes 4−6 cells wide. Leaves densely imbricate, convolute the stem when dry, widely spreading and strongly squarrose when moist, lobe broad ovate, apex rounded, margin entire; lobule ovate, ca. 2/5 as long as the lobe, with 3−5 teeth, hyaline papilla on the inner surface of the lobule at the base of the first tooth. Lobe cells thin- to slightly thicken-walled, trigones large, cordate, intermediate thickenings frequent. Cuticle smooth. Oil bodies homogeneous, usually more than 15 per median cell. Underleaves densely imbricate, 3−5 times as wide as the stem, apex undivided, slightly recurved, margins entire. Monoicous. Androecia terminal or intercalary, bracts and bracteoles similar to leaves and underleaves. Gynoecia with 1−2 *Frullania*-type innovations. Perianths obovate, with 6−10 keels, keels somewhat flexuose, beak 2 cells long. Spores globose to ellipsoidal, surface densely papillocate, with several rosettes. Elaters with 1 spiral thickening, stick to the upper 1/3 of each capsule-valve.

Habitat: Often on tree trunks and rocks, sometimes on humus, rotten logs, leaves and tree bases. Elev. 778−1829 m.

Distribution in Yachang: Widely distributed in Yachang.

Distribution in China: Widely distributed in China.

▶ A. 种群；B. 植物体（背面观）；C. 植物体一段，示蒴萼；D. 腹瓣；E. 雌苞腹叶；F. 雌苞叶；G. 侧叶；H. 腹叶；I. 茎横切面；J. 茎一段，示腹面局部植物体；K. 裂开的孢蒴和弹丝；L. 孢子；M. 背瓣中部细胞，示油体。（凭证标本：专玉梅等 201107-4）

A. Population; B. Plant (dorsal view); C. Portion of plant showing a perianth; D. Leaf lobule; E. Female bracteole; F. Female bracts; G. Leaf; H. Underleaf; I. Transverse section of stem; J. Portion of stem showing ventral merophyte; K. Dehisced capsule and elaters; L. Spores; M. Median cells of leaf lobe showing oil bodies. (All from *Wei et al. 201107-4*)

图 118 南亚顶鳞苔

Fig. 118 *Acrolejeunea sandvicensis* (Gottsche) Steph.

中华顶鳞苔

Acrolejeunea sinensis (Jian Wang bis, R.L. Zhu et Gradst.) Jian Wang bis et Gradst., Bryoph. Diversity & Evol. 36(1): 39. 2014.

植物体黄绿色，带叶宽 0.95—1.50 mm，不规则分枝，分枝多为耳叶苔型，偶尔为细鳞苔型。茎横切面约 15 个表皮细胞和 34 个内部细胞；腹面局部植物体 4—6 个细胞宽。侧叶密集覆瓦状排列，干时包裹茎，湿时平展，背瓣卵形至椭圆形，顶端圆钝，常内弯，边缘全缘；腹瓣椭圆形，长为背瓣的 1/2—3/5，具 3 个齿，齿 1—2 个细胞长，第一个齿位于腹瓣和背瓣连接处，透明疣生于腹瓣顶端第二个齿基部的内表面。侧叶细胞薄壁，有时稍厚壁，角质层平滑，三角体大，心形，中部球状加厚有时可见。油体均一型，每个细胞含 10 个以上。腹叶覆瓦状排列，宽为茎的 3—5 倍，顶端不裂，稍下弯，边缘全缘。雌雄异株？雄苞未见。雌苞具 1—2 个耳叶苔型新生枝。蒴萼未见。

生境： 石生。海拔 1240 m。

雅长分布： 蓝家湾天坑。

中国分布： 安徽，广东，广西，云南。

Plants yellowish green when fresh, 0.95−1.50 mm wide, irregularly branched, branches of the *Frullania*-type, rarely of the *Lejeunea*-type. Stem in transverse section with ca. 15 epidermal cells and ca. 34 medullary cells; ventral merophytes 4−6 cells wide. Leaves densely imbricate, clasping the stem when dry, widely spreading and convex when moist, lobe ovate to oblong, apex rounded-obtuse, usually incurved, margin entire; lobule oblong, 1/2−3/5 as long as the lobe, with 3 teeth, teeth 1−2 cells long, first tooth situated at the extreme end of the free lateral margin, hyaline papilla on the inner surface of the lobule at the proximal base of the second tooth. Lobe cells thin-walled, sometimes slightly thickened, trigones large, cordate, intermediate thickenings occasionally present. Cuticle smooth. Oil bodies homogeneous, usually more than 10 per median cell. Underleaves imbricate, 3−5 times as wide as the stem, apex undivided, slightly recurved, margins entire. Dioicous? Androecia not seen. Gynoecia with 1−2 *Frullania*-type innovations. Perianths not seen.

Habitat: On rocks. Elev. 1240 m.

Distribution in Yachang: Lanjiawan Tiankeng.

Distribution in China: Anhui, Guangdong, Guangxi, Yunnan.

▶ A. 种群；B. 植物体；C. 植物体一段；D. 茎一段，示腹面局部植物体；E. 腹瓣；F. 雌苞叶；G. 雌苞腹叶；H. 侧叶；I. 腹叶；J. 茎横切面；K. 背瓣中部细胞，示油体。（凭证标本：*韦玉梅等 191018-287*）

A. Population; B. Plant; C. Portion of plant; D. Portion of stem showing ventral merophyte; E. Leaf lobule; F. Female bracts; G. Female bracteole; H. Leaves; I. Underleaf; J. Transverse section of stem; K. Median cells of leaf lobe showing oil bodies. (All from *Wei et al. 191018-287*)

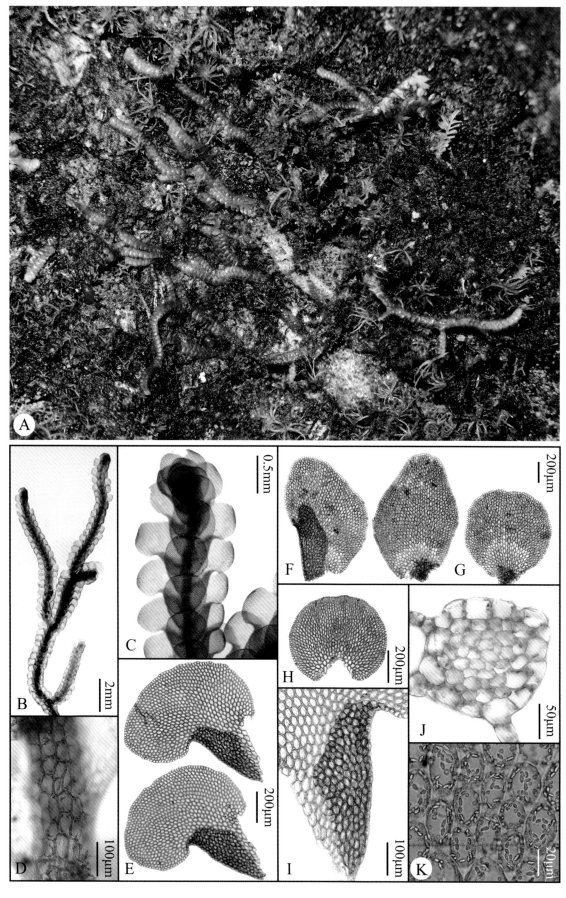

图 119 中华顶鳞苔

Fig. 119 *Acrolejeunea sinensis* (Jian Wang bis, R. L. Zhu et Gradst.) Jian Wang bis et Gradst.

大叶冠鳞苔

Lopholejeunea eulopha (Taylor) Schiffn., Hepat. (Engl.-Prantl): 129. 1893.

植物体黑绿色，带叶宽 0.73—1.25 mm，不规则分枝，分枝细鳞苔型。茎横切面约 12 个表皮细胞和 19 个内部细胞；腹面局部植物体 4—5 个细胞宽。侧叶覆瓦状排列，背瓣卵形或椭圆状卵形，顶端圆形，有时内弯，边缘全缘；腹瓣长为背瓣的 1/5—1/3，具 1—2 个齿，第一个齿位于腹瓣与背瓣连接处，有时不明显，第二个齿在近轴侧，单细胞，腹瓣顶端以 1 个细胞与背瓣相连接，透明疣生于腹瓣顶端第二个齿基部的内表面。侧叶细胞薄壁至稍厚壁，角质层平滑，三角体大，简单三角形，中部球状加厚常可见。油体未见。腹叶毗邻或覆瓦状排列，宽为茎的 4—6 倍，顶端不裂，边缘全缘。雌雄同株。雄苞常间生，雄苞叶 2—3 对，雄苞腹叶生于整个雄苞。雌苞不具新生枝，偶尔具 1—2 个假新生枝。蒴萼倒卵形，具 4—5 个带锯齿或裂片的脊，喙 2—3 个细胞长。

生境： 石生。海拔 1509 m。

雅长分布： 盘古王。

中国分布： 广东，海南，台湾，香港。首次记录于广西。

备注： 该种通常为雌雄异株，偶为雌雄同株（*Zhu & Gradstein 2005*）。

Plants blackish green when fresh, 0.73−1.25 mm wide, irregularly branched, branches of the *Lejeunea*-type. Stem in transverse section with ca. 12 epidermal cells and ca. 19 medullary cells; ventral merophytes 4−5 cells wide. Leaves imbricate, lobe ovate to oblong-ovate, apex rounded, sometimes incurved, margin entire; lobule 1/5−1/3 as long as the lobe, with 1−2 teeth, first tooth first tooth situated at the extreme end of the free lateral margin, sometimes obsolete, second tooth unicellular, apex attached to leaf lobe by only one cell, hyaline papilla on the inner surface of the lobule at the proximal base of the second tooth. Lobe cells thin- to slightly thickened-walled, trigones large, simple triangular, intermediate thickenings frequent. Cuticle smooth. Oil bodies not seen. Underleaves imbricate to contiguous, 4−6 times as wide as the stem, apex undivided, margins entire. Monoicous. Androecia usually intercalary, bracts in 2−3 pairs, bracteoles present throughout the androecium. Gynoecia without innovation, sometimes with 1−2 pseudoinnovations. Perianths obovate, with 4−5 strongly dentate or laciniate keels, beak 2−3 cells long.

Habitat: On rocks. Elev. 1509 m.

Distribution in Yachang: Panguwang.

Distribution in China: Guangdong, Hainan, Hongkong, Taiwan. New to Guangxi.

Note: The sexuality of Lopholejeunea eulopha is usually dioicous, rarely monoicous (*Zhu & Gradstein 2005*).

▶ A. 种群；B. 植物体一段；C. 茎一段，示腹面局部植物体；D. 植物体一段，示蒴萼；E. 茎横切面；F. 雄苞；G. 腹瓣；H. 侧叶；I. 腹叶；J. 雌苞叶；K. 雌苞腹叶；L. 侧叶中部细胞。（凭证标本：*唐启明 & 张仕燕 20181004-449A*）

A. Population; B. Portion of plant; C. Portion of stem showing ventral merophyte; D. Portion of plant showing a perianth; E. Transverse section of stem; F. Androeciaum; G. Leaf lobule; H. Leaves; I. Underleaf; J. Female bracts; K. Female bracteole; L. Median cells of leaf lobe. (All from *Tang & Zhang 20181004-449A*)

图 120　大叶冠鳞苔

Fig. 120　*Lopholejeunea eulopha* (Taylor) Schiffn.

黑冠鳞苔

Lopholejeunea nigricans (Lindenb.) Steph. ex Schiffn., Consp. Hepat. Arch. Ind.: 293. 1898.

植物体暗绿色，带叶宽 0.70—1.25 mm，不规则分枝，分枝细鳞苔型，偶为耳叶苔型。茎横切面约 12 个表皮细胞和 16 个内部细胞；腹面局部植物体 3—4 个细胞宽。侧叶覆瓦状排列至远生，背瓣卵形至椭圆状卵形，顶端圆，边缘全缘；腹瓣卵形，长为背瓣的 1/4—1/3，具 1 个钝的、单细胞的齿，腹瓣顶端以 1 个细胞与背瓣相连接，透明疣生于腹瓣顶端第一个齿基部的内表面。侧叶细胞薄壁至稍厚壁，角质层平滑，三角体大，简单三角形，中部球状加厚常可见。油体均一型，每个细胞含 10 个以上。腹叶毗邻至远生，宽为茎的 2—4 倍，顶端不裂，边缘全缘。雌雄同株。雄苞常间生，雄苞叶 2—4 对，雄苞腹叶生于整个雄苞。雌苞不具新生枝。蒴萼倒卵形，具 4—5 个带裂片的脊，喙 2—4 个细胞长。

生境： 土生。海拔 761—1095 m。

雅长分布： 黄猄洞天坑，拉雅沟。

中国分布： 安徽，重庆，福建，广东，广西，贵州，海南，河南，湖南，陕西，四川，台湾，香港，云南，浙江。

Plants dark green when fresh, 0.70–1.25 mm wide, irregularly branched, branches of the *Lejeunea*-type, very rarely of the *Frullania*-type. Stem in transverse section with ca.12 epidermal cells and ca. 16 medullary cells; ventral merophytes 3–4 cells wide. Leaves imbricate to distant, lobe ovate to oblong-ovate, apex rounded, margin entire, lobule ovate, 1/4–1/3 as long as the lobe, with a blunt, unicellular tooth, apex attached to leaf lobe by only one cell, hyaline papilla on the inner surface of the lobule at the proximal base of the first tooth. Lobe cells thin- to slightly thickened-walled, trigones large, simple triangular, intermediate thickenings frequent. Cuticle smooth. Oil bodies homogeneous, usually more than 10 per median cell. Underleaves distant to contiguous, 2–4 times as wide as the stem, apex undivided, margins entire. Monoicous. Androecia usually intercalary, bracts in 2–4 pairs,, bracteoles present throughout the androecium. Gynoecia without innovation. Perianths obovate, with 4–5 strongly laciniate keels, beak 2–4 cells long.

Habitat: On soil. Elev. 761–1095 m.

Distribution in Yachang: Huangjingdong Tiankeng, Layagou.

Distribution in China: Anhui, Chongqing, Fujian, Guangdong, Guangxi, Guizhou, Hainan, Henan, Hongkong, Hunan, Shaanxi, Sichuan, Taiwan, Yunnan, Zhejiang.

▶ A. 种群；B. 植物体一段带雌苞和雄苞；C. 植物体一段；D. 茎横切面；E. 茎一段，示腹面局部植物体；F. 雌苞叶；G. 雌苞腹叶；H. 腹叶；I. 侧叶；J. 腹瓣；K. 背瓣中部细胞，示油体。（凭证标本：*唐启明 & 张仕艳 20180930-47A*）

A. Population; B. Portion of plant with an androecium and gynoecia; C. Portion of plant; D. Transverse section of stem; E. Portion of stem showing ventral merophyte; F. Female bracts; G. Female bracteole; H. Underleaf; I. Leaves; J. Leaf lobule; K. Median cells of leaf lobe showing oil bodies. (All from *Tang & Zhang 20180930-47A*)

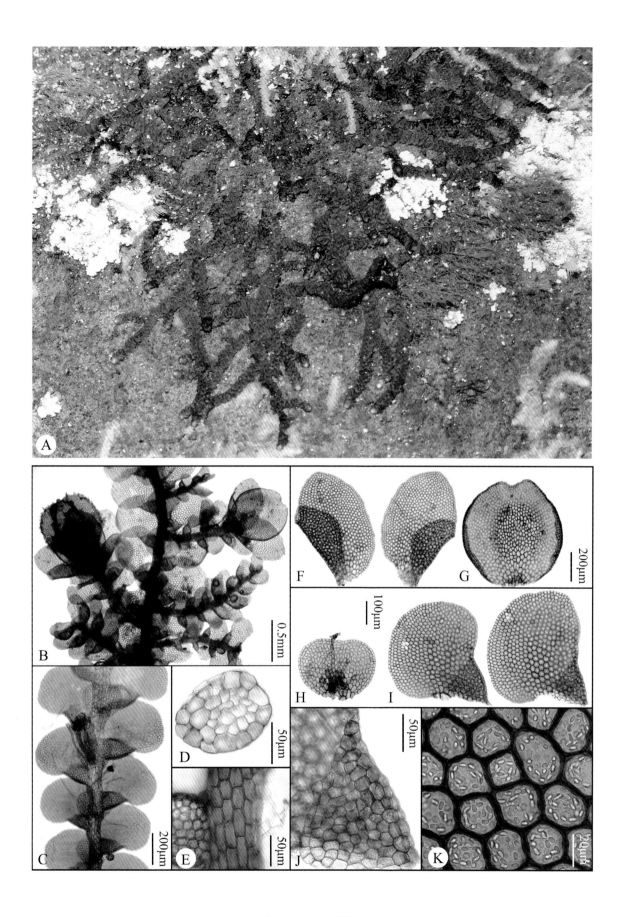

图 121 黑冠鳞苔

Fig. 121 *Lopholejeunea nigricans* (Lindenb.) Schiffn.

皱萼苔

Ptychanthus striatus (Lehm. et Lindenb.) Nees, Naturgesch. Eur. Leberm. 3: 212. 1838.

植物体深绿色，带叶宽 2.5—3.5 mm，一回或二回羽状分枝，分枝耳叶苔型，偶为细鳞苔型。茎横切面约 38 个表皮细胞和超过 100 个内部细胞；腹面局部植物体 14—18 个细胞宽。侧叶覆瓦状排列，背瓣椭圆状卵形，顶端锐尖或渐尖，边缘全缘或在顶部具不规则粗齿，腹缘常内折，背缘基部呈耳状；腹瓣椭圆状卵形，长为背瓣的 1/5—1/4，具 1—2 个齿，第一个齿 1—3 个细胞长，基部 1—3 个细胞宽，第二个齿 1—2 个细胞长，基部 1—2 个细胞宽，有时退化，透明疣位于第一个齿基部内表面。侧叶细胞稍厚壁，角质层平滑，三角体大，心形或三射形，中部球状加厚常可见。油体聚合型，每个细胞 4—10 个。腹叶覆瓦状排列至毗邻，宽为茎的 4—5 倍，顶端不裂，具不规则粗齿，基部耳状，常反折。雌雄同株。雄苞顶生或间生，雄苞叶 2—5 对，雄苞腹叶生于整个雄穗。雌苞具 1 个新生枝，新生枝叶发生顺序为细鳞苔型。蒴萼狭长倒卵形，具 10 个平滑的脊，喙长 5—6 个细胞。

生境： 石生环境常见，倒木环境也有分布。海拔 582—1342 m。

雅长分布： 二沟，老屋基天坑，逻家田屯，深洞，下棚屯。

中国分布： 除了东北以外的全国大部分省区均有分布。

Plants dark green when fresh, 2.5–3.5 mm wide, regularly pinnate or bipinnate, branching mostly of the *Frullania*-type, occasionally with *Lejeunea*-type branches. Stem in transverse section with ca. 38 epidermal cells and more than 100 medullary cells; ventral merophyte 14–18 cells wide. Leaves imbricate, lobe oblong-ovate, apex acute to acuminate, margin irregularly dentate towards apex or entire, ventral margin usually incurved, dorsal margin auriculate at base; lobule oblong-ovate, 1/5–1/4 as long as the lobe, with 1–2 teeth, first tooth 1–3 cells long, 1–3 cells wide at base, second tooth 1–2 cells long, 1–2 cells wide at base, somtimes obsolete, hyaline papilla on the inner surface of the lobule at the proximal base of the first tooth. Lobe cell slightly thicken-walled, trigones large, cordate or radiate, intermediate thickenings frequent. Cuticle smooth. Oil bodies segmented, 4–10 per median cell. Underleaves imbricate to contiguous, 4–5 times as wide as the stem, apex undivided, irregularly dentate, base auriculate, usually recurved. Monoicous. Androecia terminal or intercalary, bracts in 2–5 pairs, bracteoles present throughout the androecium. Gynoecia with 1 innovation, innovation leaf sequence lejeuneoid. Perianths narrowly obovate, with 10 smooth keels, beak 5–6 cells long.

Habitat: Often on rocks, sometimes on fallen trees. Elev. 582–1342 m.

Distribution in Yachang: Ergou, Laowuji Tiankeng, Luojiatian Tun, Shendong, Xiapeng Tun.

Distribution in China: Widely distributed in most provinces of China except Northeast Region.

▶ A. 种群；B. 植物体一段，示蒴萼；C. 雄苞；D. 侧叶；E. 雌苞腹叶；F. 雌苞叶；G. 腹叶；H. 茎横切面；I. 茎一段，示腹面局部植物体；J. 腹瓣；K. 背瓣中部细胞，示油体。（凭证标本：*韦玉梅等 191015-179*）

A. Population; B. Portion of plant showing a perianth; C. Androecium; D. Leaves; E. Female bracteole; F. Female bracts; G. Underleaves; H. Transverse section of stem; I. Portion of stem showing ventral merophyte; J. Leaf lobule; K. Median cells of leaf lobe showing oil bodies. (All from *Wei et al. 191015-179*)

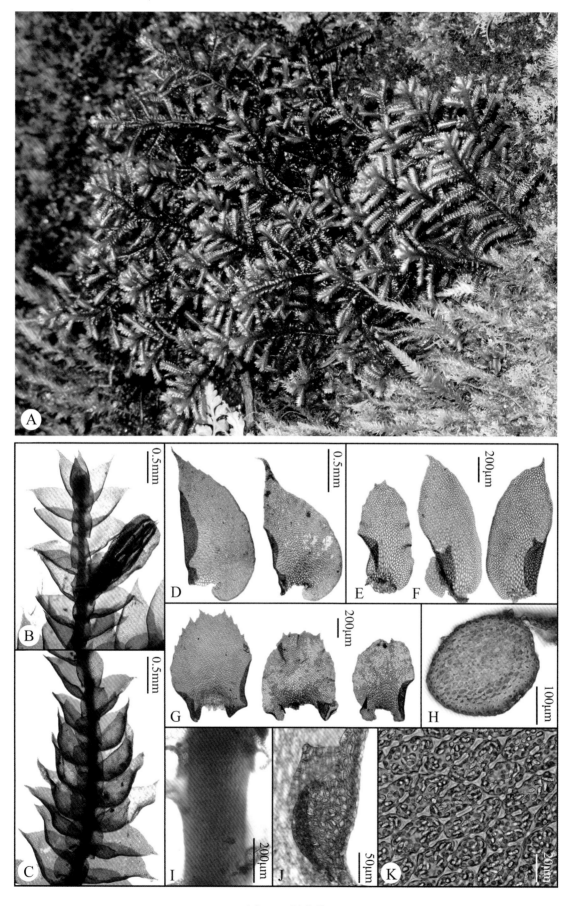

图 122 皱萼苔

Fig. 122 *Ptychanthus striatus* (Lehm. & Lindenb.) Nees

东亚多褶苔（粗齿原鳞苔）

Spruceanthus kiushianus (Horik.) X. Q. Shi, R. L. Zhu et Gradst., Taxon 64(5): 889. 2015.

Archilejeunea kiushiana (Horik.) Verd., Ann. Bryol., Suppl. 4: 46. 1934.

植物体橄榄绿色，带叶宽 0.78—1.25 mm，不规则分枝，分枝细鳞苔型。茎横切面约 12 个表皮细胞和 20 个内部细胞；腹面局部植物体 5—6 个细胞宽。侧叶覆瓦状排列，背瓣卵状椭圆形，顶端圆形，边缘全缘；腹瓣卵形或椭圆形，长为背瓣的 1/3—2/5，具 1 个齿，齿 1—3 个细胞长，基部 1—2 个细胞宽，透明疣位于腹瓣顶端第一个齿基部内表面。侧叶细胞厚壁，角质层平滑，三角体小，简单三角形，中部球状加厚少见。油体均一型，每个细胞常多于 10 个。腹叶远生至毗邻，宽为茎的 2—3 倍，顶端不裂，边缘全缘。雌雄同株。雄苞顶生或间生，雄苞叶 2—4 对，雄苞腹叶生于整个雄穗。雌苞具 1 个新生枝，新生枝叶发生顺序为细鳞苔型或密鳞苔型。蒴萼倒卵形，具 5 个平滑的脊，喙 1—2 个细胞长。

生境：石生，叶附生。海拔 1114 m。

雅长分布：黄猄洞天坑，悬崖天坑。

中国分布：安徽，福建，广东，广西，海南，江西，山东，香港，浙江。

Plants olive green when fresh, 0.78−1.25 mm wide, irregularly branched, branching of the *Lejeunea*-type. Stem in transverse section with ca. 12 epidermal cells and ca. 20 medullary cells; ventral merophytes 5−6 cells wide. Leaves imbricate, lobe ovate-oblong, apex rounded, margin entire; lobule ovate or oblong, 1/3−2/5 as long as the lobe, with 1 tooth, teeth 1−3 cells long and 1−2 cells wide at base, hyaline papilla on the inner surface of the lobule at the base of the first tooth. Lobe cell thicken-walled, trigones small, simple-triangular, intermediate thickenings scarce. Cuticle smooth. Oil bodies homogeneous, usually more than 10 per median cell. Underleaves distant to contiguous, 2−3 times as wide as the stem, apex undivided, margins entire. Monoicous. Androecia terminal or intercalary, bracts in 2−4 pairs, bracteoles present throughout the androecium. Gynoecia with 1 innovation, innovation leaf sequence lejeuneoid or pycnolejeuneoid. Perianths obovate, with 5 smooth keels, beak 1−2 cells long.

Habitat: On rocks and leaves. Elev. 1114 m.

Distribution in Yachang: Huangjingdong Tiankeng, Xuanya Tiankeng.

Distribution in China: Anhui, Fujian, Guangdong, Guangxi, Hainan, Hongkong, Jiangxi, Shandong, Zhejiang.

▶ A. 植物体一段带雄苞和雌苞；B. 植物体一段；C. 茎横切面；D. 腹瓣；E. 侧叶；F. 雌苞腹叶；G. 雌苞叶；H. 植物体一段，示腹叶；I. 茎一段，示腹面局部植物体；J. 背瓣中部细胞，示油体。（凭证标本：*唐启明等 20210721-42B*）

A. Portion of plant with an androecium and a gynoecium; B. Portion of plant; C. Transverse section of stem; D. Leaf lobule; E. Leaves; F. Female bracteole; G. Female bracts; H. Portion of plant showing underleaves; I. Portion of stem showing ventral merophyte; J. Median cells of leaf lobe showing oil bodies. (All from *Tang et al. 20210721-42B*)

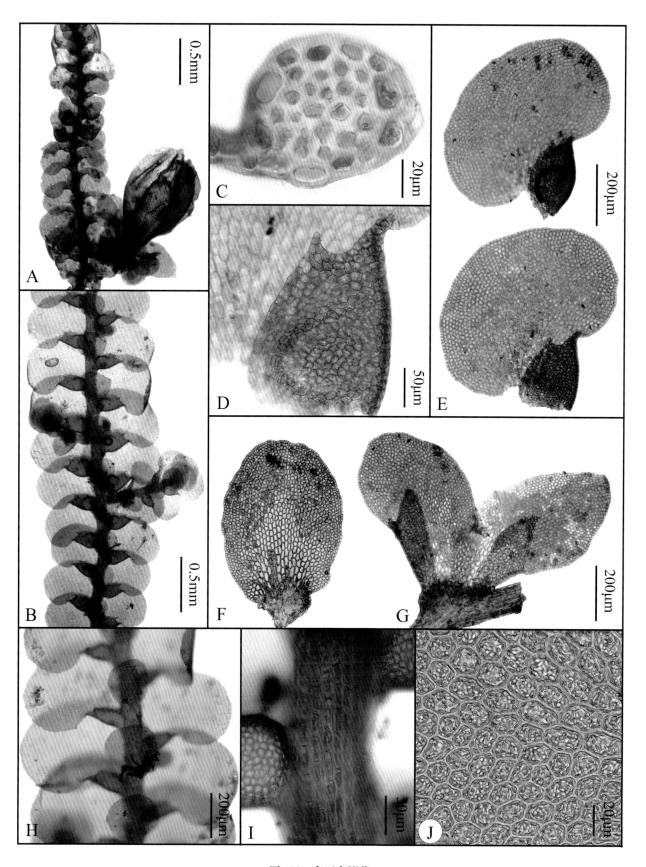

图 123 东亚多褶苔

Fig. 123 *Spruceanthus kiushianus* (Horik.) X. Q. Shi, R. L. Zhu & Gradst.

疣叶多褶苔

Spruceanthus mamillilobulus (Herzog) Verd., Hepat. Select. Crit. 9: no. 447. 1936.

植物体深绿色，带叶宽 2.1—3.8 mm，不规则分枝，分枝细鳞苔型。茎横切面约 21 个表皮细胞和 35 个内部细胞；腹面局部植物体约 9 个细胞宽。侧叶覆瓦状排列，背瓣椭圆状卵形，顶端锐尖，有时圆，常具齿，有时全缘，腹缘常内折，背缘基部呈耳状；腹瓣椭圆状卵形，长为背瓣的 1/9—1/7，有时退化，具 1—3 个单细胞齿，透明疣位于腹瓣顶端第一个齿基部内表面。侧叶细胞厚壁，角质层平滑，三角体大，简单三角形或三射形，中部球状加厚常可见。油体均一型，每个细胞常多于 20 个。腹叶覆瓦状排列，宽为茎的 2—3 倍，顶部不裂，具齿，有时全缘，顶端和侧边缘常下弯或外折，基部呈耳状。雌雄同株。雄苞常间生，雄苞叶 4—7 对，雄苞腹叶生于整个雄穗。雌苞具 1—2 个新生枝，新生枝叶发生顺序为细鳞苔型。蒴萼未见。

生境：石生。海拔 582 m。

雅长分布：二沟。

中国分布：广东，广西，贵州，四川，台湾。

Plants dark green when fresh, 2.1–3.8 mm wide, irregularly branched, branching of the *Lejeunea*-type. Stem in transverse section with ca. 21 epidermal cells and ca. 35 medullary cells; ventral merophytes ca. 9 cells wide. Leaves imbricate, lobe oblong-ovate, apex usually acute, sometimes rounded, irregularly dentate to nearly entire, ventral margin usually incurved, dorsal margin auriculate at base; lobule oblong-ovate, 1/9–1/7 as long as the lobe, occasionally reduced, with 1–3 unicellular teeth, hyaline papilla on the inner surface of the lobule at the base of the first tooth. Lobe cell thicken-walled, trigones large, simple-triangular or radiate, intermediate thickenings frequent. Cuticle smooth. Oil bodies homogeneous, usually more than 20 per median cell. Underleaves imbricate, 2–3 times as wide as the stem, apex undivided, dentate, sometimes entire, apical and lateral margin usually recurved, auriculate at base. Monoicous. Androecia usually intercalary, bracts in 4–7 pairs, bracteoles present throughout the androecium. Gynoecia with 1–2 innovations, innovation leaf sequence lejeuneoid. Perianths not seen.

Habitat: On rocks. Elev. 582 m.

Distribution in Yachang: Ergou.

Distribution in China: Guangdong, Guangxi, Guizhou, Sichuan, Taiwan.

▶ A. 种群；B. 植物体一段带雌苞；C. 茎横切面；D. 侧叶；E. 腹叶；F. 雌苞腹叶；G. 雌苞叶；H. 腹瓣；I. 茎一段，示腹面局部植物体；J. 背瓣中部细胞，示油体。（凭证标本：*韦玉梅等 201109-33*）

A. Population; B. Portion of plant with a gynoecium; C. Transverse section of stem; D. Leaves; E. Underleaves; F. Female bracteole; G. Female bracts; H. Leaf lobule; I. Portion of stem showing ventral merophyte; J. Median cells of leaf lobe showing oil bodies. (All from *Wei et al. 201109-33*)

图 124　疣叶多褶苔

Fig. 124　*Spruceanthus mamillilobulus* (Herzog) Verd.

多褶苔

Spruceanthus semirepandus (Nees) Verd., Ann. Bryol., Suppl. 4: 153. 1934.

植物体深绿色，带叶宽 3.6—5.1 mm，不规则分枝，分枝细鳞苔型。茎横切面约 35 个表皮细胞和 100 个内部细胞；腹面局部植物体 10—14 个细胞宽。侧叶覆瓦状排列，背瓣斜卵形，顶端锐尖，边缘全缘或有时顶部具细齿，腹缘常内折，背缘基部呈耳状；腹瓣椭圆状卵形，长为背瓣的 1/5—1/4，具 1—3 个单细胞齿，透明疣位于腹瓣顶端第一个齿基部内表面。侧叶细胞厚壁，角质层平滑，三角体大，三射形，中部球状加厚常可见。油体均一型，每个细胞常多于 20 个。腹叶密集覆瓦状排列，宽为茎的 3—4 倍，顶部不裂，具齿，有时全缘，侧缘常外折，基部稍呈耳状。雌雄同株。雄苞常顶生，雄苞叶 5—8 对，雄苞腹叶生于整个雄穗。雌苞具 2 个新生枝，新生枝叶发生顺序为细鳞苔型。蒴萼倒卵形，具 5—10 个平滑的脊，喙 3—4 个细胞长。

生境： 石生。海拔 1643—1755 m。

雅长分布： 大棚屯，盘古王。

中国分布： 安徽，重庆，福建，广东，广西，贵州，海南，湖南，江西，四川，台湾，西藏，香港，云南，浙江。

Plants dark green when fresh, 3.6–5.1 mm wide, irregularly branched, branching of the *Lejeunea*-type. Stem in transverse section with ca. 35 epidermal cells and ca. 100 medullary cells; ventral merophytes 10–14 cells wide. Leaves imbricate, lobe obliquely ovate, apex acute, margin entire or dentate towards apex, ventral margin usually incurved, dorsal margin auriculate at base; lobule oblong-ovate, 1/5–1/4 as long as the lobe, with 1–3 unicellular teeth, on the inner surface of the lobule at the base of the first tooth. Lobe cell thicken-walled, trigones large, radiate, intermediate thickenings frequent. Cuticle smooth. Oil bodies homogeneous, usually more than 20 per median cell. Underleaves densely imbricate, 3–4 times as wide as the stem, apex undivided, dentate, sometimes entire, lateral margins usually recurved, shortly auriculate at base. Monoicous. Androecia usually terminal, bracts in 5–8 pairs, bracteoles present throughout the androecium. Gynoecia with 2 innovations, innovation leaf sequence lejeuneoid. Perianths obovate, with 5–10 smooth keels, beak 3–4 cells long.

Habitat: On rocks. Elev. 1643–1755 m.

Distribution in Yachang: Dapeng Tun, Panguwang.

Distribution in China: Anhui, Chongqing, Fujian, Guangdong, Guangxi, Guizhou, Hainan, Hongkong, Hunan, Jiangxi, Sichuan, Taiwan, Tibet, Yunnan, Zhejiang.

▶ A. 种群；B. 植物体带雌苞和雄苞；C. 侧叶；D. 腹瓣；E. 腹叶；F. 雌苞腹叶；G. 雌苞叶；H. 茎横切面；I. 茎一段，示腹面局部植物体；J. 植物体一段，示蒴萼；K. 背瓣中部细胞，示油体。（凭证标本：*唐启明等 20190520-385*）

A. Population; B. Plant with gynoecia and an androecium; C. Leaves; D. Leaf lobule; E. Underleaves; F. Female bracteole; G. Female bracts; H. Transverse section of stem; I. Portion of stem showing ventral merophyte; J. Portion of plant showing a perianth; K. Median cells of leaf lobe showing oil bodies. (All from *Tang et al. 20190520-385*)

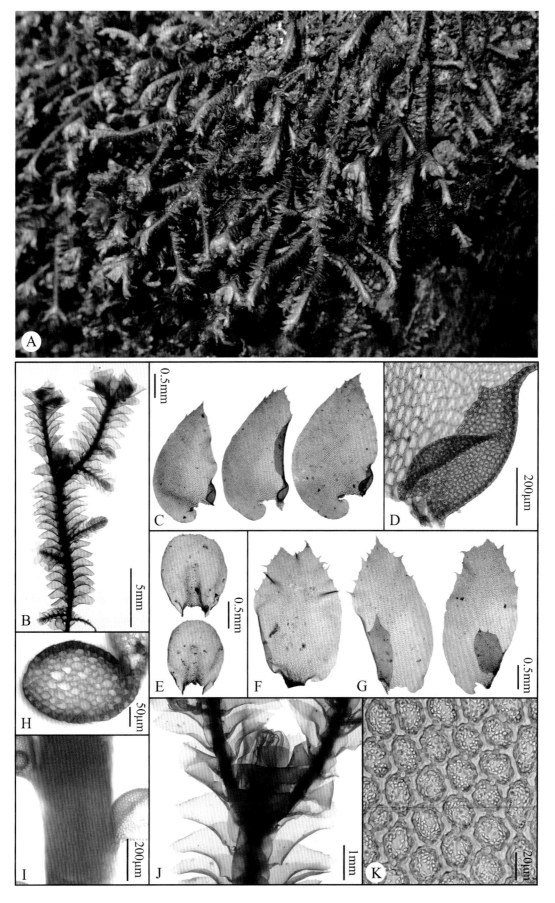

图 125 多褶苔

Fig. 125 *Spruceanthus semirepandus* (Nee.) Verd.

南亚毛鳞苔（南亚鞭鳞苔）

Thysananthus repletus (Taylor) Sukkharak et Gradst., Phytotaxa 326(2): 103. 2017.

Mastigolejeunea repleta (Taylor) A. Evans, Mem. Torrey Bot. Club 8(2): 131. 1902.

植物体深绿色至黑绿色，带叶宽 1.25—1.90 mm，不规则分枝，分枝细鳞苔型。茎横切面约 26 个表皮细胞和 80 个内部细胞；腹面局部植物体 6—12 个细胞宽。侧叶密集覆瓦状排列，背瓣长椭圆形，顶端圆钝至锐尖，边缘全缘，腹缘常内折，背缘基部稍呈耳状；腹瓣卵形，长为背瓣的 1/4—1/3，具 2—4 个粗齿，齿 3—5 个细胞长，基部 2—6 个细胞宽，透明疣生于腹瓣顶端第一个齿基部的内表面。侧叶细胞薄壁至稍厚壁，角质层平滑，三角体大，心形，中部球状加厚常可见。油体聚合型，每个细胞 1—3 个。腹叶覆瓦状排列，宽约为茎的 3 倍，顶端不裂，边缘全缘。雌雄同株。雄苞顶生或间生，雄苞叶 4—6 对，雄苞腹叶生于整个雄苞。雌苞具 1—2 个新生枝，新生枝叶发生顺序为细鳞苔型，雌苞叶背瓣长椭圆形或卵形，顶端锐尖，边缘全缘，雌苞叶腹瓣长为雌苞叶背瓣的 1/2—2/3，雌苞腹叶全缘。蒴萼倒卵形或椭圆形，具 3 个平滑的脊，喙 2—3 个细胞长。

生境：石生环境常见，树干和树基上也有分布。海拔 1048—1136 m。

雅长分布：白岩坨屯，黄猄洞天坑。

中国分布：福建，广东，广西，贵州，海南，台湾，西藏，香港，云南，浙江。

Plants dark green to blackish green when fresh, 1.25–1.90 mm wide, irregularly branched, branching of the *Lejeunea*-type. Stem in transverse section with ca. 26 epidermal cells and ca. 80 medullary cells; ventral merophytes 6–12 cells wide. Leaves densely imbricate, lobe oblong, apex obtuse-rounded to acute, margin entire, ventral margin incurved, dorsal margin auriculate at base; lobule ovate, 1/4–1/3 as long as the lobe, apex with 2–4 large teeth, teeth 3–5 cells long and 2–6 cells wide at base, hyaline papilla on the inner surface of the lobule at the base of the first tooth. Lobe cells thin- to slightly thickened-walled, trigone large, cordate, intermediate thickening frequent. Cuticle smooth. Oil bodies segmented, 1–3 per median cell. Underleaves imbricate, ca. 3 times as wide as the stem, apex undivided, margins entire. Monoicous. Androecia terminal or intercalary, bracts in 4–6 pairs, bracteoles present throughout the androecium. Gynoecia with 1–2 innovations, innovation leaf sequence lejeuneoid. Perianth obovate or oblong, with 3 smooth keels, beak 2–3 cells long.

Habitat: Often on rocks, sometimes on tree trunks and tree bases. Elev. 1048–1136 m.

Distribution in Yachang: Baiyantuo Tun, Huangjingdong Tiankeng.

Distribution in China: Fujian, Guangdong, Guangxi, Guizhou, Hainan, Hongkong, Taiwan, Tibet, Yunnan, Zhejiang.

▶ A. 种群；B. 植物体；C. 植物体一段；D. 侧叶；E. 腹叶；F. 茎一段，示腹面局部植物体；G. 茎横切面；H. 雌苞腹叶；I. 雌苞叶；J. 腹瓣；K. 背瓣中部细胞，示油体。（凭证标本：韦玉梅等 201104-43）

A. Population; B. Plant; C. Portion of plant; D. Leaf; E. Underleaves; F. Portion of stem showing ventral merophyte; G. Transverse section of stem; H. Female bracteole; I. Female bracts; J. Leaf lobule; K. Median cells of leaf lobe showing oil bodies. (All from *Wei et al. 201104-43*)

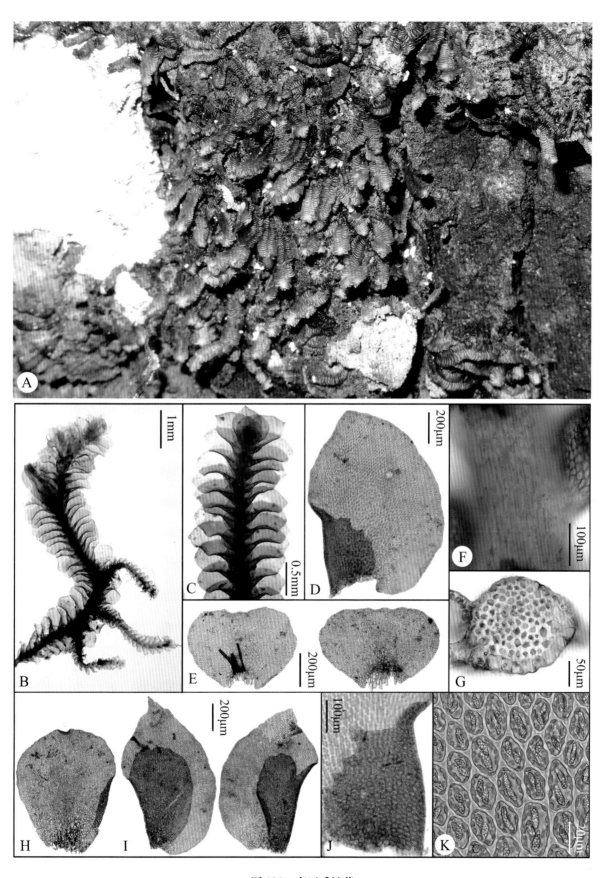

图 126 南亚毛鳞苔

Fig. 126 *Thysananthus repletus* (Taylor) Sukkharak & Gradst.

异鳞苔

Tuzibeanthus chinensis (Steph.) Mizut., J. Hattori Bot. Lab. 24: 151. 1961.

植物体橄榄绿至深绿色，带叶宽 1.60—2.53 mm，一回至二回羽状分枝，分枝耳叶苔型。茎横切面约 25 个表皮细胞和 75 个内部细胞；腹面局部植物体约 16 个细胞宽。侧叶覆瓦状排列，背瓣卵形或椭圆状卵形，顶端圆或钝圆，边缘全缘，腹缘稍内折，背缘基部呈耳状；腹瓣退化。侧叶细胞稍厚壁，角质层平滑，三角体小至大，简单三角形至三射形，中部球状加厚常见。油体聚合型，每个细胞 6—9 个。腹叶远生至毗邻，宽为茎的 3—4 倍，顶端不裂，边缘全缘，基部稍呈耳状。雌雄异株。雄苞常间生，雄苞叶 3—7 对，雄苞腹叶生于整个雄穗。雌苞具 1 个新生枝，新生枝叶发生顺序为细鳞苔型。蒴萼未见。

生境： 石生环境常见，倒木、腐木、树干、树基环境也有分布。海拔 1066—1509 m。

雅长分布： 白岩坨屯，达陇坪屯，大宴坪天坑漏斗，吊井天坑，黄猄洞天坑，蓝家湾天坑，老屋基天坑，里郎天坑，盘古王，深洞，下岩洞屯，霄罗湾洞穴，悬崖天坑，中井天坑。

中国分布： 重庆，福建，甘肃，广西，贵州，湖北，江西，陕西，四川，西藏，云南。

Plants olive green to dark green when fresh, 1.60–2.53 mm wide, regularly pinnate or bipinnate, branching of the _Frullania_-type. Stem in transverse section with ca. 25 epidermal cells and ca. 75 medullary cells; ventral merophytes ca. 16 cells wide. Leaves imbricate, lobe ovate or oblong-ovate, apex rounded or obtuse, margin entire, ventral margin slightly incurved, dorsal margin auriculate at base; lobule reduced. Lobe cell slightly thicken-walled, trigones small to large, simple-triangulate to radiate, intermediate thickenings frequent. Cuticle smooth. Oil bodies segmented, 6–9 per median cell. Underleaves distant to contiguous, 3–4 times as wide as the stem, apex undivided, margins entire, shortly auriculate at base. Dioicous. Androecia usually intercalary, bracts in 3–7 pairs, bracteoles present throughout the androecium. Gynoecia with 1 innovation, innovation leaf sequence lejeuneoid. Perianths not seen.

Habitat: Often on rocks, sometimes on fallen trees, rotten logs, tree trunks and tree bases. Elev. 1066–1509 m.

Distribution in Yachang: Baiyantuo Tun, Dalongping Tun, Dayanping Tiankeng, Diaojing Tiankeng, Huangjingdong Tiankeng, Lanjiawan Tiankeng, Laowuji Tiankeng, Lilang Tiankeng, Panguwang, Shendong, Xiayandong Tun, Xiaoluowan Cave, Xuanya Tiankeng, Zhongjing Tiankeng.

Distribution in China: Chongqing, Fujian, Gansu, Guangxi, Guizhou, Hubei, Jiangxi, Shaanxi, Sichuan, Tibet, Yunnan.

A. 种群；B. 植物体一段；C. 雄苞；D. 侧叶；E. 腹叶；F. 茎横切面；G. 雌苞叶；H. 雌苞腹叶；I. 茎一段，示腹面局部植物体；J. 腹瓣；K. 背瓣中部细胞，示油体。（凭证标本：_韦玉梅等 191012-45_）

A. Population; B. Portion of plant; C. Androeciaum; D. Leaves; E. Underleaves; F. Transverse section of stem; G. Female bracts; H. Female bracteole; I. Portion of stem showing ventral merophyte; J. Leaf lobule; K. Median cells of leaf lobe showing oil bodies. (All from _Wei et al. 191012-45_)

图 127 异鳞苔

Fig. 127 *Tuzibeanthus chinensis* (Steph.) Mizut.

圆叶唇鳞苔

Cheilolejeunea intertexta (Lindenb.) Steph., Bull. Herb. Boissier 5(2): 79. 1897.

植物体绿色至黄绿色，带叶宽 0.50—0.83 mm，不规则分枝，分枝细鳞苔型。茎横切面 7 个表皮细胞和约 5 个内部细胞，腹面局部植物体 2 个细胞宽。侧叶覆瓦状排列，背瓣椭圆形或椭圆状卵形，顶端圆，边缘全缘；腹瓣卵形，长约为背瓣的 1/3，中齿退化，角齿单细胞，透明疣位于角齿基部的远轴侧。侧叶细胞薄壁，角质层平滑，三角体简单三角形，中部球状加厚缺。油体聚合型，每个细胞 1 个。腹叶远生至毗邻，宽为茎的 2—3 倍，顶端 2 裂达 1/3—1/2 深。雌雄同株。雄苞顶生或间生，雄苞叶 2—4 对，雄苞腹叶仅生于雄穗基部。雌苞具 1 个新生枝，新生枝叶发生顺序为密鳞苔型。蒴萼倒卵形，具 4 个平滑的脊，喙 2—3 个细胞长。

生境：石生，树干。海拔 1052—1240 m。

雅长分布：黄猄洞天坑，蓝家湾天坑。

中国分布：福建，广东，广西，贵州，海南，湖北，湖南，台湾，香港，云南。

Plants green to yellowish green when fresh, 0.50−0.83 mm wide, irregularly branched, branching of the *Lejeunea*-type. Stem in transverse section with 7 epidermal cells and ca. 5 medullary cells; ventral merophyte 2 cells wide. Leaves imbricate, lobe oblong to oblong-ovate, apex rounded, margin entire; lobule ovate, ca. 1/3 as long as the lobe, second tooth unicellular, first tooth obsolete, hyaline papilla at the distal side of the second tooth. Lobe cell thin-walled, trigones simple-triangulate, intermediate thickenings absent. Cuticle smooth. Oil bodies segmented, 1 per median cell. Underleaves distant to contiguous, 2−3 times as wide as the stem, bilobed to 1/3−1/2 underleaf length. Monoicous. Androecia terminal or intercalary, bracts in 2−4 pairs, bracteoles present only at the base of the androecium. Gynoecia with 1 innovation, innovation leaf sequence pycnolejeuneoid. Perianths obovate, with 4 smooth keels, beak 2−3 cells long.

Habitat: On rocks and tree trunks. Elev. 1052−1240 m.

Distribution in Yachang: Huangjingdong Tiankeng, Lanjiawan Tiankeng.

Distribution in China: Fujian, Guangdong, Guangxi, Guizhou, Hainan, Hongkong, Hubei, Hunan, Taiwan, Yunnan.

A. 种群；B. 植物体带雌苞和雄苞；C. 植物体一段，示蒴萼；D. 植物体一段，示雄苞；E. 腹瓣；F. 侧叶；G. 雌苞叶和雌苞腹叶；H. 茎横切面；I. 茎一段，示腹叶；J. 茎一段，示腹面局部植物体；K. 背瓣中部细胞，示油体。（凭证标本：*韦玉梅等 201114-6*）

A. Population; B. Plant with gynoecia and an androecium; C. Portion of plant showing a perianth; D. Portion of plant showing an androecium; E. Leaf lobule; F. Leaves; G. Female bracts and bracteole; H. Transverse section of stem; I. Portion of stem showing underleaves; J. Portion of stem showing ventral merophyte; K. Median cells of leaf lobe showing oil bodies. (All from *Wei et al. 201114-6*)

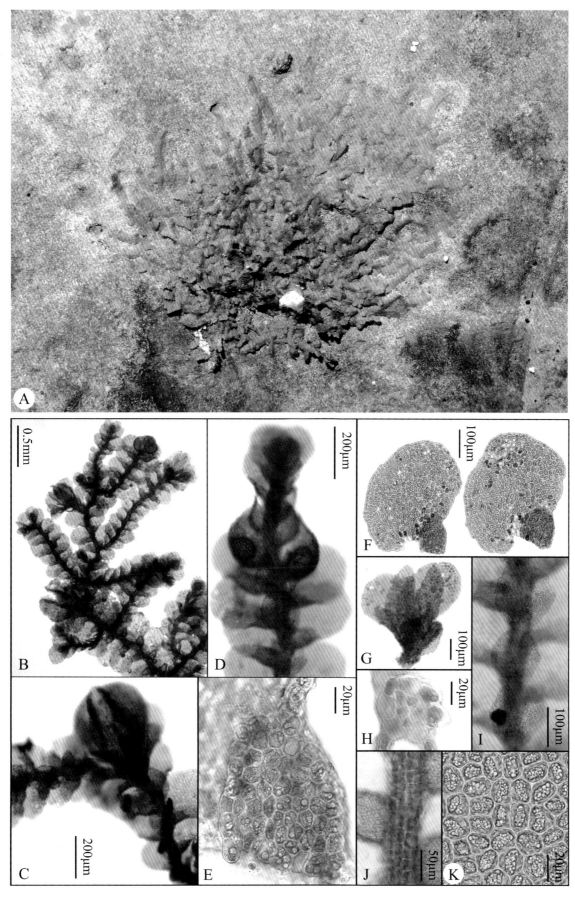

图 128 圆叶唇鳞苔

Fig. 128 *Cheilolejeunea intertexta* (Lindenb.) Steph.

粗茎唇鳞苔（瓦叶唇鳞苔、长叶唇鳞苔）

Cheilolejeunea trapezia (Nees) Kachroo et R. M. Schust., J. Linn. Soc., Bot. 56(368): 509. 1961.

Cheilolejeunea imbricata (Nees) S. Hatt., Misc. Bryol. Lichenol. 1(14): 1. 1957.

Cheilolejeunea longiloba (Steph. ex G. Hoffm.) Kachroo & R. M. Schust. ex J. J. Engel et B. C. Tan, J. Hattori Bot. Lab. 60: 294. 1986.

植物体绿色至黄绿色，带叶宽 0.97—1.25 mm，不规则分枝，分枝细鳞苔型。茎横切面 7—12 个表皮细胞和 10—17 个内部细胞；腹面局部植物体 2—4 个细胞宽。侧叶覆瓦状排列，背瓣阔卵形，顶端圆，边缘全缘；腹瓣椭圆形，长为背瓣的 1/2—3/4，中齿退化，角齿 1—5 个细胞长，透明疣位于角齿基部的远轴侧。侧叶细胞薄壁，角质层平滑，三角体小，简单三角形，中部球状加厚缺。油体聚合型，每个细胞 1 个。腹叶远生或毗邻，宽为茎的 2—3 倍，2 裂达 1/3—1/2 深。雌雄异株。雄苞未见。雌苞具 1 个新生枝，新生枝叶发生顺序为密鳞苔型。蒴萼未见。

生境：枯枝，树干，树基，树枝。海拔 1188—1862 m。

雅长分布：草王山，蓝家湾天坑。

中国分布：安徽，重庆，福建，广东，广西，贵州，海南，湖北，湖南，江西，四川，台湾，西藏，香港，云南，浙江。

Plants green to yellowish green when fresh, 0.97−1.25 mm wide, irregularly branched, branching of the *Lejeunea*-type. Stem in transverse section with 7−12 epidermal cells and 10−17 medullary cells; ventral merophyte 2−4 cells wide. Leaves imbricate, lobe broad ovate, apex rounded, margin entire; lobule oblong, 1/2−3/4 as long as the lobe, second tooth unicellular, first tooth obsolete, hyaline papilla at the distal side of the second tooth. Lobe cell thin-walled, trigones small, simple-triangulate, intermediate thickenings absent. Cuticle smooth. Oil bodies segmented, 1 per median cell. Underleaves distant to contiguous, 2−3 times as wide as the stem, bilobed to 1/3−1/2 underleaf length. Dioicous. Androecia not seen. Gynoecia with 1 innovation, innovation leaf sequence pycnolejeuneoid. Perianths not seen.

Habitat: On dead branches, tree trunks, tree bases and tree branches. Elev. 1188−1862 m.

Distribution in Yachang: Caowangshan, Lanjiawan Tiankeng.

Distribution in China: Anhui, Chongqing, Fujian, Guangdong, Guangxi, Guizhou, Hainan, Hongkong, Hubei, Hunan, Jiangxi, Sichuan, Taiwan, Tibet, Yunnan, Zhejiang.

A. 种群；B. 植物体；C. 侧叶；D. 茎横切面；E. 植物体一段，示新生枝；F. 腹瓣顶端，示透明疣；G. 茎一段，示腹面局部植物体；H. 背瓣中部细胞，示油体；I. 植物体一段；J. 腹叶；K−L. 腹瓣。（凭证标本：*韦玉梅等 191019-375*）

A. Population; B. Plant; C. Leaves; D. Transverse section of stem; E. Portion of plant showing an innovation; F. Apex of leaf lobule showing hyaline papilla; G. Portion of stem showing ventral merophyte; H. Median cells of leaf lobe showing oil bodies; I. Portion of plant; J. Underleaves; K−L. Leaf lobules. (All from *Wei et al. 191019-375*)

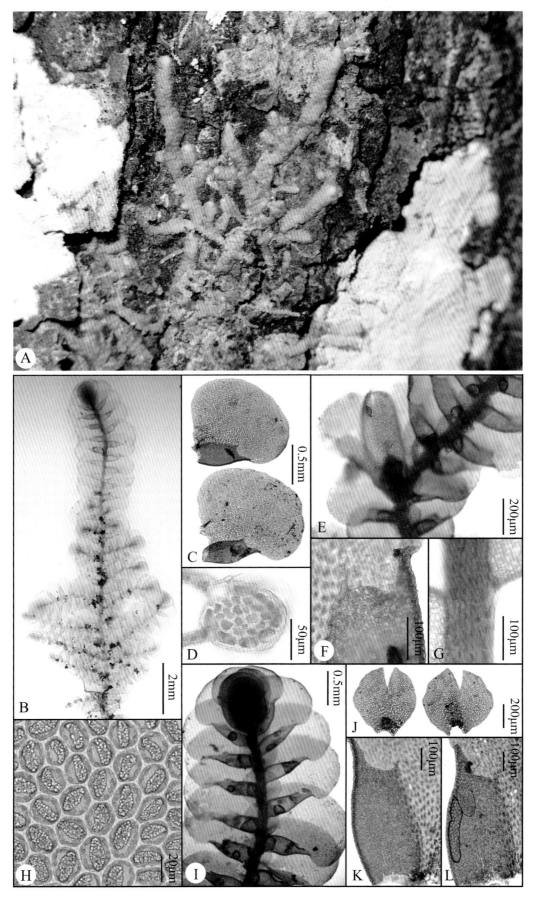

图 129 粗茎唇鳞苔

Fig. 129 *Cheilolejeunea trapezia* (Nees) Kachroo et R. M. Schust.

卷边唇鳞苔（白鳞苔）

Cheilolejeunea xanthocarpa (Lehm. et Lindenb.) Malombe, Acta Bot. Hung. 51(3/4): 326. 2009.

Leucolejeunea xanthocarpa (Lehm. et Lindenb.) A. Evans, Torreya 7: 229. 1907.

植物体绿色，带叶宽 1.06—2.10 mm，不规则分枝，分枝细鳞苔型。茎横切面 10—16 个表皮细胞和 15—22 个内部细胞；腹面局部植物体 4—6 个细胞宽。侧叶覆瓦状排列，背瓣椭圆形，顶端圆，顶部和腹缘强烈内卷；腹瓣长约为背瓣的 1/2，近轴的边缘内弯，中齿退化，角齿常单细胞，透明疣位于角齿的远轴侧。侧叶细胞薄壁，角质层平滑，三角体小，简单三角形，中部球状加厚偶可见。油体聚合型，每个细胞具 1 个油体。腹叶覆瓦状排列至毗邻，宽为茎的 4—6 倍，顶端不裂。雌雄同株。雄苞顶生或间生，雄苞叶 3—6 对，雄苞腹叶仅生于雄苞基部。雌苞具 1—2 个新生枝，新生枝叶发生顺序为密鳞苔型。蒴萼倒卵形，具 4—5 个脊，喙 7—16 个细胞长。

生境：树枝。海拔 1862 m。

雅长分布：草王山。

中国分布：福建，广东，广西，贵州，海南，湖北，江西，四川，台湾，香港，浙江。

Plants green when fresh, 1.06−2.10 mm wide, irregularly branched, branching of the *Lejeunea*-type. Stem in transverse section with 10−16 epidermal cells and 15−22 medullary cells; ventral merophyte 4−6 cells wide. Leaves imbricate, lobe oblong, apex rounded, apex and ventral margin strongly involute; lobule ca. 1/2 as long as the lobe, free lateral margin incurved, second tooth unicellular, first tooth obsolete, hyaline papilla at the distal side of the second tooth. Lobe cell thin-walled, trigones small, simple-triangulate, intermediate thickenings occasionally present. Cuticle smooth. Oil bodies segmented, 1 per median cell. Underleaves imbricate to contiguous, entire, 4−6 times as wide as the stem, apex undivided. Monoicous. Androecia terminal or intercalary, bracts in 3−6 pairs, bracteoles present only at the base of the androecium. Gynoecia with 1−2 innovations, innovation leaf sequence pycnolejeuneoid. Perianths obovate, with 4−5 keels, beak 7−16 cells long.

Habitat: On tree branches. Elev. 1862 m.

Distribution in Yachang: Caowangshan.

Distribution in China: Fujian, Guangdong, Guangxi, Guizhou, Hainan, Hongkong, Hubei, Jiangxi, Sichuan, Taiwan, Zhejiang.

▶ A. 种群；B. 植物体一段；C. 茎一段，示腹面局部植物体；D. 茎横切面；E. 侧叶；F. 腹叶；G. 腹瓣顶端；H. 雄苞；I. 植物体一段，示蒴萼；J. 背瓣中部细胞，示油体。（凭证标本：*韦玉梅等 201108-31A*）

A. Population; B. Portion of plant; C. Portion of stem showing ventral merophyte; D. Transverse section of stem; E. Leaves; F. Underleaf; G. Apex of leaf lobule; H. Androecium; I. Portion of plant showing perianth; J. Median cells of leaf lobe showing oil bodies. (All from *Wei et al. 201108-31A*)

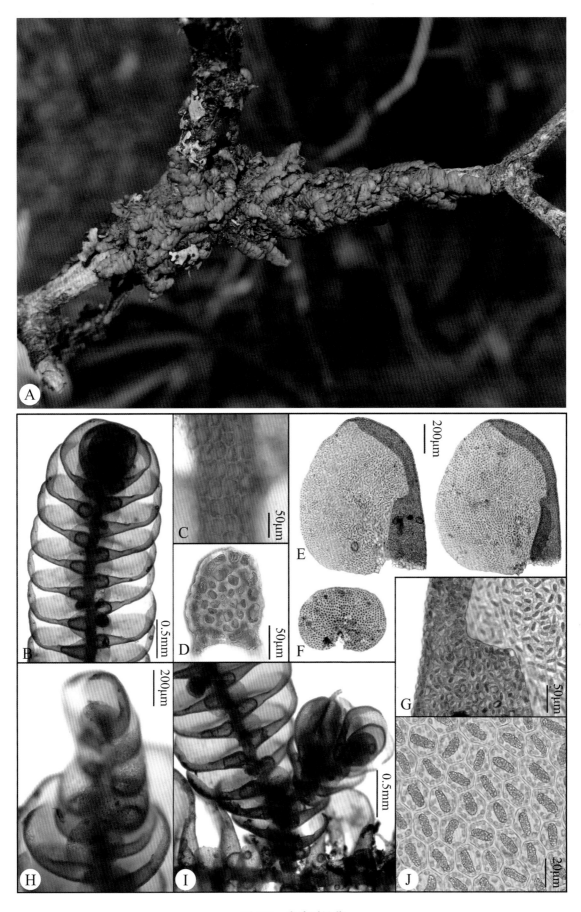

图 130 卷边唇鳞苔

Fig. 130 *Cheilolejeunea xanthocarpa* (Lehm. & Lindenb.) Malombe

单胞疣鳞苔

Cololejeunea kodamae Kamim., Feddes Repert. Spec. Nov. Regni Veg. 58: 55. 1955.

植物体浅绿色，带叶宽 0.45—0.95 mm，不规则分枝，分枝细鳞苔型。茎横切面 5 个表皮细胞和 1 个内部细胞；腹面局部植物体 1（—2）个细胞宽。侧叶覆瓦状排列，背瓣卵形，顶端锐尖，边缘具刺疣；腹瓣卵形，长约为背瓣的 1/2，腹面常或多或少具刺状的突起（靠近近轴边缘平滑），中齿 2—3 个细胞长，基部 1—3 个细胞宽，角齿钝，单细胞，透明疣位于中齿基部内表面；副体 1 个细胞。侧叶细胞薄壁，背部表面每个细胞具一个刺疣，腹部表面平滑，三角体小，简单三角形，中部球状加厚缺。油体未见。腹叶缺。雄苞和雌苞未见。芽胞圆盘状，由 15—16 个细胞构成，生于背瓣腹面。

生境： 叶附生。海拔 1240 m。

雅长分布： 蓝家湾天坑。

中国分布： 福建，广西，贵州，台湾。

Plants light green when fresh, 0.45–0.95 mm wide, irregularly branched, branching of the *Lejeunea*-type. Stem in transverse section with 5 epidermal cells and 1 medullary cell; ventral merophyte 1(–2) cells wide. Leaves imbricate, lobe ovate, apex acute, margin spinose throughout; lobule ovate, ca. 1/2 as long as the lobe, spinose on ventral surface (except for proximal portion), first tooth 2–3 cells long, 1–3 cells wide at base, second tooth blunt, unicellular, hyaline papilla on the inner surface of lobule at the base of the first tooth. Stylus unicellular. Lobe cell thin-walled, spinose on dorsal surface, smooth on ventral surface, trigones small, simple-triangulate, intermediate thickenings absent. Oil bodies not seen. Underleaves absent. Androecia and Gynoecia not seen. Gemmae discoid, 15–16-celled, on ventral surface of leaf lobe.

Habitat: Epiphyllous. Elev. 1240 m.

Distribution in Yachang: Lanjiawan Tiankeng.

Distribution in China: Fujian, Guangxi, Guizhou, Taiwan.

▶ A. 种群；B. 植物体；C. 侧叶（腹面观）；D. 侧叶（背面观）；E–F. 腹瓣；G. 茎横切面；H. 茎一段，示腹面局部植物体；I. 背瓣中部细胞；J. 芽胞。（凭证标本：*韦玉梅等 191019-398A*）

A. Population; B. Plant; C. Leaf (ventral view); D. Leaf (dorsal view); E–F. Leaf lobules; G. Transverse section of stem; H. Portion of stem showing ventral merophyte; I. Median cells of leaf lobe; J.: Gemmae. (All from *Wei et al. 191019-398A*)

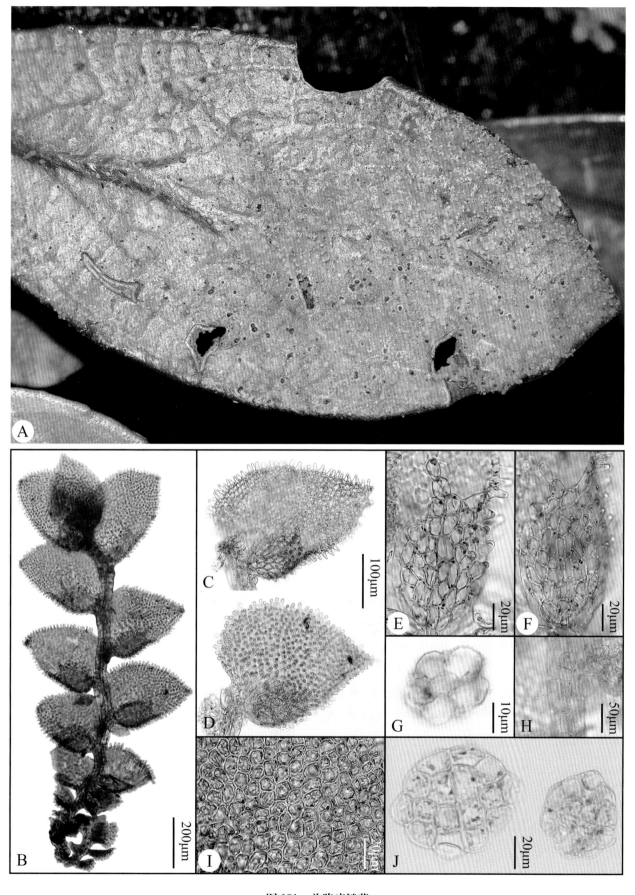

图 131 单胞疣鳞苔

Fig. 131 *Cololejeunea kodamae* Kamim.

狭瓣疣鳞苔

Cololejeunea lanciloba Steph., Hedwigia 34(5): 250. 1895.

植物体浅绿色，带叶宽 0.87—1.50 mm，不规则分枝，分枝细鳞苔型。茎横切面 5—7 个表皮细胞和 1 个内部细胞；腹面局部植物体 2 个细胞宽。侧叶覆瓦状排列，背瓣近椭圆形或卵形，顶端圆，边缘具 1—3 列透明细胞；腹瓣狭长舌状，长约为背瓣的 1/3，中齿位于腹瓣顶端，角齿位于近轴边缘，有时缺，透明疣位于中齿基部近轴侧；副体 1 个细胞。侧叶细胞薄壁，角质层密被细疣，三角体小，简单三角形，中部球状加厚缺。油体聚合型，每个细胞具 6—10 个油体。腹叶缺。雌雄同株。雄苞顶生或间生，雄苞叶 3—7 对，雄苞腹叶缺。雌苞具 1 个新生枝，雌苞腹叶缺。蒴萼倒卵形或倒心形，具 2—4 个平滑的脊，喙 1—2 个细胞长。芽胞圆盘状，由 32—35 个细胞构成，生于背瓣腹面。

生境： 叶附生。海拔 620 m。

雅长分布： 二沟。

中国分布： 福建，广东，广西，贵州，海南，江西，台湾，香港，云南。

Plants light green when fresh, 0.87−1.50 mm wide, irregularly branched, branching of the *Lejeunea*-type. Stem in transverse section with 5−7 epidermal cells and 1 medullary cell; ventral merophyte 2 cells wide. Leaves imbricate, lobe suboblong to ovate, apex rounded, margin entire, bordered by 1−3 rows of hyaline cells; lobule narrowly ligulate, ca. 1/3 as long as the lobe, first tooth at apex of the lobule, second tooth at proximal margin, sometimes absent, hyaline papilla at the proximal side of basal cell of the first tooth. Stylus unicellular. Lobe cell thin-walled, trigones small, simple-triangulate, intermediate thickenings absent. Cuticle finely punatate. Oil bodies segmented, 6−10 per median cell. Underleaves absent. Monoicous. Androecia terminal or intercalary, bracts in 3−7 pairs, bracteoles absent. Gynoecia with 1 innovation, bracteole absent. Perianths obovate or obcordate, with 2−4 smooth keels, beak 1−2 cells long. Gemmae discoid, 32−35-celled, on ventral surface of leaf lobe.

Habitat: Epiphyllous. Elev. 620 m.

Distribution in Yachang: Ergou.

Distribution in China: Fujian, Guangdong, Guangxi, Guizhou, Hainan, Hongkong, Jiangxi, Taiwan, Yunnan.

▶ A. 种群；B. 植物体一段；C. 植物体一段，示蒴萼；D. 雌苞；E. 侧叶；F. 腹瓣；G. 茎一段，示腹面局部植物体；H. 茎横切面；I. 芽胞；J. 背瓣中部细胞，示角质层细疣；K. 背瓣中部细胞，示油体。（凭证标本：韦玉梅等 201109-24）

A. Population; B. Portion of plant; C. Portion of plant showing perianths; D. Gynoecium; E. Leaves; F. Leaf lobule; G. Portion of stem showing ventral merophyte; H. Transverse section of stem; I. Gemma; J. Median cells of leaf lobe showing dorsal cuticle; K. Median cells of leaf lobe showing oil bodies. (All from *Wei et al. 201109-24*)

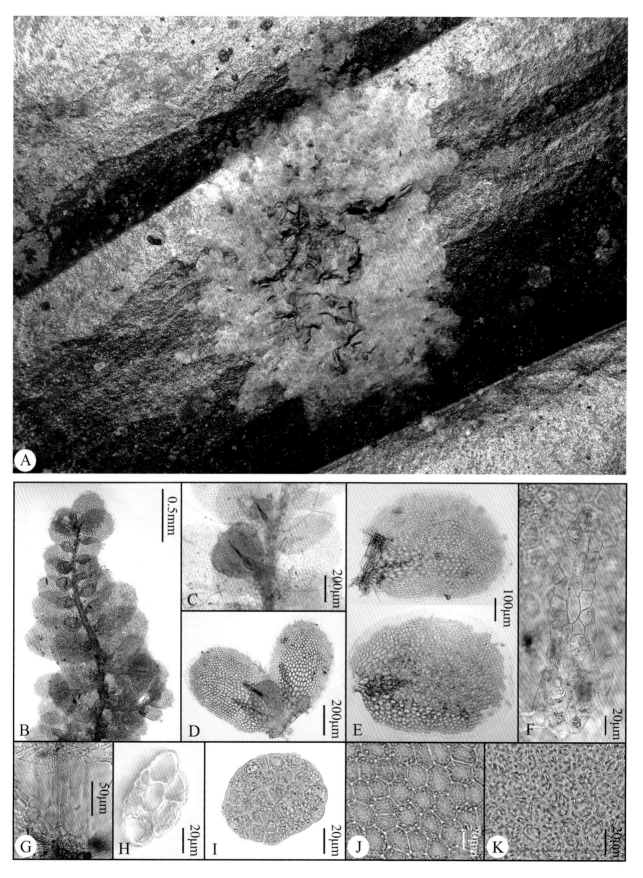

图 132　狭瓣疣鳞苔

Fig. 132　*Cololejeunea lanciloba* Steph.

阔瓣疣鳞苔

Cololejeunea latilobula (Herzog) Tixier, Bryophyt. Biblioth. 27: 156. 1985.

植物体浅绿色，带叶宽 1.70—2.53 mm，不规则分枝，分枝细鳞苔型。茎横切面 6—7 个表皮细胞和 1 个内部细胞；腹面局部植物体 2—3 个细胞宽。侧叶覆瓦状排列，背瓣卵形，顶端圆，顶部边缘和背缘常具 1—4 列透明细胞；腹瓣阔舌状，长约为背瓣的 1/3，齿不明显，透明疣位于腹瓣顶端；副体 1 个细胞。侧叶细胞薄壁，角质层密被细疣，三角体小，简单三角形，中部球状加厚缺。油体聚合型，每个细胞具 9—13 个油体。腹叶缺。雌雄同株。雄苞顶生或间生，雄苞叶 2—6 对，雄苞腹叶缺。雌苞具 1 个新生枝，雌苞腹叶缺。蒴萼倒卵形，具 2—4 个平滑的脊，喙 1 个细胞长。芽胞未见。

生境：石生，树干，树基，藤茎，枯叶，叶附生。海拔 765—1240 m。

雅长分布：黄猄洞天坑，拉雅沟，蓝家湾天坑，瞭望台，旁墙屯。

中国分布：澳门，广东，广西，贵州，海南，湖北，湖南，江西，台湾，西藏，香港，云南，浙江。

Plants light green when fresh, 1.70−2.53 mm wide, irregularly branched, branching of the *Lejeunea*-type. Stem in transverse section with 6−7 epidermal cells and 1 medullary cell; ventral merophyte 2−3 cells wide. Leaves imbricate, lobe ovate, apex rounded, margin entire, apical and dorsal margin usually bordered by 1−4 rows of hyaline cells; lobule broad ligulate, ca. 1/3 as long as the lobe, tooth indistinct, hyaline papilla at apex of the lobule. Stylus unicellular. Lobe cell thin-walled, trigones small, simple-triangulate, intermediate thickenings absent. Cuticle finely punatate. Oil bodies segmented, 9−13 per median cell. Underleaves absent. Monoicous. Androecia terminal or intercalary, bracts in 2−6 pairs, bracteoles absent. Gynoecia with 1 innovation, bracteole absent. Perianths obovate, with 2−4 smooth keels, beak 1 cell long. Gemmae not seen.

Habitat: On rocks, tree trunks, tree bases, liana and leaves. Elev. 765−1240 m.

Distribution in Yachang: Huangjingdong Tiankeng, Layagou, Lanjiawan Tiankeng, Liaowangtai, Pangqiang Tun.

Distribution in China: Guangdong, Guangxi, Guizhou, Hainan, Hongkong, Hubei, Hunan, Jiangxi, Macao, Taiwan, Tibet, Yunnan, Zhejiang.

▶ A. 种群；B. 植物体带雄苞和雌苞；C. 植物体一段；D. 叶顶端；E. 雌苞；F. 侧叶；G. 腹瓣；H. 茎横切面；I. 背瓣中部细胞，示角质层细疣；J. 背瓣中部细胞，示油体。（凭证标本：*韦玉梅等 201115-1B*）

A. Population; B. Plant with androecium and gynoecium; C. Portion of plant; D. Apex of leaf lobe; E. Gynoecium; F. leaves; G. Leaf lobule; H. Transverse section of stem; I. Median cells of leaf lobe showing dorsal cuticle; J. Median cells of leaf lobe showing oil bodies. (All from *Wei et al. 201115-1B*)

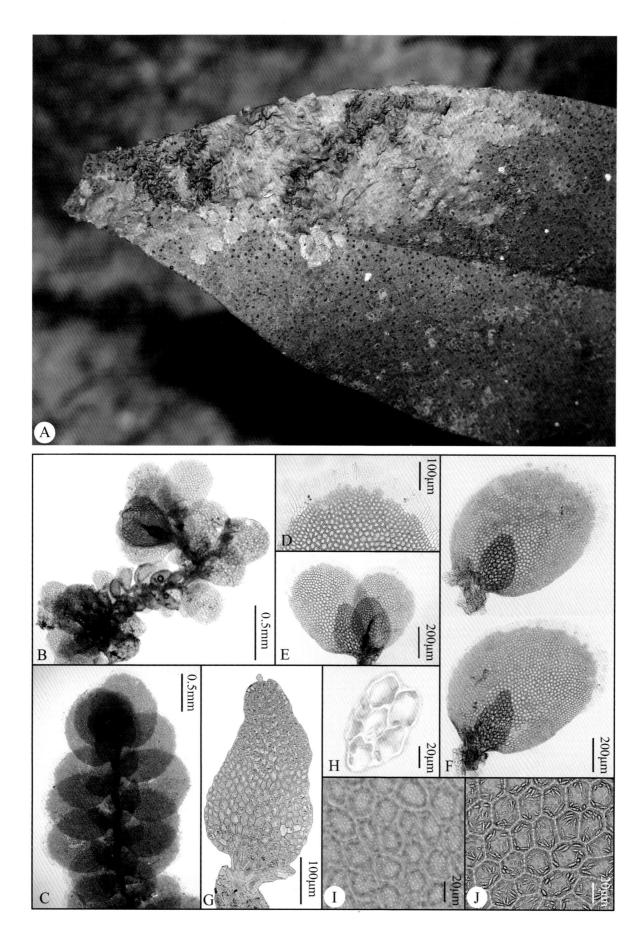

图 133　阔瓣疣鳞苔

Fig. 133　*Cololejeunea latilobula* (Herzog) Tixier

阔体疣鳞苔

Cololejeunea latistyla R. L. Zhu, Hikobia 11: 544. 1994.

植物体浅绿色，带叶宽 0.77—1.25 mm，不规则分枝，分枝细鳞苔型。茎横切面 5 个表皮细胞和 1 个内部细胞；腹面局部植物体 1—2 个细胞宽。侧叶覆瓦状排列，背瓣卵形或斜卵形，顶端圆或钝圆，顶部边缘和背缘常具 1—3 列透明细胞；腹瓣卵形或椭圆状卵形，长约为背瓣的 1/2，具 2 个齿，中齿 2—3 个细胞长，基部 1—2 个细胞宽，角齿 2—3 个细胞长，基部 1—3 个细胞宽，透明疣位于中齿基部近轴侧，有时在内表面；副体大，6—12 个细胞长，2—6 个细胞宽。侧叶细胞薄壁，角质层平滑，三角体小，简单三角形，中部球状加厚缺。油体聚合型。腹叶缺。雄苞和雌苞未见。芽胞圆盘状，由 17—22 个细胞构成，生于背瓣腹面。

生境：树干。海拔 1126—1376 m。

雅长分布：黄猄洞天坑，盘古王，下棚屯。

中国分布：福建，云南，浙江。首次记录于广西。

Plants light green when fresh, 0.77–1.25 mm wide, irregularly branched, branching of the *Lejeunea*-type. Stem in transverse section with 5 epidermal cells and 1 medullary cell; ventral merophyte 1–2 cells wide. Leaves imbricate, lobe ovate or obliquely ovate, apex rounded to obtuse, margin entire, apical and dorsal margin usually bordered by 1–3 rows of hyaline cells; lobule ovate or oblong-ovate, ca. 1/2 as long as the lobe, first tooth 2–3 cells long, 1–2 cells wide at base, second tooth 2–3 cells long, 1–3 cells wide at base, hyaline papilla at the proximal side of basal cell of the first tooth, sometimes on the inner surface of lobule at the base of the first tooth. Stylus large, 6–12 cells long, 2–6 cells wide. Lobe cell thin-walled, trigones small, simple-triangulate, intermediate thickenings absent. Cuticle smooth. Oil bodies segmented. Underleaves absent. Androecia and gynoecia not seen. Gemmae discoid, 17–22-celled, on ventral surface of leaf lobe.

Habitat: On tree trunks. Elev. 1126–1376 m.

Distribution in Yachang: Huangjingdong Tiankeng, Panguwang, Xiapeng Tun.

Distribution in China: Fujian, Yunnan, Zhejiang. New to Guangxi.

▶ A. 种群；B. 植物体；C. 植物体一段，示副体；D. 侧叶和副体；E. 背瓣中部细胞，示油体；F. 茎一段，示腹面局部植物体；G. 芽胞；H. 副体；I. 茎横切面；J. 腹瓣。（凭证标本：*韦玉梅等 201110-4*）

A. Population; B. Plant; C. Portion of plant showing styli; D. Leaves and styli; E. Median cells of leaf lobe showing oil bodies; F. Portion of stem showing ventral merophyte; G. Gemmae; H. Stylus; I. Transverse section of stem; J. Leaf lobule. (All from *Wei et al. 201110-4*)

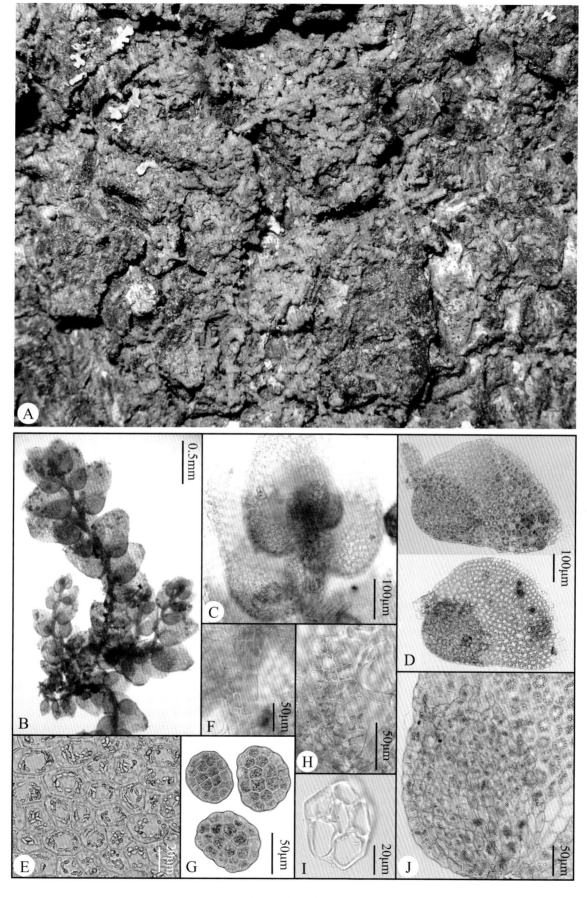

图 134　阔体疣鳞苔

Fig. 134　*Cololejeunea latistyla* R. L. Zhu

鳞叶疣鳞苔

Cololejeunea longifolia (Mitt.) Benedix ex Mizut., J. Hattori Bot. Lab. 26: 184. 1963.

植物体浅绿色或黄绿色，带叶宽 0.90—1.60 mm，不规则分枝，分枝细鳞苔型。茎横切面 5—6 个表皮细胞和 1 个内部细胞；腹面局部植物体 1 个细胞宽。侧叶远生至毗邻，背瓣披针形，顶端近锐尖或渐尖，边缘全缘；腹瓣椭圆形，长为背瓣的 1/4—1/3，具 2 个齿，中齿 2 个细胞长，基部 1 个细胞宽，角齿单细胞，常退化，透明疣位于中齿基部内表面；副体 1 个细胞。侧叶细胞薄壁，角质层平滑，三角体小至大，简单三角形，中部球状加厚常见。油体聚合型，每个细胞具 7—15 个油体。腹叶缺。雌雄同株同苞。生殖苞具 1 个新生枝，苞叶略小于侧叶，苞叶腹瓣长为苞叶背瓣的 1/2—2/3，苞叶腹叶缺。蒴萼倒卵形，顶端常平截，具 5 个脊，喙 1—2 个细胞长。芽胞未见。

生境： 叶附生环境常见，石生、树干、树基、树枝环境也有分布。海拔 931—1755 m。

雅长分布： 大宴坪天坑漏斗，吊井天坑，黄猄洞天坑，蓝家湾天坑，李家坨屯，盘古王，旁墙屯，山干屯，深洞，下棚屯，中井屯竖井，中井屯。

中国分布： 除了东北以外的全国大部分省区均有分布。

Plants light green or yellowish green when fresh, 0.90–1.60 mm wide, irregularly branched, branching of the *Lejeunea*-type. Stem in transverse section with 5–6 epidermal cells and 1 medullary cell; ventral merophyte 1 cell wide. Leaves distant to contiguous, lobe lanceolate, apex subacute or acuminate, margin entire; lobule oblong, 1/4–1/3 as long as the lobe, apex with two teeth, first tooth 2 cells long, 1 cell wide at base, second tooth unicellular, usually obsolete, hyaline papilla on the inner surface of lobule at the base of the first tooth. Stylus unicellular. Lobe cell thin-walled, trigones small to large, simple-triangulate, intermediate thickenings frequent. Oil bodies segmented, 7–15 per median cell. Underleaves absent. Synoicous. Inflorescences with 1 innovation, bracts slightly smaller than leaves, bract lobule 1/2–2/3 as long as the bract lobe, bracteole absent. Perianths obovate, with 5 smooth keels, apex truncate, beak 1–2 cells long. Gemmae not seen.

Habitat: Often epiphyllous, sometimes on rocks, tree trunks, tree bases and tree branches. Elev. 931–1755 m.

Distribution in Yachang: Dayanping Tiankeng, Diaojing Tiankeng, Huangjingdong Tiankeng, Lanjiawan Tiankeng, Lijiatuo Tun, Panguwang, Pangqiang Tun, Shangan Tun, Shendong, Xiapeng Tun, Zhongjing Tun.

Distribution in China: Widely distributed in most provinces of China except Northeast Region.

► A. 种群；B–C. 植物体一段具蒴萼；D. 苞叶；E. 侧叶；F. 蒴萼横切面；G. 茎横切面；H. 腹瓣；I. 茎一段，示腹面局部植物体；J. 背瓣中部细胞，示油体。（凭证标本：韦玉梅等 201105-29）

A. Population; B–C. Portions of plants with perianths; D. Bracts; E. Leaves; F. Transverse section of perianth; G. Transverse section of stem; H. Leaf lobule; I. Portion of stem showing ventral merophyte; J. Median cells of leaf lobe showing oil bodies. (All from *Wei et al. 201105-29*)

图 135　鳞叶疣鳞苔

Fig. 135　*Cololejeunea longifolia* (Mitt.) Benedix

大瓣疣鳞苔

Cololejeunea magnilobula (Horik.) S. Hatt., Bull. Tokyo Sci. Mus. 11: 99. 1944.

植物体浅绿色，带叶宽 0.62—0.75 mm，不规则分枝，分枝细鳞苔型。茎横切面 5 个表皮细胞和 1 个内部细胞；腹面局部植物体 1 个细胞宽。侧叶远生至毗邻，背瓣近圆形，顶端圆，边缘稍具圆齿；腹瓣椭圆形，与背瓣近于等长，近轴边缘强烈内卷，中齿 2 个细胞长，基部 1 个细胞宽，角齿常退化，透明疣位于中齿基部内表面；副体 1 个细胞。侧叶细胞薄壁，角质层平滑，三角体小，简单三角形，中部球状加厚缺。油体聚合型，每个细胞具 4—8 个油体。腹叶缺。雌雄异株？雄苞常顶生，雄苞叶 1—3 对，雄苞腹叶缺。雌苞，蒴萼未见。芽胞圆盘状，由 24—29 个细胞构成，生于背瓣腹面。

生境： 树枝。海拔 1862 m。

雅长分布： 草王山。

中国分布： 福建，贵州，海南，台湾，浙江。首次记录于广西。

Plants light green when fresh, 0.62−0.75 mm wide, irregularly branched, branching of the *Lejeunea*-type. Stem in transverse section with 5 epidermal cells and 1 medullary cell; ventral merophyte 1 cell wide. Leaves distant to contiguous, lobe suborbicular, apex rounded, margin slightly crenulate; lobule oblong, as long as the lobe, free margin strongly involute, first tooth 2 cells long, 1 cell wide at base, second tooth obsolete, hyaline papilla on the inner surface of lobule at the base of the first tooth. Stylus unicellular. Lobe cell thin-walled, trigones small, simple-triangulate, intermediate thickenings absent. Oil bodies segmented, 4−8 per median cell. Underleaves absent. Dioicous? Androecia usually terminal, bracts in 1−3 pairs, bracteoles absent. Gynoecia and perianths not seen. Gemmae discoid, 24−29-celled, on ventral surface of leaf lobe.

Habitat: On tree branches. Elev. 1862 m.

Distribution in Yachang: Caowangshan.

Distribution in China: Fujian, Guizhou, Hainan, Taiwan, Zhejiang. New to Guangxi.

▶ A. 种群；B. 植物体具雄苞；C. 植物体一段；D. 侧叶；E. 背瓣中部细胞，示油体；F. 茎横切面；G. 茎一段，示腹面局部植物体；H. 芽胞；I. 腹瓣。（凭证标本：*韦玉梅等 201108-31C*）

A. Population; B. Plant with androecia; C. Portion of plant; D. Leaves; E. Median cells of leaf lobe showing oil bodies; F. Transverse section of stem; G. Portion of stem showing ventral merophyte; H. Gemmae; I. Leaf lobule. (All from *Wei et al. 201108-31C*)

图 136 大瓣疣鳞苔

Fig. 136 *Cololejeunea magnilobula* (Horik.) S. Hatt.

粗柱疣鳞苔

Cololejeunea ornata A. Evans, Bryologist 41(4): 73. 1938.

植物体浅绿色，带叶宽 0.35—0.47 mm，不规则分枝，分枝细鳞苔型。茎横切面 5—6 个表皮细胞和 1 个内部细胞；腹面局部植物体 1 个细胞宽。侧叶覆瓦状排列，背瓣卵形，顶端锐尖，边缘具刺疣；腹瓣卵形，长为背瓣的 1/2—2/3，腹面具刺状的突起，中齿 2 个细胞长，基部 1 个细胞宽，有时不明显，角齿常退化，透明疣位于中齿基部内表面；副体大，3—6 个细胞长，1—2 个细胞宽。侧叶细胞薄壁，背部表面每个细胞具一个刺疣，腹部表面平滑，三角体小，简单三角形，中部球状加厚缺。油体未见。腹叶缺。雌雄同株。雄苞常顶生，雄苞叶 2—4 对，雄苞腹叶缺。雌苞具 1 个新生枝，雌苞腹叶缺。蒴萼椭圆，表面密被刺突（除了基部），具 5 个脊，喙 1—2 个细胞长。芽胞未见。

生境：土生，树基，枯叶，叶附生。海拔 1069—1361 m。

雅长分布：大宴坪天坑漏斗，黄猄洞天坑，蓝家湾天坑。

中国分布：安徽，广西，贵州，黑龙江，江西，四川，浙江。

Plants light green when fresh, 0.35–0.47 mm wide, irregularly branched, branching of the *Lejeunea*-type. Stem in transverse section with 5–6 epidermal cells and 1 medullary cell; ventral merophyte 1 cell wide. Leaves imbricate, lobe ovate, apex acute, margin spinose throughout; lobule ovate, 1/2–2/3 as long as the lobe, spinose on ventral surface, first tooth 2 cells long, 1 cells wide at base, sometimes indistinct, second tooth usually obsolete, hyaline papilla on the inner surface of lobule at the base of the first tooth. Stylus large, 3–6 cells long, 1–2 cells wide. Lobe cell thin-walled, spinose on dorsal surface, smooth on ventral surface, trigones small, simple-triangulate, intermediate thickenings absent. Oil bodies not seen. Underleaves absent. Monoicous. Androecia usually terminal, bracts in 2–4 pairs, bracteoles absent. Gynoecia with 1 innovation, bracteole absent. Perianths oblong, densely spinose on the surface except for base, with 5 keels, beak 1–2 cells long. Gemmae not seen.

Habitat: On soil, tree bases and leaves. Elev. 1069–1361 m.

Distribution in Yachang: Dayanping Tiankeng, Huangjingdong Tiankeng, Lanjiawan Tiankeng.

Distribution in China: Anhui, Guangxi, Guizhou, Heilongjiang, Jiangxi, Sichuan, Zhejiang.

▶ A. 种群；B. 植物体带蒴萼；C. 植物体一段；D. 侧叶（腹面观）；E. 侧叶（腹面观）；F. 茎横切面；G. 腹瓣；H. 茎一段，示副体。（凭证标本：*韦玉梅等 201114-4*）

A. Population; B. Plant with perianths; C. Portion of plant; D. Leaf (ventral view); E. Leaf (dorsal view); F. Transverse section of stem; G. Leaf lobule; H. Portion of stem showing styli. (All from *Wei et al. 201114-4*)

图 137 粗柱疣鳞苔

Fig. 137 *Cololejeunea ornata* A. Evans

粗齿疣鳞苔

Cololejeunea planissima (Mitt.) Abeyw., Ceylon J. Sci., Biol. Sci. 2(1): 73. 1959.

植物体浅绿色，带叶宽 1.20—1.63 mm，不规则分枝，分枝细鳞苔型。茎横切面 5—6 个表皮细胞和 1 个内部细胞；腹面局部植物体 2 个细胞宽。侧叶覆瓦状排列，背瓣椭圆形，顶端圆，顶部边缘和背缘常具 1—3 列透明细胞；腹瓣三角形，三角状舌形，舌形，或三角状卵形，长约为背瓣的 1/3，齿不明显，或有时在近轴侧边缘具 1 个钝齿，透明疣位于腹瓣顶端；副体 1 个细胞。侧叶细胞薄壁，角质层近于平滑，三角体小，简单三角形，中部球状加厚缺。油体聚合型，每个细胞具 6—11 个油体。腹叶缺。雌雄同株。雄苞常间生，雄苞叶 1—5 对，雄苞腹叶缺。雌苞具 1 个新生枝，雌苞腹叶缺。蒴萼倒卵形或倒心形，具 2—4 个平滑的脊，喙 1 个细胞长。芽胞未见。

生境：树干，树基，树枝，叶附生。海拔 503—1240 m。

雅长分布：二沟，黄猄洞天坑，蓝家湾天坑，旁墙屯，深洞，一沟。

中国分布：安徽，福建，广东，广西，贵州，海南，湖南，江西，四川，台湾，西藏，香港，云南，浙江。

Plants light green when fresh, 1.20–1.63 mm wide, irregularly branched, branching of the *Lejeunea*-type. Stem in transverse section with 5–6 epidermal cells and 1 medullary cell; ventral merophyte 2 cells wide. Leaves imbricate, lobe oblong, apex rounded, apical and dorsal margin usually bordered by 1–3 rows of hyaline cells; lobule triangular, triangular-ligulate, ligulate, or triangular-ovate, ca. 1/3 as long as the lobe, tooth indistinct, sometimes with a blunt tooth on proximal margin, hyaline papilla at apex of the lobule. Stylus unicellular. Lobe cell thin-walled, trigones small, simple-triangulate, intermediate thickenings absent. Cuticle nearly smooth. Oil bodies segmented, 6–11 per median cell. Underleaves absent. Monoicous. Androecia usually intercalary, bracts in 1–5 pairs, bracteoles absent. Gynoecia with 1 innovation, bracteole absent. Perianths obovate or obcordate, with 2–4 smooth keels, beak 1 cell long. Gemmae not seen.

Habitat: On tree trunks, tree bases, tree branches and leaves. Elev. 503–1240 m.

Distribution in Yachang: Ergou, Huangjingdong Tiankeng, Lanjiawan Tiankeng, Pangqiang Tun, Shendong, Yigou.

Distribution in China: Anhui, Fujian, Guangdong, Guangxi, Guizhou, Hainan, Hongkong, Hunan, Jiangxi, Sichuan, Taiwan, Tibet, Yunnan, Zhejiang.

▶ A. 种群；B. 植物体一段；C. 植物体一段，示雌苞和雄苞；D. 侧叶；E–H. 腹瓣；I. 叶顶部；J. 背瓣中部细胞，示油体；K. 雌苞叶；L. 茎横切面；M. 茎一段，示腹面局部植物体。（凭证标本：*韦玉梅等 201109-36A*）

A. Population; B. Portion of plant; C. Portion of plant showing an androecium and gynoecia; D. Leaves; E–H. Leaf lobules; I. Apex of leaf lobe; J. Median cells of leaf lobe showing oil bodies; K. Female bracts; L. Transverse section of stem; M. Portion of stem showing ventral merophyte. (All from *Wei et al. 201109-36A*)

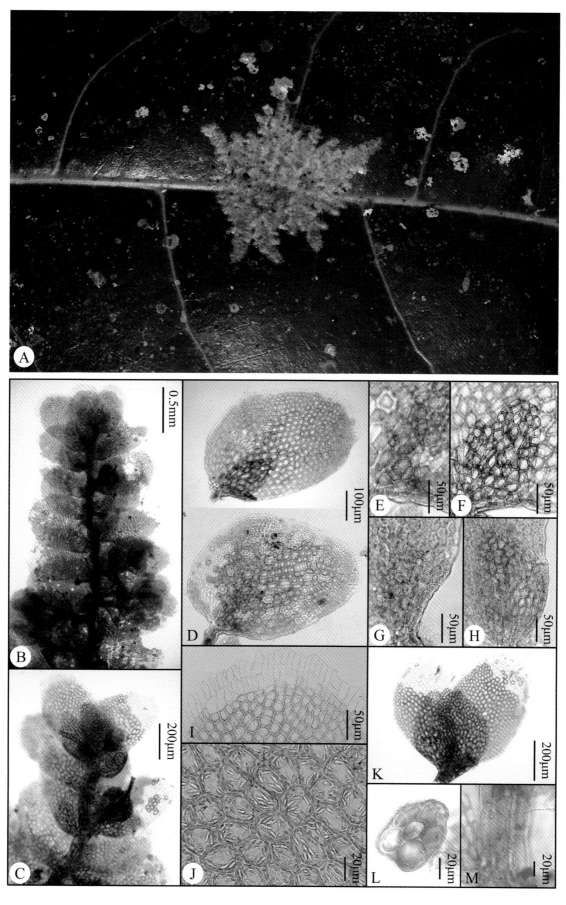

图 138　粗齿疣鳞苔

Fig. 138　*Cololejeunea planissima* (Mitt.) Abeyw.

尖叶疣鳞苔

Cololejeunea pseudocristallina P. C. Chen et P. C. Wu, Acta Phytotax. Sin. 9(3): 257. 1964.

植物体浅绿色，带叶宽 0.25—0.36 mm，不规则分枝，分枝细鳞苔型。茎横切面 6—7 个表皮细胞和 1 个内部细胞；腹面局部植物体 2 个细胞宽。侧叶覆瓦状排列，背瓣卵形，顶端钝尖，边缘具刺疣；腹瓣方形或长方形，长为背瓣的 1/2—2/3，腹面具刺状的突起，齿不明显，透明疣位于腹瓣顶端，但常见不到；副体 1 个细胞。侧叶细胞薄壁，背面每个细胞具一个刺疣，腹面表面平滑，三角体小，简单三角形，中部球状加厚缺。油体聚合型。腹叶缺。雌雄同株。雄苞顶生或间生，雄苞叶 2—3 对，雄苞腹叶缺。雌苞具 1 个新生枝，雌苞腹叶缺。蒴萼椭圆形，表面密被刺突（除了基部），具 5 个脊，喙 1—2 个细胞长。芽胞未见。

生境： 石生，树干，苔类上附生，叶附生。海拔 1188—1324 m。

雅长分布： 大宴坪天坑漏斗，蓝家湾天坑。

中国分布： 重庆，广东，广西，贵州，湖北，江西，香港，云南。

Plants light green when fresh, 0.25–0.36 mm wide, irregularly branched, branching of the *Lejeunea*-type. Stem in transverse section with 6–7 epidermal cells and 1 medullary cell; ventral merophyte 2 cell wide. Leaves imbricate, lobe ovate, apex obtuse, margin spinose throughout; lobule quadrate or retangular, 1/2–2/3 as long as the lobe, spinose on ventral surface, tooth indistinct, hyaline papilla at apex of the lobule, seldom observed. Stylus unicellular. Lobe cell thin-walled, spinose on dorsal surface, smooth on ventral surface, trigones small, simple-triangulate, intermediate thickenings absent. Oil bodies segmented. Underleaves absent. Monoicous. Androecia usually terminal, bracts in 2–3 pairs, bracteoles absent. Gynoecia with 1 innovation, bracteole absent. Perianths oblong, densely spinose on the surface except for base, with 5 keels, beak 1–2 cells long. Gemmae not seen.

Habitat: On rocks, tree trunks and leaves. Elev. 1188–1324 m.

Distribution in Yachang: Dayanping Tiankeng, Lanjiawan Tiankeng.

Distribution in China: Chongqing, Guangdong, Guangxi, Guizhou, Hongkong, Hubei, Jiangxi, Yunnan.

▶ A. 种群；B. 植物体带雄苞和雌苞；C. 植物体一段；D. 腹瓣；E. 茎一段，示腹面局部植物体；F. 茎一段，示单细胞副体；G. 茎横切面。（凭证标本：韦玉梅等 *201105-41*）

A. Population; B. Plant with an androecium and a gynoecium; C. Portion of plant; D. Leaf lobule; E. Portion of stem showing ventral merophyte; F. Portion of stem showing unicellular stylus; G. Transverse section of stem. (All from *Wei et al. 201105-41*)

图 139　尖叶疣鳞苔

Fig. 139　*Cololejeunea pseudocristallina* P. C. Chen & P. C. Wu

拟疣鳞苔

Cololejeunea raduliloba Steph., Hedwigia 34(5): 251. 1895.

植物体浅绿色，带叶宽 1.00—1.30 mm，不规则分枝，分枝细鳞苔型。茎横切面 5—6 个表皮细胞和 1 个内部细胞；腹面局部植物体 1（—2）个细胞宽。侧叶毗邻，偶覆瓦状排列，背瓣椭圆形，顶端圆，边缘全缘；腹瓣舌形，与茎平行，腹面具刺状的突起，中齿不明显，角齿位于近轴侧边缘，透明疣位于腹瓣顶端；副体 1 个细胞。侧叶细胞薄壁，角质层平滑，三角体小，简单三角形，中部球状加厚缺。油体聚合型，每个细胞具 6—11 个油体。腹叶缺。雌雄同株。雄苞顶生，雄苞叶 1—3 对，雄苞腹叶缺。雌苞具 1 个新生枝，雌苞腹叶缺。蒴萼倒卵形，具 2—4 个脊，喙 1 个细胞长。芽胞未见。

生境：倒木，叶附生。海拔 503—590 m。

雅长分布：二沟，一沟。

中国分布：安徽，福建，广东，广西，贵州，海南，湖南，江西，台湾，香港，云南，浙江。

Plants light green when fresh, 1.00–1.30 mm wide, irregularly branched, branching of the *Lejeunea*-type. Stem in transverse section with 5–6 epidermal cells and 1 medullary cell; ventral merophyte 1(–2) cell wide. Leaves contiguous, rarely imbricate, lobe oblong, apex rounded, margin entire; lobule ligulate, parallel to stem, first tooth indistinct, second tooth on proximal margin, hyaline papilla at apex of the lobule. Stylus unicellular. Lobe cell thin-walled, trigones small, simple-triangulate, intermediate thickenings absent. Cuticle smooth. Oil bodies segmented, 6–11 per median cell. Underleaves absent. Monoicous. Androecia usually terminal, bracts in 1–3 pairs, bracteoles absent. Gynoecia with 1 innovation, bracteole absent. Perianths obovate, with 2–4 keels, beak 1 cell long. Gemmae not seen.

Habitat: On fallen tree and leaves. Elev. 503–590 m.

Distribution in Yachang: Ergou, Yigou.

Distribution in China: Anhui, Fujian, Guangdong, Guangxi, Guizhou, Hainan, Hongkong, Hunan, Jiangxi, Taiwan, Yunnan, Zhejiang.

▶ A. 种群；B. 植物体一段；C. 植物体一段，示雄苞；D. 植物体一段，示蒴萼；E. 侧叶；F. 背瓣中部细胞，示油体；G. 茎横切面；H. 茎一段，示腹面局部植物体；I–J. 腹瓣。（凭证标本：*韦玉梅等 201109-30*）

A. Population; B. Portion of plant; C. Portion of plant showing androecia; D. Portion of plant showing perianths; E. Leaves; F. Median cells of leaf lobe showing oil bodies; G. Transverse section of stem; H. Portion of stem showing ventral merophyte; I–J. Leaf lobules. (All from *Wei et al. 201109-30*)

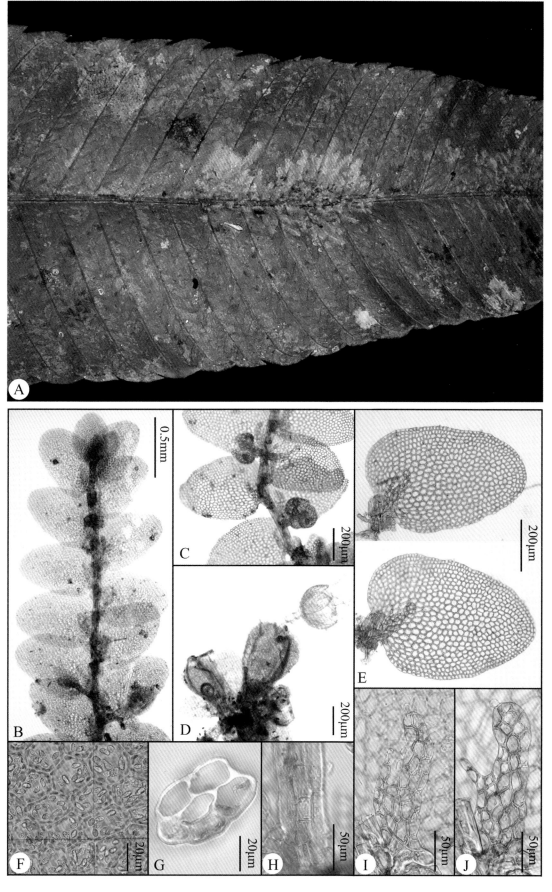

图 140 拟疣鳞苔

Fig. 140 *Cololejeunea raduliloba* Steph.

全缘疣鳞苔

Cololejeunea schwabei Herzog, J. Hattori Bot. Lab. 14: 54. 1955.

植物体浅绿色，带叶宽 0.90—1.27 mm，不规则分枝，分枝细鳞苔型。茎横切面 5 个表皮细胞和 1 个内部细胞；腹面局部植物体 1 个细胞宽。侧叶覆瓦状排列，背瓣卵形，顶端圆，边缘全缘；腹瓣舌形或三角状舌形，与茎平行，齿不明显，透明疣位于腹瓣顶端；副体大，3—6 个细胞长，1—2 个细胞宽。侧叶细胞薄壁，角质层平滑，三角体小，简单三角形，中部球状加厚缺。油体未见。腹叶缺。雌雄同株。雄苞常间生，雄苞叶 3—5 对，雄苞腹叶缺。雌苞具 1 个新生枝，雌苞腹叶缺。蒴萼倒卵形或梨形，具 5 个脊，喙 1 个细胞长。芽胞未见。

生境：石生。海拔 582—765 m。

雅长分布：二沟，拉雅沟。

中国分布：广东，贵州，海南，台湾，香港。首次记录于广西。

Plants light green when fresh, 0.90–1.27 mm wide, irregularly branched, branching of the *Lejeunea*-type. Stem in transverse section with 5 epidermal cells and 1 medullary cell; ventral merophyte 1 cells wide. Leaves imbricate, lobe ovate, apex rounded, margin entire; lobule ligulate or triangular-ligulate, parallel to stem, tooth indistinct, hyaline papilla at apex of the lobule. Stylus large, 3–6 cells long, 1–2 cells wide. Lobe cell thin-walled, trigones small, simple-triangulate, intermediate thickenings absent. Cuticle smooth. Oil bodies not seen. Underleaves absent. Monoicous. Androecia usually intercalary, bracts in 3–5 pairs, bracteoles absent. Gynoecia with 1 innovation, bracteole absent. Perianths obovate or pyriform, with 5 keels, beak 1 cell long. Gemmae not seen.

Habitat: On rocks. Elev. 582–765 m.

Distribution in Yachang: Ergou, Layagou.

Distribution in China: Guangdong, Guizhou, Hainan, Hongkong, Taiwan. New to Guangxi.

▶ A. 种群；B. 植物体一段；C. 茎一段，示腹瓣和副体；D. 侧叶；E. 雌苞叶；F. 茎横切面；G. 蒴萼；H–I. 腹瓣和副体。（凭证标本：*韦玉梅等 201109-34*）

A. Population; B. Portion of plant; C. Portion of stem showing lobules and styli; D. Leaves; E. Female bracts; F. Transverse section of stem; G. Perianth; H–I. Leaf lobules and styli. (All from *Wei et al. 201109-34*)

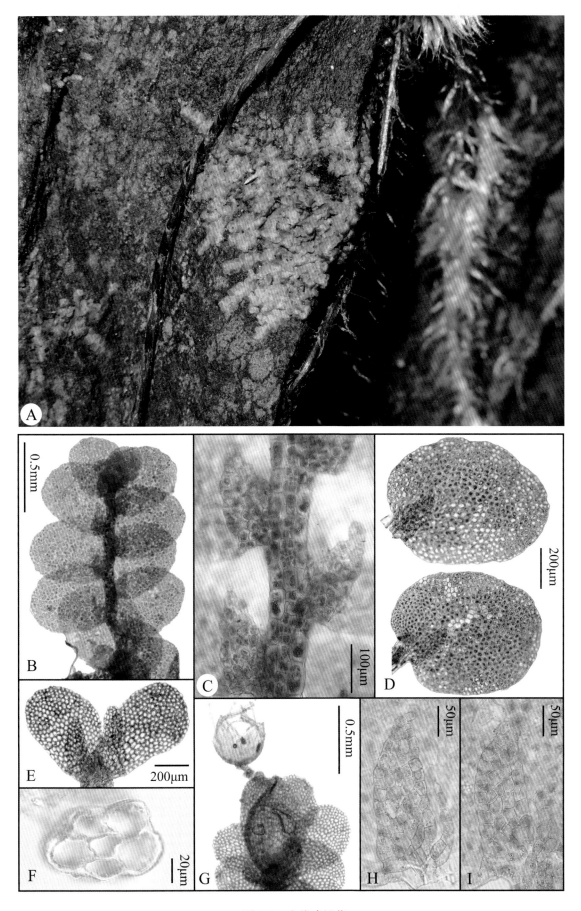

图 141　全缘疣鳞苔

Fig. 141　*Cololejeunea schwabei* Herzog

卵叶疣鳞苔

Cololejeunea shibiensis Mizut., J. Hattori Bot. Lab. 57: 437. 1984.

植物体淡绿色，带叶宽 1.00—1.40 mm，不规则分枝，分枝细鳞苔型。茎横切面 5 个表皮细胞和 1 个内部细胞；腹面局部植物体 2 个细胞宽。侧叶覆瓦状排列，背瓣卵形，顶端圆至钝圆，顶部边缘和背缘常具 1—5 列透明细胞；腹瓣卵形，长约为背瓣的 1/2，中齿 2—3 个细胞长，基部 1—2 个细胞宽，角齿 2—3 个细胞长，基部 1—2 个细胞宽，透明疣位于中齿基部内表面；副体大，2—7 个细胞长，1 个细胞宽。侧叶细胞薄壁，角质层平滑，三角体小，简单三角形，中部球状加厚缺。油体未见。腹叶缺。雄苞和雌苞未见。芽胞圆盘状，由 20 个细胞构成，生于背瓣腹面。

生境：石生，叶附生。海拔 1215—1533 m。

雅长分布：逻家田屯，香当天坑。

中国分布：福建，台湾。首次记录于广西。

Plants pale green when fresh, 1.00–1.40 mm wide, irregularly branched, branching of the *Lejeunea*-type. Stem in transverse section with 5 epidermal cells and 1 medullary cell; ventral merophyte 2 cells wide. Leaves imbricate, lobe ovate, apex rounded to obtuse, apical and dorsal margin usually bordered by 1–5 rows of hyaline cells; lobule ovate, ca. 1/2 as long as the lobe, first tooth 2–3 cells long, 1–2 cells wide at base, second tooth 2–3 cells long, 1–2 cells wide at base, hyaline papilla on the inner surface of lobule at the base of the first tooth. Stylus large, 2–7 cells long, 1 cell wide. Lobe cell thin-walled, trigones small, simple-triangulate, intermediate thickenings absent. Cuticle smooth. Oil bodies not seen. Underleaves absent. Androecia and gynoecia not seen. Gemmae discoid, 20-celled, on ventral surface of leaf lobe.

Habitat: On rocks and leaves. Elev. 1215–1533 m.

Distribution in Yachang: Luojiatian Tun, Xiangdang Tiankeng.

Distribution in China: Fujian, Taiwan. New to Guangxi.

▶ A. 种群；B. 植物体一段；C. 侧叶；D–E. 腹瓣和副体；F. 茎横切面；G. 叶顶部边缘；H. 芽胞。（凭证标本：*唐启明等 20190521-433B*）

A. Population; B. Portion of plant; C. Leaves; D–E. Leaf lobules and styli; F. Transverse section of stem; G. Apex of leaf lobe; H. Gemmae. (All from *Tang 20190521-433B*)

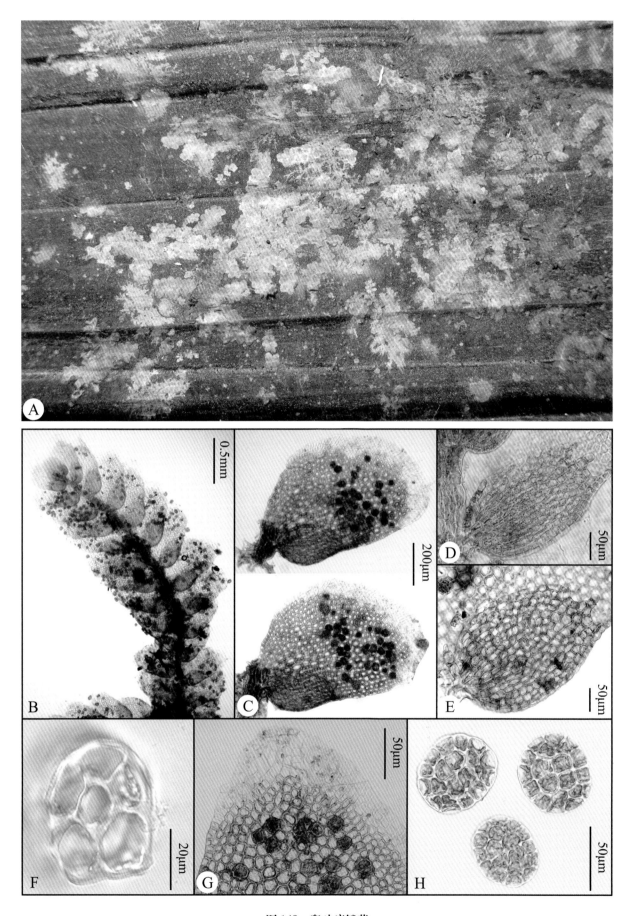

图 142 卵叶疣鳞苔

Fig. 142 *Cololejeunea shibiensis* Mizut.

刺疣鳞苔

Cololejeunea spinosa (Horik.) Pandé et R. N. Misra, J. Indian Bot. Soc. 22(2/4): 166. 1943.

植物体绿色，带叶宽 0.30—0.35 mm，不规则分枝，分枝细鳞苔型。茎横切面 5 个表皮细胞和 1 个内部细胞；腹面局部植物体 1 个细胞宽。侧叶远生至毗邻，背瓣卵形至椭圆状卵形，顶端圆钝，边缘具刺疣；腹瓣卵形，长约为背瓣的 1/2，中齿 2 个细胞长，基部 1 个细胞宽，角齿单细胞，透明疣位于中齿基部内表面；副体 1 个细胞。侧叶细胞薄壁，背面每个细胞具一个刺疣，腹面表面平滑，三角体小，简单三角形，中部球状加厚缺。油体未见。腹叶缺。雄苞和雌苞未见。芽胞圆盘状，由 16 个细胞构成，生于背瓣腹面。

生境： 树干，叶附生。海拔 1239—1240 m。

雅长分布： 蓝家湾天坑。

中国分布： 安徽，福建，广东，广西，贵州，海南，河南，湖南，江西，四川，台湾，西藏，香港，云南，浙江。

Plants green when fresh, 0.30−0.35 mm wide, irregularly branched, branching of the *Lejeunea*-type. Stem in transverse section with 5 epidermal cells and 1 medullary cell; ventral merophyte 1 cell wide. Leaves distant to contiguous, lobe ovate to oblong-ovate, apex obtuse, margin spinose throughout; lobule ovate, ca. 1/2 as long as the lobe, first tooth 2 cells long, 1 cells wide at base, second tooth unicellular, hyaline papilla on the inner surface of lobule at the base of the first tooth. Stylus unicellular. Lobe cell thin-walled, spinose on dorsal surface, smooth on ventral surface, trigones small, simple-triangulate, intermediate thickenings absent. Oil bodies not seen. Underleaves absent. Androecia and gynoecia not seen. Gemmae discoid, 16-celled, on ventral surface of leaf lobe.

Habitat: On tree trunks and leaves. Elev. 1239−1240 m.

Distribution in Yachang: Lanjiawan Tiankeng.

Distribution in China: Anhui, Fujian, Guangdong, Guangxi, Guizhou, Hainan, Henan, Hongkong, Hunan, Jiangxi, Sichuan, Taiwan, Tibet, Yunnan, Zhejiang.

▶ A–B. 种群；C. 植物体；D. 植物体一段带蒴萼；E. 侧叶（腹面观和背面观）；F. 植物体一段；G. 腹瓣；H. 茎一段，示腹面局部植物体；I. 茎横切面；J. 芽胞。（凭证标本：韦玉梅等 201114-5）

A–B. Populations; C. Plant; D. Portion of plant with a perianth; E. Leaf (dorsal view and ventral view); F. Portion of plant; G. Leaf lobule; H. Portion of stem showing ventral merophyte; I. Transverse section of stem; J. Gemma. (All from *Wei et al. 201114-5*)

图 143　刺疣鳞苔

Fig. 143　*Cololejeunea spinosa* (Horik.) Pandé & R. N. Misra

疣瓣疣鳞苔

Cololejeunea subkodamae Mizut., J. Hattori Bot. Lab. 60: 448, 1986.

植物体浅绿色至黄绿色，带叶宽 0.32—0.55 mm，不规则分枝，分枝细鳞苔型。茎横切面 5 个表皮细胞和 1 个内部细胞；腹面局部植物体 1 个细胞宽。侧叶覆瓦状排列至毗邻，背瓣卵形，顶端圆钝至钝尖，边缘具刺疣；腹瓣卵形，长为背瓣的 1/2—2/3，腹面常或多或少具刺状的突起（近轴部分平滑），中齿 1—2 个细胞长，基部 1 个细胞宽，角齿单细胞，中齿和角齿常交叉在一起，透明疣位于中齿基部内表面；副体 1 个细胞。侧叶细胞薄壁，背部表面每个细胞具一个刺疣，腹部表面平滑，三角体小，简单三角形，中部球状加厚缺。油体聚合型。腹叶缺。雌雄同株。雄苞顶生或间生，雄苞叶 2—3 对，与侧叶在形状和大小上相似，雄苞腹叶缺。雌苞具 1—2 个新生枝，雌苞腹叶缺。蒴萼椭圆形，表面密被刺突（除了基部），具 3—4 个脊，喙 1—2 个细胞长。芽胞未见。

生境： 枯枝，叶附生。海拔 1240—1244 m。

雅长分布： 蓝家湾天坑。

中国分布： 安徽，福建，广西，贵州，湖南，台湾，浙江。

Plants light green to yellowish green when fresh, 0.32−0.55 mm wide, irregularly branched, branching of the *Lejeunea*-type. Stem in transverse section with 5 epidermal cells and 1 medullary cell; ventral merophyte 1 cell wide. Leaves imbricate to contiguous, lobe ovate, apex obtuse to obtuse-acute, margin spinose throughout; lobule ovate, 1/2−2/3 as long as the lobe, spinose on ventral surface (except for proximal portion), first tooth 1−2 cells long, 1 cells wide at base, second tooth unicellular, two teeth usually crossing each other, hyaline papilla on the inner surface of lobule at the base of the first tooth. Stylus unicellular. Lobe cell thin-walled, spinose on dorsal surface, smooth on ventral surface, trigones small, simple-triangulate, intermediate thickenings absent. Oil bodies segmented. Underleaves absent. Monoicous. Androecia terminal or intercalary, bracts in 2−3 pairs, resembling leaves in size and shape, bracteoles absent. Gynoecia with 1−2 innovation, bracteole absent. Perianths oblong, densely spinose on the surface except for base, with 3−4 keels, beak 1−2 cells long. Gemmae not seen.

Habitat: On dead branches and leaves. Elev. 1240−1244 m.

Distribution in Yachang: Lanjiawan Tiankeng.

Distribution in China: Anhui, Fujian, Guangxi, Guizhou, Hunan, Taiwan, Zhejiang.

▶ A. 种群；B. 植物体；C. 植物体一段，示蒴萼；D. 茎一段，示腹瓣；E. 侧叶（背面观）；F. 侧叶（腹面观）；G. 茎横切面。（凭证标本：韦玉梅等 *201114-3B*）

A. Population; B. Plant; C. Portion of plant showing a perianth; D. Portion of stem showing lobules; E. Leaf (dorsal view); F. Leaf (ventral view); G. Transverse section of stem. (All from *Wei et al. 201114-3B*)

图 144 疣瓣疣鳞苔
Fig. 144 *Cololejeunea subkodamae* Mizut.

管叶苔

Colura calyptrifolia (Hook.) Dumort., Recueil Observ. Jungerm.: 12. 1835.

植物体浅绿色，带叶宽 1.55—2.30 mm，不规则分枝，分枝细鳞苔型。茎横切面 7 个表皮细胞和 3 个内部细胞；腹面局部植物体 2 个细胞宽。侧叶远生至毗邻，背瓣披针形，边缘全缘；腹瓣大，长约为背瓣的 2 倍，基部圆柱状，向上强烈鼓起形成一个囊，囊纺锤形，顶端形成长管状，达叶长的 1/4—1/3，活瓣位于囊的口部，舌状，由 1 圈 16 个透明的边缘细胞和约 12 个中部细胞组成。叶细胞薄壁，角质层平滑，三角体和中部球状加厚缺。油体聚合型，每个细胞具 8—30 个油体。腹叶远生，顶端深 2 裂，裂瓣线状披针形，长 6—9 个细胞，基部宽 3—4 个细胞。雌雄同株。雄苞顶生，雄苞叶 2—3 对，雄苞腹叶未见。雌苞具 1 个新生枝，新生枝叶发生顺序为密鳞苔型。蒴萼倒卵形，具 5 个角状的脊，脊短，占蒴萼长度的 1/5—1/4，喙长 1 个细胞。芽胞盘状，由 26—27 个细胞组成，通常生于叶顶端。

生境： 枯枝。海拔 1747—1829 m。

雅长分布： 草王山。

中国分布： 广西，台湾。

Plants light green when fresh, 1.55–2.30 mm wide, irregularly branched, branching of the *Lejeunea*-type. Stem in transverse section with 7 epidermal cells and 3 medullary cells; ventral merophyte 2 cells wide. Leaves distant to contiguous, lobe lanceolate, margin entire; lobule large, 2 times as long as the lobe, cylindrical at base, flaring toward sac, sac strongly inflated, fusiform, forming a long tubular beak towards the apex, lobular beak 1/4–1/3 of leaf lenght; valve at the mouth of the sac, ligulate, composed of one circleof ca. 16 hyaline marginal cells and ca. 12 median cells. Lobe cell thin-walled, trigones and intermediate thickenings absent. Cuticle smooth. Oil bodies segmented, 8–30 per median cell. Underleaves distant, deeply bilobed, lobes linear-lanceolate, 6–9 cells long, 3–4 cells wide at base. Monoicous. Androecia terminal, bracts in 2–3 pairs, bracteoles not seen. Gynoecia with 1 innovation, innovation leaf sequence pycnolejeuneoid. Perianths obovate, with 5 horn-like keels, keels short, 1/5–1/4 as long as the perianth, beak 1 cell long. Gemmae discoid, 26–27-celled, usually at the top of leaf.

Habitat: On dead branches. Elev. 1747–1829 m.

Distribution in Yachang: Caowangshan.

Distribution in China: Guangxi, Taiwan.

▶ A. 种群；B. 植物体带蒴萼；C. 蒴萼；D. 侧叶；E. 活瓣；F. 腹叶；G. 芽胞；H. 茎横切面。（凭证标本：*韦玉梅等 191013-123*）

A. Population; B. Plant with a perianth; C. Perianths; D. Leaves; E. Valve; F. Underleaf; G. Gemmae; H. Transverse section of stem. (All from *Wei et al. 191013-123*)

图 145 管叶苔

Fig. 145 *Colura calyptrifolia* (Hook.) Dumort.

细角管叶苔

Colura tenuicornis (A. Evans) Steph., Sp. Hepat. (Stephani) 5: 942. 1916.

植物体浅绿色，带叶宽 1.8—2.5 mm，不规则分枝，分枝细鳞苔型。茎横切面 7 个表皮细胞和 3 个内部细胞；腹面局部植物体 2 个细胞宽。侧叶远生至毗邻，背瓣披针形，边缘全缘；腹瓣大，长约为背瓣的 2 倍，基部圆柱状，向上鼓起形成一个囊，囊披针形，顶端形成长管状，达叶长的 1/2，活瓣位于囊的口部，舌状，由 1 圈 16 个透明的边缘细胞和约 12 个中部细胞组成。叶细胞薄壁，角质层平滑，三角体和中部球状加厚缺。油体未见。腹叶远生，顶端深 2 裂，裂瓣线状披针形，长 5—7 个细胞，基部宽 2—3 个细胞。雄苞和雌苞。芽胞盘状，由 24—26 个细胞组成，通常生于叶顶端。

生境：树基生。海拔 1625 m。

雅长分布：草王山。

中国分布：福建，广东，广西，贵州，海南，四川，台湾，西藏，云南，浙江。

备注：细角管叶苔与管叶苔极为相似，但两者可通过腹瓣的囊形状（前者纺锤形，后者披针形）、叶先端管状部分的长短（前者为叶长的 1/4—1/3，后者一般占叶长的 1/2）以及蒴萼脊的长短（前者占蒴萼长度的 1/5—1/4，后者占 1/3—1/2）加以区分。

Plants light green when fresh, 1.8–2.5 mm wide, irregularly branched, branching of the *Lejeunea*-type. Stem in transverse section with 7 epidermal cells and 3 medullary cells; ventral merophyte 2 cells wide. Leaves distant to contiguous, lobe lanceolate, margin entire; lobule large, 2 times as long as the lobe, cylindrical at base, flaring toward sac, sac inflated, lanceolate, forming a long tubular beak towards the apex, lobular beak ca. 1/2 of leaf lenght; valve at the mouth of the sac, ligulate, composed of one circleof ca. 16 hyaline marginal cells and ca. 12 median cells. Lobe cell thin-walled, trigones and intermediate thickenings absent. Cuticle smooth. Oil bodies not seen, 8–30 per median cell. Underleaves distant, deeply bilobed, lobes linear-lanceolate, 5 -7 cells long, 2–3 cells wide at base. Androecia and gynoecia not seen. Gemmae discoid, 24–26-celled, uasually at the tip of leaf.

Habitat: On tree base. Elev. 1625 m.

Distribution in Yachang: Caowangshan.

Distribution in China: Fujian, Guangdong, Guangxi, Guizhou, Hainan, Sichuan, Taiwan, Tibet, Yunnan, Zhejiang.

Note: *Colura tenuicornis* is most closely related to *C. calyptrifolia*, the later deffers in its fusiform lobule sac (lanceolate in *C. tenuicornis*), the sac with a short beak (1/4–1/3 of total leaf length vs. ca. 1/2 in *C. tenuicornis*), and the short horns of perianth (1/5–1/4 of perianth length vs. 1/3–1/2 in *C. tenuicornis*).

▶ A–D. 植物体；E. 侧叶；F. 芽胞；G. 茎横切面 H. 茎一段，示腹叶。（凭证标本：*韦玉梅等 221114-7B*）

A–D. Plants; E. Leaves; F. Gemmae; G. Transverse section of stem; H. Portion of stem showing underleaves. (All from *Wei et al. 221114-7B*)

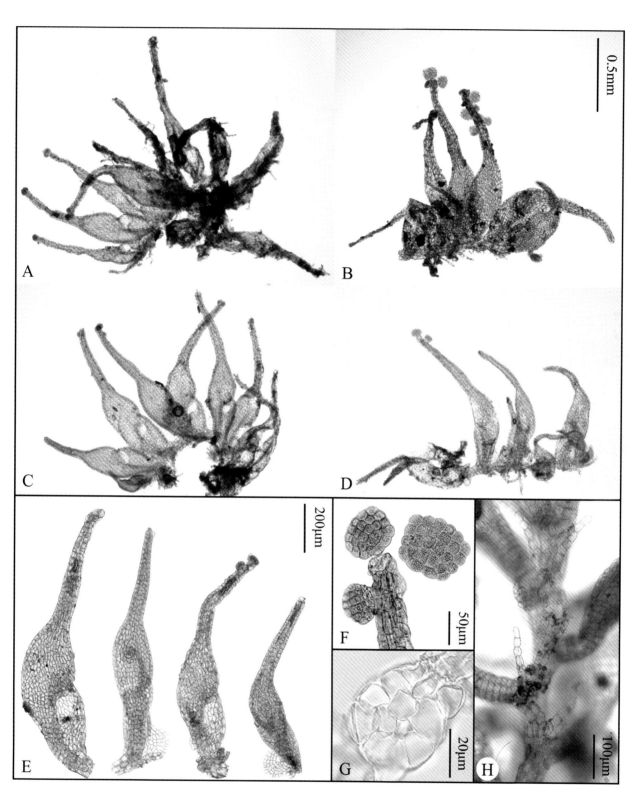

图 146　细角管叶苔

Fig. 146　*Colura tenuicornis* (A. Evans) Steph.

狭叶角鳞苔

Drepanolejeunea angustifolia (Mitt.) Grolle, J. Jap. Bot. 40(7): 206. 1965.

植物体黄绿色，带叶宽 0.28—0.45 mm，不规则分枝，分枝细鳞苔型。茎横切面 7 个表皮细胞和 3 个内部细胞；腹面局部植物体 2 个细胞宽。侧叶远生至毗邻，背瓣狭卵形或披针形，顶端锐尖或细尖，有时稍内弯，边缘全缘或具细圆齿；腹瓣长为背瓣的 1/3—1/2，近轴边缘由 4 个加长细胞组成，中齿为 1 个长的弯曲的细胞，弯向脊，角齿不明显，透明疣位于中齿基部近轴侧。叶细胞稍厚壁，角质层平滑，三角体小至大，简单三角形，中部球状加厚常见。油体聚合型，每个细胞具 1—2 个油体。油胞（1—）2 个，生于背瓣基部。腹叶远生，顶端深 2 裂，裂瓣常 3—4 个细胞，基部 1（—2）个细胞宽。雄苞未见。雌苞通常具 1 个新生枝，新生枝叶发生顺序密鳞苔型。

生境：藤茎，树干，倒木，石生。海拔 1625—1878 m。

雅长分布：草王山。

中国分布：安徽，重庆，福建，广东，广西，贵州，海南，湖南，江西，四川，台湾，西藏，香港，云南，浙江。

Plants yellowish green when fresh, 0.28–0.45 mm wide, irregularly branched, branching of the *Lejeunea*-type. Stem in transverse section with 7 epidermal cells and 3 medullary cell; ventral merophyte 2 cells wide. Leaves remote to contiguous, lobe narrowly ovate to lanceolate, apex acute to long-acuminate, sometimes incurved, margin entire to crenulate; lobule 1/3–1/2 as long as the lobe, free lateral margin bordered by 4 rectangular or rectangular-linear marginal cells, first tooth unicellular, elongate and curved, second tooth obsolete, hyaline papilla at the proximal side of first tooth. Lobe cell slightly thick-walled, trigones small to large, simple-triangulate, intermediate thickenings frequent. Cuticle smooth. Oil bodies segmented, 1–2 per median cell. Ocelli (1–)2 at base of leaf lobe. Underleaves distant, deeply bilobed, lobes 3–4 cells long, 1(–2) cells wide at base. Androecia not seen. Gynoecia with an innovation, innovation leaf sequence pycnolejeuneoid.

Habitat: On liana, tree trunk, fallen tree, and rock. Elev. 1625–1878 m.

Distribution in Yachang: Caowangshan.

Distribution in China: Anhui, Chongqing, Fujian, Guangdong, Guangxi, Guizhou, Hainan, Hongkong, Hunan, Jiangxi, Sichuan, Taiwan, Tibet, Yunnan, Zhejiang.

▶ A. 种群；B. 植物体；C. 植物体一段；D. 植物体一段带雌苞；E. 茎横切面；F. 侧叶（腹面观）；G. 侧叶（背面观）；H. 植物体一段（背面观），示油胞；I. 茎一段，示腹叶；J. 腹瓣。（凭证标本：*韦玉梅等 221114-11*）

A. Population; B. Plant; C. Portion of plant; D. Portion of plant with a gynoecium; E. Transverse section of stem; F. Leaves (ventral view); G. Leaf (dorsal view); H. Portion of plant (dorsal view) showing ocelli; I. Portion of stem showing underleaves; J. Leaf lobule. (All from *Wei et al. 221114-11*)

图 147 狭叶角鳞苔

Fig. 147 *Drepanolejeunea angustifolia* (Mitt.) Grolle

日本角鳞苔

Drepanolejeunea erecta (Steph.) Mizut., J. Hattori Bot. Lab. 40: 442. 1976.

植物体绿色或黄绿色，带叶宽 0.50—1.20 mm，不规则分枝，分枝细鳞苔型。茎横切面 7 个表皮细胞和 3 个内部细胞；腹面局部植物体 2 个细胞宽。侧叶覆瓦状排列至相接，背瓣卵形，顶端圆钝至钝尖，边缘具细齿；腹瓣长为背瓣的 1/3—1/2，近轴边缘由 4 个加长细胞组成，中齿为 1 个长的弯曲的细胞，弯向脊，角齿不明显，透明疣位于中齿基部近轴侧。叶细胞厚壁，角质层平滑，三角体大，简单三角形，中部球状加厚有时可见。油体聚合型，每个细胞具 2—6 个油体。油胞 1—2 个，生于背瓣基部。腹叶远生，宽为茎的 2—3 倍，顶端 2 裂至 1/3-2/3 深，裂瓣三角形或三角状披针形，基部宽 4—7 个细胞。雄苞和雌苞未见。

生境： 石生，倒木，树干，树基，树枝。海拔 837—1823 m。

雅长分布： 草王山，二沟，黄猄洞天坑，蓝家湾天坑，盘古王，全达村。

中国分布： 安徽，福建，甘肃，广东，广西，贵州，海南，湖北，湖南，江西，台湾，西藏，香港，云南，浙江。

Plants light green to yellowish green when fresh, 0.50−1.20 mm wide, irregularly branched, branching of the *Lejeunea*-type. Stem in transverse section with 7 epidermal cells and 3 medullary cell; ventral merophyte 2 cells wide. Leaves imbricate to contiguous, lobe ovate, apex obtuse to obtuse-acute, margin slightly denticulate; lobule 1/3−1/2 as long as the lobe, free lateral margin bordered by 4 rectangular or rectangular-linear marginal cells, first tooth unicellular, elongate and curved, second tooth obsolete, hyaline papilla at the proximal side of first tooth. Lobe cell thick-walled, trigones large, simple-triangulate, intermediate thickenings occasionally present. Cuticle smooth. Oil bodies segmented, 2−6 per median cell. Ocelli 1−2 at base of leaf lobe. Underleaves distant, 2−3 times as wide as the stem, bilobed to 1/3−2/3 underleaf length, lobes triangular to triangular-lanceolate, 4−7 cells wide at base. Androecia and gynoecia not seen.

Habitat: On rocks, fallen trees, tree trunks, tree bases and tree branches. Elev. 837−1823 m.

Distribution in Yachang: Caowangshan, Ergou, Huangjingdong Tiankeng, Lanjiawan Tiankeng, Panguwang, Quanda Village.

Distribution in China: Anhui, Fujian, Gansu, Guangdong, Guangxi, Guizhou, Hainan, Hongkong, Hubei, Hunan, Jiangxi, Taiwan, Tibet, Yunnan, Zhejiang.

▶ A. 种群；B. 植物体一段；C. 茎一段，示腹叶；D. 侧叶；E. 腹瓣；F. 叶基部细胞，示油胞；G. 腹叶；H. 茎横切面；I. 背瓣中部细胞，示油体。（凭证标本：*韦玉梅等 201104-29*）

A. Population; B. Portion of plant; C. Portion of stem showing underleaves; D. Leaves; E. Leaf lobule; F. Basal cells of leaf lobe showing a ocellus; G. Underleaf; H. Transverse section of stem; I. Median cells of leaf lobe showing oil bodies. (All from *Wei et al. 201104-29*)

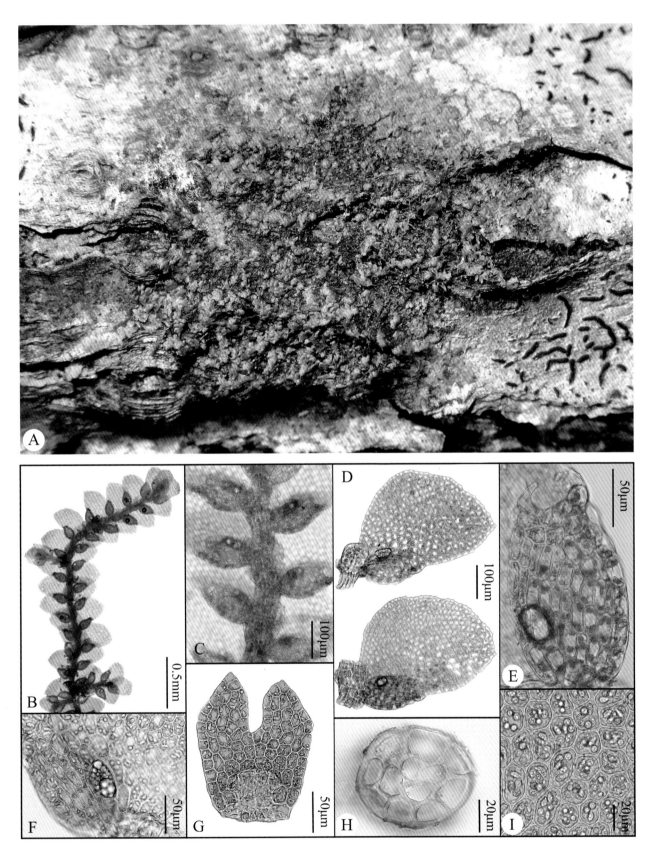

图 148 日本角鳞苔

Fig. 148 *Drepanolejeunea erecta* (Steph.) Mizut.

单齿角鳞苔

Drepanolejeunea ternatensis (Gottsche) Schiffn., Hepat. (Engl.-Prantl): 126. 1893.

植物体黄绿色，带叶宽 0.32—0.50 mm，不规则分枝，分枝细鳞苔型。茎横切面 7 个表皮细胞和 3 个内部细胞；腹面局部植物体 2 个细胞宽。侧叶毗邻至远生，易脱落，背瓣斜卵状披针形，顶端锐尖，常内弯，背侧边缘具不规则齿，背面每个细胞具 1 个疣或平滑；腹瓣长约为背瓣的 1/2，近轴边缘由 4 个加长细胞组成，中齿为 1 个长的稍弯曲的细胞，角齿不明显，透明疣位于中齿基部近轴侧。叶细胞稍厚壁，三角体小至大，简单三角形，中部球状加厚常见。油体聚合型，每个细胞具 1—2 个油体。油胞 1—2 个，生于背瓣基部。腹叶远生，顶端深 2 裂，裂瓣常（2—）3 个细胞，基部 1—2 个细胞宽。雄苞和雌苞未见。

生境： 树基，倒木生。海拔 1625—1869 m。

雅长分布： 草王山。

中国分布： 安徽，福建，广东，广西，贵州，海南，台湾，香港，浙江。

Plants yellowish green when fresh, 0.32−0.50 mm wide, irregularly branched, branching of the *Lejeunea*-type. Stem in transverse section with 7 epidermal cells and 3 medullary cell; ventral merophyte 2 cells wide. Leaves contiguous to remote, usually caducous, lobe obliquely ovate-lanceolate, apex acute, usually incurved, margin irregularly denticulate, dorsal surface smooth or with 1 papilla per cell; lobule ca. 1/2 as long as the lobe, free lateral margin bordered by 4 rectangular or rectangular-linear marginal cells, first tooth unicellular, elongate and slightly curved, second tooth obsolete, hyaline papilla at the proximal side of first tooth. Lobe cell slightly thick-walled, trigones small to large, simple-triangulate, intermediate thickenings frequent. Oil bodies segmented, 1−2 per median cell. Ocelli 1−2 at base of leaf lobe. Underleaves distant, deeply bilobed, lobes (2−)3 cells long, 1−2 cells wide at base. Androecia and gynoecia not seen.

Habitat: On tree bases and fallen trees. Elev. 1625−1869 m.

Distribution in Yachang: Caowangshan.

Distribution in China: Anhui, Fujian, Guangdong, Guangxi, Guizhou, Hainan, Hongkong, Taiwan, Zhejiang.

A. 种群；B. 植物体；C. 植物体一段（腹面观）；D. 植物体一段（背面观）；E. 侧叶（腹面观）；F. 侧叶（背面观）；G. 植物体一段（背面观），示油胞；H. 茎一段，示腹叶；I. 茎横切面；J. 腹瓣。（凭证标本：*韦玉梅等 221114-7A*）

A. Population; B. Plant; C. Portion of plant (ventral view); D. Portion of plant (dorsal view); E. Leaves (ventral view); F. Leaves (dorsal view); G. Portion of plant (dorsal view) showing ocelli; H. Portion of stem showing underleaves; I. Transverse section of stem; J. Leaf lobule. (All from *Wei et al. 221114-7A*)

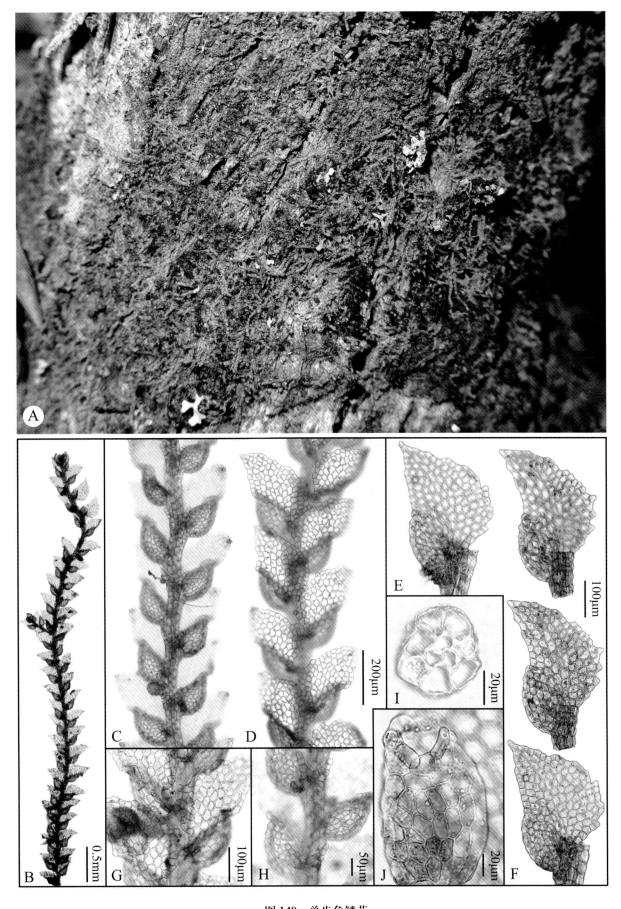

图 149 单齿角鳞苔

Fig. 149 *Drepanolejeunea ternatensis* (Gottsche) Schiffn.

短叶角鳞苔

Drepanolejeunea vesiculosa (Mitt.) Steph., Sp. Hepat. (Stephani) 5: 356. 1913.

植物体绿色至黄绿色，带叶宽 0.28—0.45 mm，不规则分枝，分枝细鳞苔型。茎横切面 7 个表皮细胞和 3 个内部细胞；腹面局部植物体 2 个细胞宽。侧叶疏松覆瓦状排列，易脱落，背瓣斜卵状披针形，顶端锐尖，常内弯，背侧边缘全缘或具细圆齿，背面每个细胞常具 1 个疣；腹瓣长约为背瓣的 1/2，近轴边缘由 4 个加长细胞组成，中齿为 1 个长的弯曲的细胞，角齿不明显，透明疣位于中齿基部近轴侧。叶细胞稍厚壁，三角体小至大，简单三角形，中部球状加厚常见。油体聚合型，每个细胞具 1—2 个油体。油胞（1—）2 个，生于背瓣基部。腹叶远生，顶端深 2 裂，裂瓣常 3—4 个细胞，基部 2 个细胞宽。雄苞未见。雌苞通常具 1 个新生枝，新生枝叶发生顺序密鳞苔型。

生境：树干，倒木生。海拔 1869—1902 m。

雅长分布：草王山。

中国分布：福建，广东，广西，贵州，海南，台湾，云南。

Plants green to yellowish green when fresh, 0.28–0.45 mm wide, irregularly branched, branching of the *Lejeunea*-type. Stem in transverse section with 7 epidermal cells and 3 medullary cell; ventral merophyte 2 cells wide. Leaves loosely imbricate, usually caducous, lobe obliquely ovate-lanceolate, apex acute, usually incurved, margin entire to crenulate, dorsal surface usually with 1 papilla per cell; lobule ca. 1/2 as long as the lobe, free lateral margin bordered by 4 rectangular or rectangular-linear marginal cells, first tooth unicellular, elongate and curved, second tooth obsolete, hyaline papilla at the proximal side of first tooth. Lobe cell slightly thick-walled, trigones small to large, simple-triangulate, intermediate thickenings frequent. Oil bodies segmented, 1–2 per median cell. Ocelli (1–)2 at base of leaf lobe. Underleaves distant, deeply bilobed, lobes 3–4 cells long, 2 cells wide at base. Androecia not seen. Gynoecia with an innovation, innovation leaf sequence pycnolejeuneoid.

Habitat: On tree trunks and fallen trees. Elev. 1869–1902 m.

Distribution in Yachang: Caowangshan.

Distribution in China: Fujian, Guangdong, Guangxi, Guizhou, Hainan, Taiwan, Yunnan.

▶ A. 种群；B. 植物体一段（腹面观）；C. 植物体一段（背面观）；D. 侧叶（腹面观）；E. 侧叶（背面观）；F. 植物体一段（背面观），示油胞；G. 茎一段，示腹叶；H. 腹瓣；I. 茎横切面。（凭证标本：*韦玉梅等 221112-71*）

A. Population; B. Portions of plants (ventral view); C. Portion of plant (dorsal view); D. Leaves (ventral view); E. Leaf (dorsal view); F. Portion of plant (dorsal view) showing ocelli; G. Portion of stem showing underleaves; H. Leaf lobule; I. Transverse section of stem. (All from *Wei et al. 221112-71*)

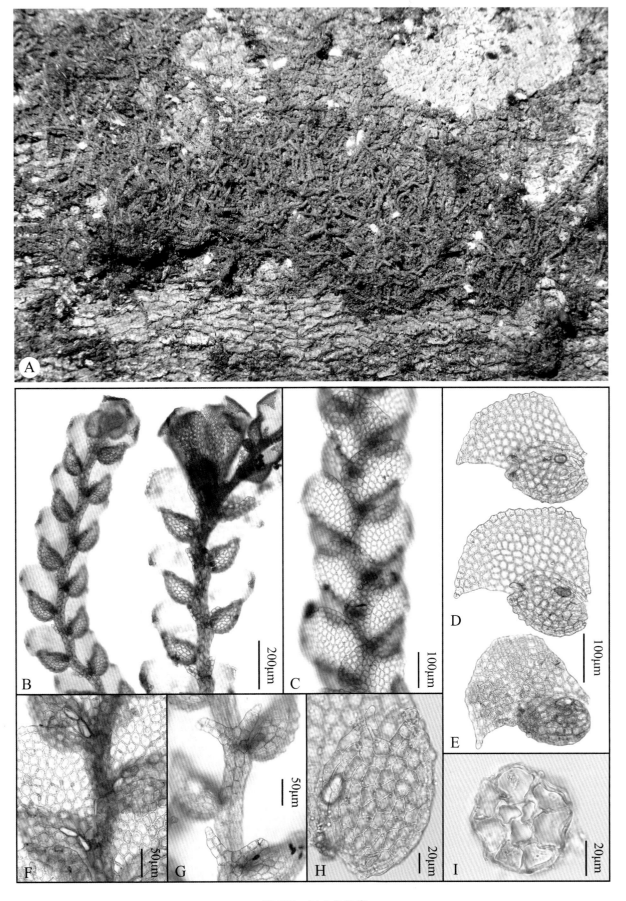

图 150 短叶角鳞苔

Fig. 150 *Drepanolejeunea vesiculosa* (Mitt.) Steph.

狭瓣细鳞苔

Lejeunea anisophylla Mont., Ann. Sci. Nat. Bot. (sér. 2) 19: 263. 1843.

植物体黄绿色，带叶宽 0.67—1.05 mm，不规则分枝，分枝细鳞苔型。茎横切面 7 个表皮细胞和约 9 个内部细胞，腹面局部植物体 2 个细胞宽。侧叶覆瓦状排列至毗邻，背瓣卵形至椭圆形，顶端圆，边缘全缘；腹瓣卵形，长约为背瓣的 1/4，有时退化，中齿单细胞，角齿退化，透明疣位于中齿基部的近轴侧。侧叶细胞薄壁，角质层平滑，三角体大，简单三角形，中部球状加厚常见。油体聚合型，每个细胞 6—10 个。腹叶远生，宽约为茎的 2 倍，顶端 2 裂达 1/2—2/3 深，裂瓣外侧边缘常具 1 个钝齿。雌雄同株。雄苞常顶生，雄苞叶 2—6 对，雄苞腹叶仅生于雄苞基部。雌苞具 1 个新生枝，新生枝叶发生顺序细鳞苔型。蒴萼倒心形或倒卵形，具 4—5 个平滑的脊，喙 1—2 个细胞长。

生境：石生，枯枝，树基，钙华基质，叶附生。海拔 491—1320 m。

雅长分布：大宴坪天坑漏斗，黄猄洞天坑，深洞，一沟，中井屯。

中国分布：除了东北以外的全国大部分省区均有分布。

Plants yellowish green when fresh, 0.67–1.05 mm wide, irregularly branched, branching of the *Lejeunea*-type. Stem in transverse section with 7 epidermal cells and ca. 9 medullary cells; ventral merophyte 2 cells wide. Leaves imbricate to contiguous, lobe ovate to oblong, apex rounded, margin entire; lobule ovate, ca. 1/4 as long as the lobe, sometimes reduced, first tooth unicellular, second tooth obsolete, hyaline papilla at the proximal side of the first tooth. Lobe cell thin-walled, trigones large, simple-triangulate, intermediate thickenings frequent. Cuticle smooth. Oil bodies segmented, 6–10 per median cell. Underleaves distant, ca. 2 times as wide as the stem, bilobed to 1/2–2/3 underleaf length, usually with a blunt tooth at outer margins of lobes. Monoicous. Androecia usually terminal, bracts in 2–6 pairs, bracteoles present only at the base of the androecium. Gynoecia with 1 innovation, innovation leaf sequence lejeuneoid. Perianths obcordate or obovate, with 4–5 smooth keels, beak 1–2 cells long.

Habitat: On rocks, dead branches, tree bases, calcareous substrates and leaves. Elev. 491–1320 m.

Distribution in Yachang: Dayanping Tiankeng, Huangjingdong Tiankeng, Shendong, Yigou, Zhongjing Tun.

Distribution in China: Widely distributed in most provinces of China except Northeast Region.

A. 种群；B. 植物体一段；C. 植物体一段，示蒴萼；D. 雄苞；E. 侧叶；F. 腹叶；G. 茎横切面；H. 腹瓣；I. 茎一段，示腹面局部植物体；J. 背瓣中部细胞，示油体。（凭证标本：黄萍等 210717-13）

A. Population; B. Portion of plant; C. Porttion of plant showing a perianth; D. Androecium; E. Leaves; F. Underleaves; G. Transverse section of stem; H. Leaf lobule; I. Portion of stem showing ventral merophyte; J. Median cells of leaf lobe showing oil bodies. (All from *Huang et al. 210717-13*)

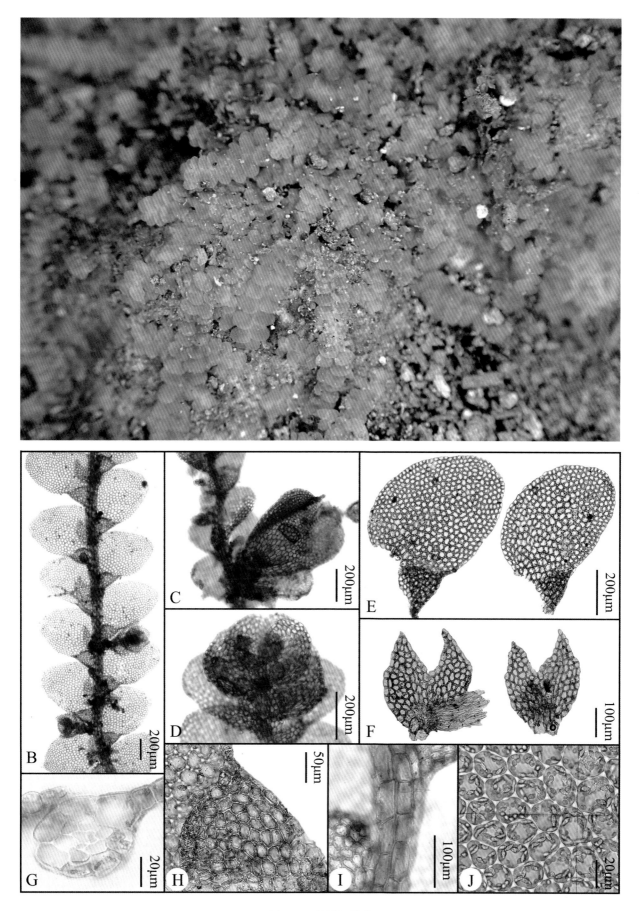

图 151　狭瓣细鳞苔

Fig. 151　*Lejeunea anisophylla* Mont.

拟细鳞苔（狭鳞苔）

Lejeunea apiculata Sande Lac., Ned. Kruidk. Arch. 3(4): 421. 1854.

Stenolejeunea apiculata (Sande Lac.) R. M. Schust., Beih. Nova Hedwigia 9: 144. 1963.

植物体淡绿色至黄绿色，带叶宽 0.48—0.95 mm，不规则分枝，分枝细鳞苔型。茎横切面 7 个表皮细胞和约 6 个内部细胞，腹面局部植物体 2 个细胞宽。侧叶覆瓦状排列至毗邻，背瓣卵形，顶端锐尖至细尖，边缘全缘；腹瓣卵形，长为背瓣的 1/4—1/3，有时退化，中齿单细胞，角齿退化，透明疣位于中齿基部的近轴侧。侧叶细胞薄壁，角质层平滑，三角体小，简单三角形，中部球状加厚有时可见。油体未见。腹叶远生，宽为茎的 1.5—2.0 倍，顶端 2 裂达 1/2—2/3 深，有时裂瓣外侧边缘具 1 个钝齿。雌雄同株。雄苞常顶生，雄苞叶 2—5 对，雄苞腹叶仅生于雄苞基部。雌苞具 1 个新生枝，新生枝叶发生顺序细鳞苔型。蒴萼未见。

生境：树干。海拔 1747 m。

雅长分布：草王山。

中国分布：广东，海南，台湾，香港。首次记录于广西。

Plants pale green to yellowish green when fresh, 0.48−0.95 mm wide, irregularly branched, branching of the *Lejeunea*-type. Stem in transverse section with 7 epidermal cells and ca. 6 medullary cells; ventral merophyte 2 cells wide. Leaves imbricate to contiguous, lobe ovate, apex acute to apiculate, margin entire; lobule ovate, 1/4−1/3 as long as the lobe, sometimes reduced, first tooth unicellular, second tooth obsolete, hyaline papilla at the proximal side of the first tooth. Lobe cell thin-walled, trigones large, simple-triangulate, intermediate thickenings occasionally present. Cuticle smooth. Oil bodies not seen. Underleaves distant, 1.5−2.0 times as wide as the stem, bilobed to 1/2−2/3 underleaf length, sometimes with a blunt tooth at outer margins of lobes. Monoicous. Androecia usually terminal, bracts in 2−5 pairs, bracteoles present only at the base of the androecium. Gynoecia with 1 innovation, innovation leaf sequence lejeuneoid. Perianths not seen.

Habitat: On tree trunks. Elev. 1747 m.

Distribution in Yachang: Caowangshan.

Distribution in China: Guangdong, Hainan, Hongkong, Taiwan. New to Guangxi.

▶ A. 植物体；B. 植物体一段；C. 侧叶；D. 植物体一段，示腹叶；E. 植物体一段，示雌苞；F. 腹瓣；G. 雌苞腹叶；H. 雌苞叶；I. 茎横切面。（凭证标本：*唐启明等 20190518-292C*）

A. Plant; B. Portion of plant; C. Leaves; D. Portion of plant showing underleaves; E. Portion of plant showing a gynoecium; F. Leaf lobule; G. Female bracteole; H. Female bracts; I. Transverse section of stem. (All from *Tang et al. 20190518-292C*)

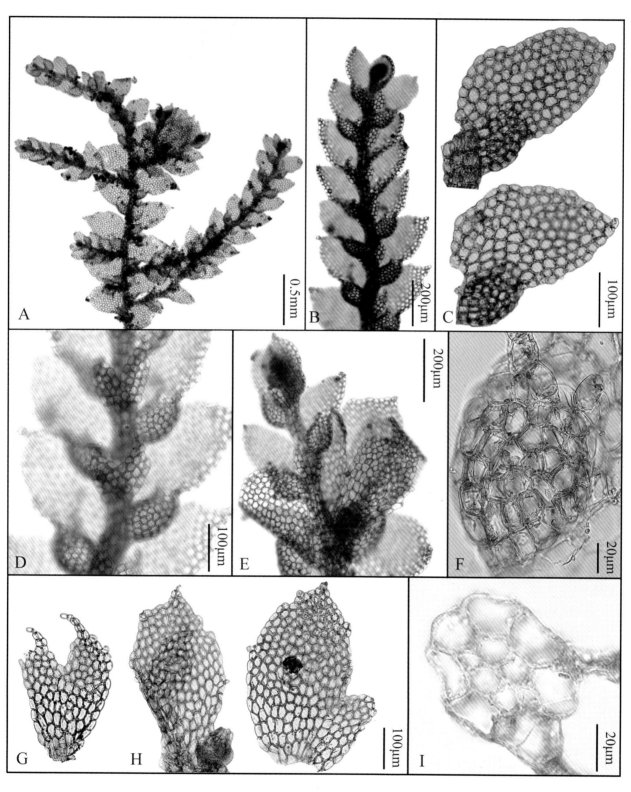

图 152 拟细鳞苔

Fig. 152 *Lejeunea apiculata* Sande Lac.

双齿细鳞苔

Lejeunea bidentula Herzog, Symb. Sin. 5: 51. 1930.

植物体浅绿色至黄绿色，带叶宽 1.15—1.60 mm，不规则分枝，分枝细鳞苔型。茎横切面 7 个表皮细胞和 6—8 个内部细胞，腹面局部植物体 2 个细胞宽。侧叶覆瓦状排列，背瓣宽卵形，顶端圆，边缘全缘；腹瓣卵形，长约为背瓣的 1/2，中齿大，3—5 个细胞长，基部 2—4 个细胞宽，角齿 1–2 个细胞，透明疣位于中齿基部的近轴侧。侧叶细胞薄壁，角质层平滑，三角体小，简单三角形，中部球状加厚缺。油体聚合型，每个细胞 4—9 个。腹叶远生至毗邻，宽为茎的 3—4 倍，顶端 2 裂达 1/2 深。雌雄同株。雄苞常顶生，雄苞叶 2—3 对，雄苞腹叶仅生于雄苞基部。雌苞具 1 个新生枝，新生枝叶发生顺序细鳞苔型。蒴萼倒卵形，具 5 个脊，喙 5—8 个细胞长。

生境： 树干，树枝。海拔 1747—1834 m。

雅长分布： 草王山。

中国分布： 甘肃，贵州，海南，四川，台湾，云南。首次记录于广西。

Plants light green to yellowish green when fresh, 1.15–1.60 mm wide, irregularly branched, branching of the *Lejeunea*-type. Stem in transverse section with 7 epidermal cells and 6–8 medullary cells; ventral merophyte 2 cells wide. Leaves imbricate, lobe ovate, apex rounded, margin entire; lobule ovate, ca. 1/2 as long as the lobe, first tooth 3–5 cells long, 2–4 cells wide at base, second tooth with 1–2 cells, hyaline papilla at the proximal side of the first tooth. Lobe cell thin-walled, trigones small, simple-triangulate, intermediate thickenings absent. Cuticle smooth. Oil bodies segmented, 4–9 per median cell. Underleaves distant, 3–4 times as wide as the stem, bilobed to 1/2 underleaf length. Monoicous. Androecia usually terminal, bracts in 2–3 pairs, bracteoles present only at the base of the androecium. Gynoecia with 1 innovation, innovation leaf sequence lejeuneoid. Perianths obovate, with 5 keels, beak 5–8 cells long.

Habitat: On tree trunks and tree branches. Elev. 1747–1834 m.

Distribution in Yachang: Caowangshan.

Distribution in China: Gansu, Guizhou, Hainan, Sichuan, Taiwan, Yunnan. New to Guangxi.

▶ A. 种群；B. 植物体一段带蒴萼；C. 茎一段，示腹叶；D. 侧叶；E. 腹叶；F. 茎横切面；G. 腹瓣；H. 背瓣中部细胞，示油体。（凭证标本：*韦玉梅等 191013-127*）

A. Population; B. Portion of plant with perianths; C. Portion of stem showing underleaves; D. Leaf; E. Underleaves; F. Transverse section of stem; G. Leaf lobule; H. Median cells of leaf lobe showing oil bodies. (All from *Wei et al. 191013-127*)

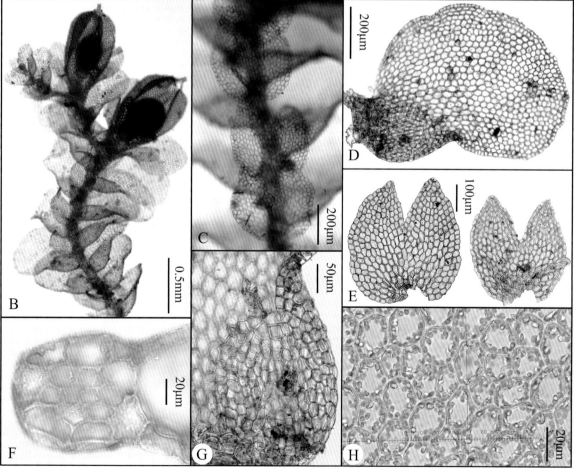

图 153 双齿细鳞苔

Fig. 153 *Lejeunea bidentula* Herzog

瓣叶细鳞苔

Lejeunea cocoes Mitt., J. Proc. Linn. Soc., Bot. 5(18): 114. 1860.

植物体浅绿色至黄绿色，带叶宽 0.35—0.48 mm，不规则分枝，分枝细鳞苔型。茎横切面 7 个表皮细胞和 4—5 个内部细胞，腹面局部植物体 2 个细胞宽。侧叶远生，背瓣卵形，顶端圆钝至钝尖，边缘全缘；腹瓣常退化，卵形（发育良好时），长约为背瓣的 1/2，中齿单细胞，角齿退化，透明疣位于中齿基部的近轴侧。侧叶细胞薄壁，角质层具细疣，三角体小，简单三角形，中部球状加厚缺。油体聚合型，每个细胞 5—11 个。腹叶远生，宽为茎的 1.0—1.5 倍，顶端 2 裂约达 2/3 深。雌雄异株。雄苞未见。雌苞具 1—2 个新生枝，新生枝叶发生顺序细鳞苔型。蒴萼未见。

生境： 石生和树干环境常见，钙华基质、腐木、树基、岩面薄土生环境也有分布。海拔 514—1823 m。

雅长分布： 草王山，达陇坪屯，二沟，黄猄洞天坑，拉雅沟，蓝家湾天坑，李家坨屯，龙坪村，全达村，深洞，一沟。

中国分布： 重庆，福建，广东，广西，贵州，海南，湖北，江西，台湾，香港，云南，浙江。

Plants light green to yellowish green when fresh, 0.35−0.48 mm wide, irregularly branched, branching of the *Lejeunea*-type. Stem in transverse section with 7 epidermal cells and 4−5 medullary cells; ventral merophyte 2 cells wide. Leaves distant, lobe ovate, apex rounded-obtuse to acute-obtuse, margin entire; lobule usually reduced, ovate (when well developed), ca. 1/2 as long as the lobe, first tooth unicellular, second tooth obsolete, hyaline papilla at the proximal side of the first tooth. Lobe cell thin-walled, trigones small, simple-triangulate, intermediate thickenings absent. Cuticle finely punctate. Oil bodies segmented, 5−11 per median cell. Underleaves distant, 1.0−1.5 times as wide as the stem, bilobed to ca. 2/3 underleaf length. Dioicous. Androecia not seen. Gynoecia with 1−2 innovations, innovation leaf sequence lejeuneoid. Perianths not seen.

Habitat: Often on tree trunks and rocks, sometimes on calcareous substrates, rotten logs, tree bases and on rocks with a thin layer of soil. Elev. 514−1823 m.

Distribution in Yachang: Caowangshan, Dalongping Tun, Ergou, Huangjingdong Tiankeng, Layagou, Lanjiawan Tiankeng, Lijiatuo Tun, Longping Village, Quanda Village, Shendong, Yigou.

Distribution in China: Chongqing, Fujian, Guangdong, Guangxi, Guizhou, Hainan, Hongkong, Hubei, Jiangxi, Taiwan, Yunnan, Zhejiang.

A. 种群；B–C. 植物体一段；D. 植物体一段带雌苞；E–F. 腹叶；G. 茎横切面；H. 腹瓣；I. 背瓣中部细胞，示角质层细疣；J. 背瓣中部细胞，示油体（凭证标本：韦玉梅等 201109-35）

A. Population; B–C. Portions of plants; D. Portion of plant with a gynoecium; E–F. Underleaves; G. Transverse section of stem; H. Leaf lobule; I. Median cells of leaf lobe showing finely punctate cuticle; J. Median cells of leaf lobe showing oil bodies. (All from *Wei et al. 201109-35*)

图 154 瓣叶细鳞苔

Fig. 154 *Lejeunea cocoes* Mitt.

弯叶细鳞苔

Lejeunea curviloba Steph., Sp. Hepat. (Stephani) 5: 774. 1915.

植物体黄绿色，带叶宽 0.45—0.85 mm，不规则分枝，分枝细鳞苔型。茎横切面 7 个表皮细胞和约 4 个内部细胞，腹面局部植物体 2 个细胞宽。侧叶覆瓦状排列至毗邻，背瓣卵圆形，顶端圆，常内弯，边缘全缘；腹瓣卵形，长为背瓣的 1/4—2/5，中齿单细胞，常指向叶顶端，角齿退化，透明疣位于中齿基部的近轴侧。侧叶细胞薄壁，角质层平滑，三角体小至大，简单三角形，中部球状加厚有时可见。油体聚合型，每个细胞 3—6 个。腹叶远生，宽为茎的 1.5—2.0 倍，顶端 2 裂约达 1/2 深。雌雄异株。雄苞未见。雌苞具 1 个新生枝，新生枝叶发生顺序细鳞苔型。蒴萼未见。

生境：石生，土生，树干，叶附生。海拔 1030—1747 m。

雅长分布：草王山，白岩垱漏斗。

中国分布：安徽，重庆，福建，广东，广西，贵州，海南，湖北，湖南，江西，四川，台湾，西藏，香港，云南，浙江。

Plants yellowish green when fresh, 0.45–0.85 mm wide, irregularly branched, branching of the *Lejeunea*-type. Stem in transverse section with 7 epidermal cells and ca. 4 medullary cells; ventral merophyte 2 cells wide. Leaves imbricate to contiguous, lobe rounded-ovate, apex rounded, usually incurved, margin entire; lobule ovate, 1/4–2/5 as long as the lobe, first tooth unicellular, spreading toward leaf apex, second tooth obsolete, hyaline papilla at the proximal side of the first tooth. Lobe cell thin-walled, trigones small to large, simple-triangulate, intermediate thickenings occasionally present. Cuticle smooth. Oil bodies segmented, 3–6 per median cell. Underleaves distant, 1.5–2.0 times as wide as the stem, bilobed to ca. 1/2 underleaf length. Dioicous. Androecia not seen. Gynoecia with 1 innovation, innovation leaf sequence lejeuneoid. Perianths not seen.

Habitat: On rocks, soil, tree trunks and leaves. Elev. 1030–1747 m.

Distribution in Yachang: Caowangshan, Baiyandang.

Distribution in China: Anhui, Chongqing, Fujian, Guangdong, Guangxi, Guizhou, Hainan, Hongkong, Hubei, Hunan, Jiangxi, Sichuan, Taiwan, Tibet, Yunnan, Zhejiang.

▶ A. 种群；B. 植物体一段；C. 植物体一段带雌苞；D. 侧叶；E. 腹瓣；F–G. 腹叶；H. 茎横切面；I. 背瓣中部细胞，示油体。（凭证标本：*黄萍等 210718-36*）

A. Population; B. Portion of plant; C. Portion of plant with a gynoecium; D. Leaves; E. Leaf lobule; F–G. Underleaves; H. Transverse section of stem; I. Median cells of leaf lobe showing oil bodies. (All from *Huang 210718-36*)

图 155 弯叶细鳞苔

Fig. 155 *Lejeunea curviloba* Steph.

神山细鳞苔

Lejeunea eifrigii Mizut., J. Hattori Bot. Lab. 33: 244. 1970.

植物体绿色，带叶宽 0.90—1.20 mm，不规则分枝，分枝细鳞苔型。茎横切面 7 个表皮细胞和 8—10 个内部细胞，腹面局部植物体 2 个细胞宽。侧叶覆瓦状排列至毗邻，背瓣阔卵形至三角状卵形，顶端细尖，边缘全缘；腹瓣常退化，卵形（发育良好时），长约为背瓣的 1/5，中齿单细胞，角齿退化，透明疣位于中齿基部的近轴侧。侧叶细胞薄壁，角质层平滑，三角体小，简单三角形，中部球状加厚缺。油体聚合型，每个细胞 5—9 个。腹叶远生，宽为茎的 2—3 倍，顶端 2 裂达 1/3—1/2 深。雌雄同株。雄苞常顶生，雄苞叶 2—3 对，雄苞腹叶仅生于雄苞基部。雌苞具 1 个新生枝，新生枝叶发生顺序细鳞苔型。蒴萼未见。

生境：石生。海拔 931 m。

雅长分布：深洞。

中国分布：福建，甘肃，广东，广西，贵州，海南，台湾，香港。

Plants yellowish green when fresh, 0.90−1.20 mm wide, irregularly branched, branching of the *Lejeunea*-type. Stem in transverse section with 7 epidermal cells and ca. 8−10 medullary cells; ventral merophyte 2 cells wide. Leaves imbricate to contiguous, lobe broad ovate to triangular-ovate, apex apiculate, margin entire; lobule usually reduced, ovate (when well developed), ca. 1/5 as long as the lobe, first tooth unicellular, second tooth obsolete, hyaline papilla at the proximal side of the first tooth. Lobe cell thin-walled, trigones small, simple-triangulate, intermediate thickenings absent. Cuticle smooth. Oil bodies segmented, 5−9 per median cell. Underleaves distant, 2−3 times as wide as the stem, bilobed to 1/3−1/2 underleaf length. Monoicous. Androecia usually terminal, bracts in 2−3 pairs, bracteoles present only at the base of the androecium. Gynoecia with 1 innovation, innovation leaf sequence lejeuneoid. Perianths not seen.

Habitat: On rocks. Elev. 931 m.

Distribution in Yachang: Shendong.

Distribution in China: Fujian, Gansu, Guangdong, Guangxi, Guizhou, Hainan, Hongkong, Taiwan.

▶ A. 种群；B. 植物体一段；C. 植物体一段，示雌苞和雄苞；D. 侧叶；E. 植物体一段，示腹叶；F. 背瓣中部细胞，示油体；G. 雌苞腹叶；H. 雌苞叶；I. 腹瓣；J. 茎横切面。（凭证标本：*韦玉梅等 191017-260*）

A. Population; B. Portion of plant; C. Portion of plant showing a androecium and a gynoecium; D. Leaves; E. Portion of plant showing underleaves; F. Median cells of leaf lobe showing oil bodies; G. Female bracteole; H. Female bracts; I. Leaf lobule; J. Transverse section of stem. (All from *Wei et al. 191017-260*)

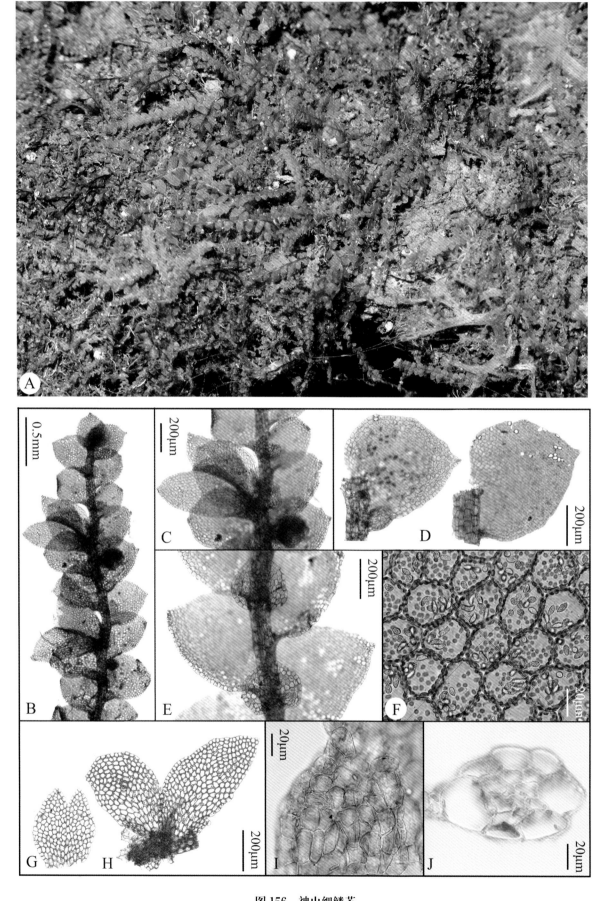

图 156 神山细鳞苔

Fig. 156 *Lejeunea eifrigii* Mizut.

黄色细鳞苔

Lejeunea flava (Sw.) Nees, Naturgesch. Eur. Leberm. 3: 277. 1838.

植物体黄绿色，带叶宽 0.95—1.52 mm，不规则分枝，分枝细鳞苔型。茎横切面 7 个表皮细胞和约 9 个内部细胞，腹面局部植物体 2 个细胞宽。侧叶覆瓦状排列，背瓣卵形至椭圆状卵形，顶端圆或偶尔钝，边缘全缘；腹瓣椭圆状卵形，长约为背瓣的 1/4，中齿单细胞，角齿退化，透明疣位于中齿基部的近轴侧。侧叶细胞薄壁，角质层具细疣，三角体小至大，简单三角形，中部球状加厚常见。油体聚合型，每个细胞 2—5 个。腹叶疏松覆瓦状排列，宽为茎的 4—6 倍，顶端 2 裂达 1/3—1/2 深。雌雄同株。雄苞常顶生，雄苞叶 2—4 对，雄苞腹叶仅生于雄苞基部。雌苞具 1 个新生枝，新生枝叶发生顺序细鳞苔型。蒴萼椭圆形，具 5 个脊，喙 2—4 个细胞长。

生境： 树干环境常见，枯枝、树枝环境也有分布。海拔 1048—1862 m。

雅长分布： 草王山，黄猄洞天坑，盘古王，悬崖天坑，中井屯。

中国分布： 除了东北以外的全国大部分省区均有分布。

Plants yellowish green when fresh, 0.95–1.52 mm wide, irregularly branched, branching of the *Lejeunea*-type. Stem in transverse section with 7 epidermal cells and ca. 9 medullary cells; ventral merophyte 2 cells wide. Leaves imbricate, lobe ovate to oblong-ovate, apex rounded or occasionally obtuse, margin entire; lobule oblong-ovate, ca. 1/4 as long as the lobe, first tooth unicellular, second tooth obsolete, hyaline papilla at the proximal side of the first tooth. Lobe cell thin-walled, trigones small to large, simple-triangulate, intermediate thickenings frequent. Cuticle finely punctate. Oil bodies segmented, 2–5 per median cell. Underleaves loosely imbricate, 4–6 times as wide as the stem, bilobed to 1/3–1/2 underleaf length. Monoicous. Androecia usually terminal, bracts in 2–4 pairs, bracteoles present only at the base of the androecium. Gynoecia with 1 innovation, innovation leaf sequence lejeuneoid. Perianths oblong, with 5 keels, beak 2–4 cells long.

Habitat: Often on tree trunks, sometimes on (dead) branches. Elev. 1048–1862 m.

Distribution in Yachang: Caowangshan, Huangjingdong Tiankeng, Panguwang, Xuanya Tiankeng, Zhongjing Tun.

Distribution in China: Widely distributed in most provinces of China except Northeast Region.

▶ A. 种群；B. 植物体一段；C. 植物体一段，示蒴萼；D. 雄苞；E. 侧叶；F. 腹叶；G. 腹瓣；H. 茎横切面；I. 腹瓣茎一段，示腹面局部植物体；J. 背瓣中部细胞，示角质层细疣；K. 背瓣中部细胞，示油体。（凭证标本：*韦玉梅等 201108-32*）

A. Population; B. Portion of plant; C. Portion of plant with a perianth; D. Androecium; E. Leaves; F. Underleaves; G. Leaf lobule; H. Transverse section of stem; I. Portion of stem showing ventral merophyte; J. Median cells of leaf lobe showing dorsal cuticle; K. Median cells of leaf lobe showing oil bodies. (All from *Wei et al. 201108-32*)

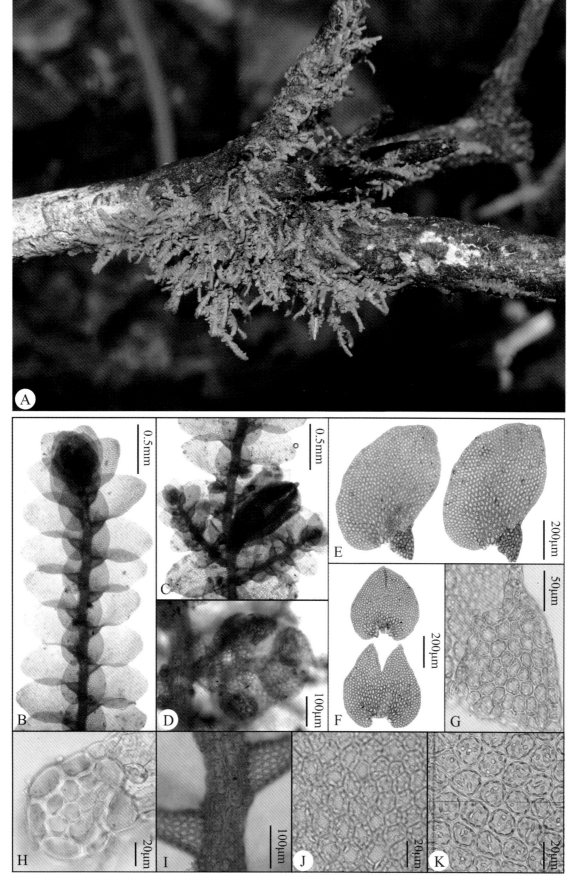

图 157 黄色细鳞苔

Fig. 157 *Lejeunea flava* (Sw.) Nees

巨齿细鳞苔

Lejeunea kodamae Ikegami et Inoue, J. Jap. Bot. 36(1): 7. 1961.

植物体浅绿色，带叶宽 0.50—0.82 mm，不规则分枝，分枝细鳞苔型。茎横切面 7 个表皮细胞和 3—6 个内部细胞，腹面局部植物体 2 个细胞宽。侧叶覆瓦状排列，背瓣卵形至近圆形，顶端圆或偶尔钝，边缘全缘；腹瓣常退化，长方形，顶部常突出 2—5 个细胞长，基部 1—2 个细胞宽，卵形（发育良好时），长约为背瓣的 1/3，中齿 2—3 个细胞长，基部 1—2 个细胞宽，角齿 1—2 个细胞，透明疣位于中齿基部的近轴侧。侧叶细胞薄壁，角质层平滑，三角体小，简单三角形，中部球状加厚缺。油体聚合型，每个细胞 4—7 个。腹叶远生，宽为茎的 2—3 倍，顶端 2 裂达 1/2 深。雌雄同株。雄苞常顶生，雄苞叶 2—3 对，雄苞腹叶仅生于雄苞基部。雌苞具 1 个新生枝，新生枝叶发生顺序细鳞苔型。蒴萼倒卵形，具 5 个弱的脊，喙 2—3 个细胞长。

生境： 树干和树基环境常见，石生、腐木、树枝、叶附生环境也有分布。海拔 1099—1755 m。

雅长分布： 达陇坪屯，大棚屯，大宴坪天坑漏斗，吊井天坑，黄猄洞天坑，老屋基天坑，李家坨屯，里郎天坑，逻家田屯，盘古王，旁墙屯，深洞，塘英村，下棚屯，悬崖天坑，中井屯。

中国分布： 贵州。首次记录于广西。

Plants light green when fresh, 0.50−0.82 mm wide, irregularly branched, branching of the *Lejeunea*-type. Stem in transverse section with 7 epidermal cells and 3−6 medullary cells; ventral merophyte 2 cells wide. Leaves imbricate, lobe ovate to suborbicular, apex rounded or occasionally obtuse, margin entire; lobule usually reduced, retangular, apex forming a projection, 2−5 cells long, 1−2 cells wide at base, ovate (when well developed), ca. 1/3 as long as the lobe, first tooth 2−3 cells long, 1−2 cells wide at base, second tooth with 1−2 cells, hyaline papilla at the proximal side of the first tooth. Lobe cell thin-walled, trigones small, simple-triangulate, intermediate thickenings absent. Cuticle smooth. Oil bodies segmented, 4−7 per median cell. Underleaves distant, 2−3 times as wide as the stem, bilobed to ca. 1/2 underleaf length. Monoicous. Androecia usually terminal, bracts in 2−3 pairs, bracteoles present only at the base of the androecium. Gynoecia with 1 innovation, innovation leaf sequence lejeuneoid. Perianths obovate, with 5 weak keels, beak 2−3 cells long.

Habitat: Often on tree trunks and tree bases, sometimes on rocks, rotten logs, tree branches and leaves. Elev. 1099−1755 m.

Distribution in Yachang: Dalongping Tun, Dapeng Tun, Dayanping Tiankeng, Diaojing Tiankeng, Huangjingdong Tiankeng, Laowuji Tiankeng, Lijiatuo Tun, Lilang Tiankeng, Luojiatian Tun, Panguwang, Pangqiang Tun, Shendong, Tangying Village, Xiapeng Tun, Xuanya Tiankeng, Zhongjing Tun.

Distribution in China: Guizhou. New to Guangxi.

A. 种群；B. 植物体；C. 植物体一段，示蒴萼；D. 雄苞；E. 发育良好的腹瓣；F. 茎一段，示退化的腹瓣；G. 侧叶；H. 茎横切面；I. 雌苞；J. 背瓣中部细胞，示油体；K. 茎一段，示腹叶。（凭证标本：韦玉梅等 201105-19）

A. Population; B. Plant; C. Portion of plant showing a perianth; D. Androecium; E. Well developed lobule; F. Portion of stem showing reduced lobules; G. Leaves; H. Transverse section of stem; I. Gynoecium; J. Median cells of leaf lobe showing oil bodies; K. Portion of stem showing underleaves. (All from *Wei et al. 201105-19*)

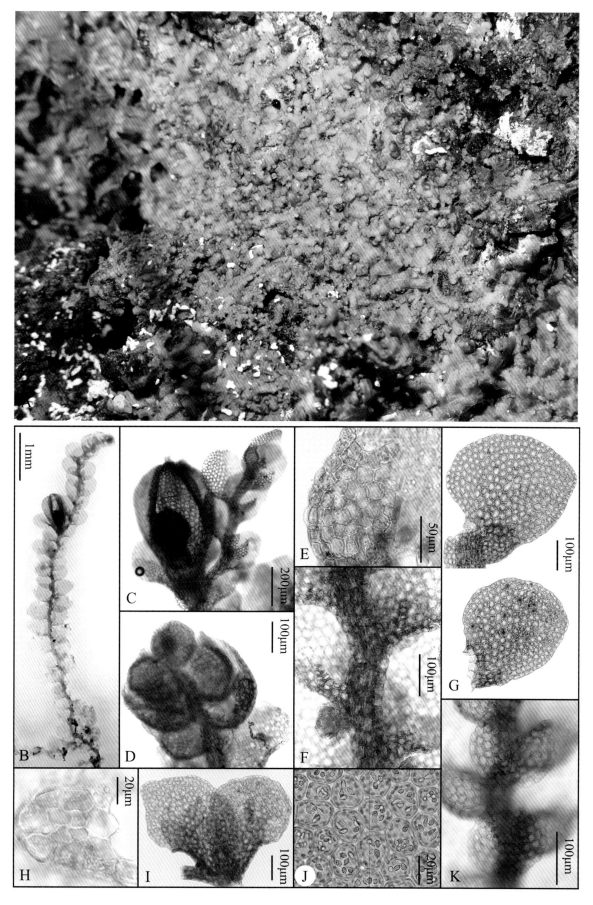

图 158 巨齿细鳞苔

Fig. 158 *Lejeunea kodamae* Ikegami & Inoue

科诺细鳞苔

Lejeunea konosensis Mizut., J. Hattori Bot. Lab. 71: 127. 1992.

植物体黄绿色，带叶宽 0.45—0.70 mm，不规则分枝，分枝细鳞苔型。茎横切面 7 个表皮细胞和约 6 个内部细胞，腹面局部植物体 2 个细胞宽。侧叶覆瓦状排列至远生，背瓣卵形至椭圆状卵形，顶端圆或偶尔圆钝，边缘全缘；腹瓣卵形，长为背瓣的 1/3—2/5，中齿单细胞，角齿退化，透明疣位于中齿基部的近轴侧。侧叶细胞薄壁，角质层具细疣，三角体小至大，简单三角形，中部球状加厚常见。油体聚合型，每个细胞 3—6 个。腹叶远生，宽为茎的 3—4 倍，顶端 2 裂达 1/3—1/2 深。雌雄同株。雄苞常间生，雄苞叶 2—3 对，雄苞腹叶生于整个雄苞。雌苞具 1 个新生枝，新生枝叶发生顺序细鳞苔型。蒴萼未见。

生境： 石生环境常见，土生、树基、叶附生、岩面薄土生环境也有分布。海拔 1048—1252 m。

雅长分布： 大宴坪天坑漏斗，黄猄洞天坑，蓝家湾天坑，下岩洞屯，悬崖天坑。

中国分布： 广东，广西，贵州，浙江。

Plants yellowish green when fresh, 0.45–0.70 mm wide, irregularly branched, branching of the *Lejeunea*-type. Stem in transverse section with 7 epidermal cells and ca. 6 medullary cells; ventral merophyte 2 cells wide. Leaves imbricate to distant, lobe ovate to oblong-ovate, apex rounded or occasionally rounded-obtuse, margin entire; lobule ovate, 1/3–2/5 as long as the lobe, first tooth unicellular, second tooth obsolete, hyaline papilla at the proximal side of the first tooth. Lobe cell thin-walled, trigones small to large, simple-triangulate, intermediate thickenings frequent. Cuticle finely punctate. Oil bodies segmented, 3–6 per median cell. Underleaves distant, 3–4 times as wide as the stem, bilobed to 1/3–1/2 underleaf length. Monoicous. Androecia usually intercalary, bracts in 2–3 pairs, bracteoles present throughout the androecium. Gynoecia with 1 innovation, innovation leaf sequence lejeuneoid. Perianths not seen.

Habitat: Often on rocks, sometimes on soil, tree bases, leaves and on rocks with a thin layer of soil. Elev. 1048–1252 m.

Distribution in Yachang: Dayanping Tiankeng, Huangjingdong Tiankeng, Lanjiawan Tiankeng, Xiayandong Tun, Xuanya Tiankeng.

Distribution in China: Guangdong, Guangxi, Guizhou, Zhejiang.

A. 种群；B. 植物体一段；C. 植物体一段带雄苞；D. 植物体一段，示腹叶；E. 侧叶；F. 腹瓣；G. 茎横切面；H. 背瓣中部细胞，示油体；I. 背瓣中部细胞，示角质层细疣。（凭证标本：*韦玉梅等 201104-41*）

A. Population; B. Portion of plant; C. Portion of plant with androecium; D. Portion of plant showing underleaves; E. Leaves; F. Leaf lobule; G. Transverse section of stem; H. Median cells of leaf lobe showing oil bodies; I. Median cells of leaf lobe showing dorsal cuticle. (All from *Wei et al. 201104-41*)

图 159 科诺细鳞苔

Fig. 159 *Lejeunea konosensis* Mizut.

麦氏细鳞苔

Lejeunea micholitzii Mizut., J. Hattori Bot. Lab. 33: 236. 1970.

植物体深绿色，带叶宽 0.82—1.15 mm，不规则分枝，分枝细鳞苔型。茎横切面 7 个表皮细胞和约 6 个内部细胞，腹面局部植物体 2 个细胞宽。侧叶毗邻至远生，背瓣卵形，顶端圆或偶尔圆钝，边缘全缘；腹瓣常强烈退化，卵形（发育良好时），长约为背瓣的 1/5，中齿单细胞，角齿退化，透明疣位于中齿基部的近轴侧。侧叶细胞薄壁，角质层平滑，三角体小至大，简单三角形，中部球状加厚常见。油体聚合型，每个细胞含 15 个以上。腹叶远生，宽为茎的 2—3 倍，顶端 2 裂达 1/2—2/3 深。雄苞和雌苞未见。

生境： 石生。海拔 1198 m。

雅长分布： 蓝家湾天坑。

中国分布： 广西，海南，台湾。

Plants dark green when fresh, 0.82–1.15 mm wide, irregularly branched, branching of the *Lejeunea*-type. Stem in transverse section with 7 epidermal cells and ca. 6 medullary cells; ventral merophyte 2 cells wide. Leaves contiguous to distant, lobe ovate, apex rounded or occasionally rounded-obtuse, margin entire; lobule often strongly reduced, sometimes ovate (when well developed), ca. 1/5 as long as the lobe, first tooth unicellular, second tooth obsolete, hyaline papilla at the proximal side of the first tooth. Lobe cell thin-walled, trigones small to large, simple-triangulate, intermediate thickenings frequent. Cuticle smooth. Oil bodies segmented, usually more than 15 per median cell. Underleaves distant, 2–3 times as wide as the stem, bilobed to 1/2–2/3 underleaf length. Androecia and gynoecia not seen.

Habitat: On rocks. Elev. 1198 m.

Distribution in Yachang: Lanjiawan Tiankeng.

Distribution in China: Guangxi, Hainan, Taiwan.

▶ A. 种群；B. 植物体一段；C. 茎一段，示腹叶；D. 侧叶；E. 发育良好的腹瓣；F. 退化的腹瓣；G. 茎横切面；H. 茎一段，示腹面局部植物体；I. 背瓣中部细胞，示油体。（凭证标本：*韦玉梅等 201114-7*）

A. Population; B. Portion of plant; C. Portion of stem showing underleaves; D. Leaves; E. Well developed lobule; F. Reduced lobule; G. Transverse section of stem; H. Portion of stem showing ventral merophyte; I. Median cells of leaf lobe showing oil bodies. (All from *Wei et al. 201114-7*)

图 160　麦氏细鳞苔

Fig. 160　*Lejeunea micholitzii* Mizut.

暗绿细鳞苔

Lejeunea obscura Mitt., J. Proc. Linn. Soc., Bot. 5(18): 112. 1860[1861].

植物体黄绿色，带叶宽 0.90—1.20 mm，不规则分枝，分枝细鳞苔型。茎横切面 7 个表皮细胞和约 8 个内部细胞，腹面局部植物体 2 个细胞宽。侧叶覆瓦状排列至毗邻，背瓣阔卵形，顶端圆，边缘全缘；腹瓣常强烈退化，长方形，长约为背瓣的 1/6，中齿单细胞，角齿退化，透明疣位于中齿基部的近轴侧。侧叶细胞薄壁，角质层平滑，三角体小，简单三角形，中部球状加厚缺。油体均一型，每个细胞含 20 个以上。腹叶常远生，宽为茎的 2—4 倍，顶端 2 裂达 1/3—1/2 深。雌雄同株。雄苞常顶生，雄苞叶 2—4 对，雄苞腹叶仅生于雄苞基部。雌苞具 1 个新生枝，新生枝叶发生顺序细鳞苔型。蒴萼倒卵形，具 5 个脊，喙 1—2 个细胞长。

生境： 叶附生和石生环境常见，土生、倒木、腐殖质、腐木、枯枝、树干、树基、树枝环境也有分布。海拔 507—1829 m。

雅长分布： 草王山，大宴坪天坑漏斗，吊井天坑，二沟，黄猄洞天坑，九十九堡，蓝家湾天坑，逻家田屯，盘古王，旁墙屯，深洞，下岩洞屯，香当天坑，一沟，中井天坑，中井屯竖井，中井屯。

中国分布： 除了东北以外的全国大部分省区均有分布。

Plants yellowish green when fresh, 0.90–1.20 mm wide, irregularly branched, branching of the *Lejeunea*-type. Stem in transverse section with 7 epidermal cells and ca. 8 medullary cells; ventral merophyte 2 cells wide. Leaves imbricate to contiguous, lobe broad ovate, apex rounded, margin entire; lobule usually reduced, retangular, ca. 1/6 as long as the lobe, first tooth unicellular, second tooth obsolete, hyaline papilla at the proximal side of the first tooth. Lobe cell thin-walled, trigones small, simple-triangulate, intermediate thickenings absent. Cuticle smooth. Oil bodies homogeneous, usually more than 20 per median cell. Underleaves usually distant, 2–4 times as wide as the stem, bilobed to 1/3–1/2 underleaf length. Monoicous. Androecia usually terminal, bracts in 2–4 pairs, bracteoles present only at the base of the androecium. Gynoecia with 1 innovation, innovation leaf sequence lejeuneoid. Perianths obovate, with 5 keels, beak 1–2 cells long.

Habitat: Often on rock and leaves, sometimes on soil, fallen trees, humus, rotten logs, dead branches, tree trunks, tree bases and tree branches. Elev. 507–1829 m.

Distribution in Yachang: Caowangshan, Dayanping Tiankeng, Diaojing Tiankeng, Ergou, Huangjingdong Tiankeng, Jiushijiubao, Lanjiawan Tiankeng, Luojiatian Tun, Panguwang, Pangqiang Tun, Shendong, Xiayandong Tun, Xiangdang Tiankeng, Yigou, Zhongjing Tiankeng, Zhongjing Tun.

Distribution in China: Widely distributed in most provinces of China except Northeast Region.

▶ A. 种群；B. 植物体一段带雌苞和雄苞；C. 侧叶；D. 腹叶；E. 茎一段，示退化的腹瓣；F. 茎横切面；G. 雄苞；H. 背瓣中部细胞，示油体。（凭证标本：韦玉梅等 201105-14A）

A. Population; B. Portion of plant with gynoecia and a androecium; C. Leaves; D. Underleaves; E. Portion of stem showing reduced lobules; F. Transverse section of stem; G. Androecium; H. Median cells of leaf lobe showing oil bodies. (All from *Wei et al.* 201105-14A)

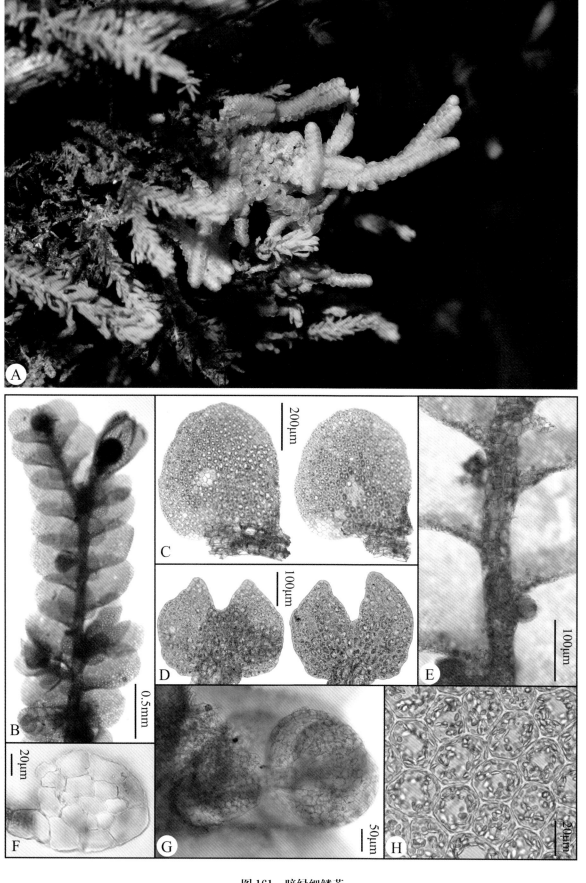

图 161 暗绿细鳞苔

Fig. 161 *Lejeunea obscura* Mitt.

角齿细鳞苔

Lejeunea otiana S. Hatt., Bot. Mag. (Tokyo) 65(763/764): 15. 1952.

植物体浅绿色，带叶宽 0.90—1.32 mm，不规则分枝，分枝细鳞苔型。茎横切面 7 个表皮细胞和约 14 个内部细胞，腹面局部植物体 2 个细胞宽。侧叶覆瓦状排列，背瓣卵形至宽卵形，顶端圆，边缘全缘；腹瓣常退化，长方形，顶部常突出 2—4 个细胞长，基部 2—4 个细胞宽，卵形（发育良好时），长为背瓣的 1/4—1/3，中齿 2—3 个细胞长，基部 1—2 个细胞宽，角齿单细胞或不明显，透明疣位于中齿基部的近轴侧。侧叶细胞薄壁，角质层平滑，三角体小至大，简单三角形，中部球状加厚常可见。油体聚合型，每个细胞 7—13 个。腹叶远生，宽为茎的 2—3 倍，顶端 2 裂约达 2/3 深。雌雄同株。雄苞常顶生，雄苞叶 2—4 对，雄苞腹叶仅生于雄苞基部。雌苞具 1 个新生枝，新生枝叶发生顺序细鳞苔型。蒴萼倒卵形，具 5 个脊，喙 2—3 个细胞长。

生境：石生，腐木，树干，叶附生。海拔 491—931 m。

雅长分布：二沟，拉雅沟，深洞，一沟。

中国分布：贵州，香港。首次记录于广西。

Plants light green when fresh, 0.90–1.32 mm wide, irregularly branched, branching of the *Lejeunea*-type. Stem in transverse section with 7 epidermal cells and ca. 14 medullary cells; ventral merophyte 2 cells wide. Leaves imbricate, lobe ovate to broad ovate, apex rounded, margin entire; lobule usually reduced, retangular, apex often forming a projection, 2–4 cells long, 2–4 cells wide at base, ovate (when well developed), 1/4–1/3 as long as the lobe, first tooth 2–3 cells long, 1–2 cells wide at base, second tooth unicellular or indistinct, hyaline papilla at the proximal side of the first tooth. Lobe cell thin-walled, trigones small to large, simple-triangulate, intermediate thickenings frequent. Cuticle smooth. Oil bodies segmented, 7–13 per median cell. Underleaves distant, 2–3 times as wide as the stem, bilobed to ca. 2/3 underleaf length. Monoicous. Androecia usually terminal, bracts in 2–4 pairs, bracteoles present only at the base of the androecium. Gynoecia with 1 innovation, innovation leaf sequence lejeuneoid. Perianths obovate, with 5 keels, beak 2–3 cells long.

Habitat: On rocks, rotten logs, tree trunks and Leaves. Elev. 491–931 m.

Distribution in Yachang: Ergou, Layagou, Shendong, Yigou.

Distribution in China: Guizhou, Hongkong. New to Guangxi.

▶ A. 种群；B. 植物体一段；C. 侧叶；D. 茎横切面；E. 茎一段，示发育良好的腹瓣；F. 茎一段，示腹叶；G–H. 茎一段，示退化的腹瓣；I. 雄苞；J. 背瓣中部细胞，示油体。（凭证标本：*韦玉梅等 191017-249*）

A. Population; B. Portion of plant; C. Leaves; D. Transverse section of stem; E. Portion of stem showing well developed lobules; F. Portion of stem showing underleaves; G–H. Portions of stems showing reduced lobules; I. Androecium; J. Median cells of leaf lobe showing oil bodies. (All from *Wei et al. 191017-249*)

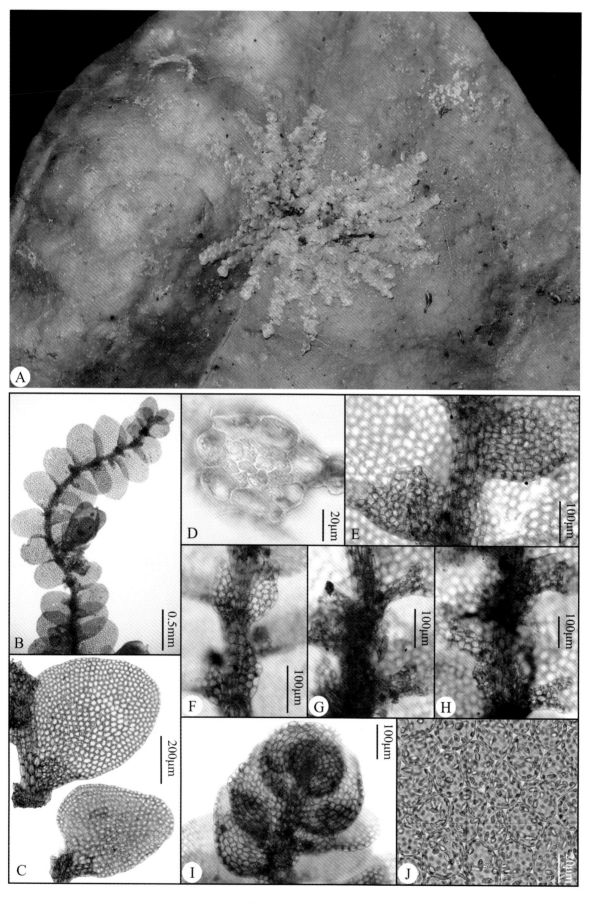

图 162 角齿细鳞苔

Fig. 162 *Lejeunea otiana* S. Hatt.

疣萼细鳞苔

Lejeunea tuberculosa Steph., Sp. Hepat. (Stephani) 5: 790. 1915.

植物体绿色，带叶宽 0.49—0.65 mm，不规则分枝，分枝细鳞苔型。茎横切面 7 个表皮细胞和 10—12 个内部细胞，腹面局部植物体 2 个细胞宽。侧叶覆瓦状排列至远生，背瓣卵形，顶端圆或偶尔圆钝，边缘全缘；腹瓣卵形，长约为背瓣的 1/3，中齿单细胞，角齿退化，透明疣位于中齿基部的近轴侧。侧叶细胞薄壁，角质层具细疣，三角体小至大，简单三角形，中部球状加厚有时可见。油体聚合型，每个细胞 2—5 个。腹叶远生，宽为茎的 1.5—2.0 倍，顶端 2 裂达 1/2—2/3 深。雌雄异株。雄苞未见。雌苞具 1 个新生枝，新生枝叶发生顺序细鳞苔型。蒴萼未见。

生境：石生环境常见，岩面薄土生环境也有分布。海拔 1338—1774 m。

雅长分布：草王山，大棚屯，盘古王，全达村，下棚屯。

中国分布：安徽，广东，广西，贵州，海南，陕西，台湾，香港，云南。

Plants green when fresh, 0.49−0.65 mm wide, irregularly branched, branching of the *Lejeunea*-type. Stem in transverse section with 7 epidermal cells and 10−12 medullary cells; ventral merophyte 2 cells wide. Leaves imbricate to distant, lobe ovate, apex rounded or occasionally rounded-obtuse, margin entire; lobule ovate, ca. 1/3 as long as the lobe, first tooth unicellular, second tooth obsolete, hyaline papilla at the proximal side of the first tooth. Lobe cell thin-walled, trigones small to large, simple-triangulate, intermediate thickenings occasionally present. Cuticle finely punctate. Oil bodies segmented, 2−5 per median cell. Underleaves distant, 1.5−2.0 times as wide as the stem, bilobed to 1/2−2/3 underleaf length. Dioicous. Androecia not seen. Gynoecia with 1 innovation, innovation leaf sequence lejeuneoid. Perianths not seen.

Habitat: Often on rocks, sometimes on rocks with a thin layer of soil. Elev. 1338−1774 m.

Distribution in Yachang: Caowangshan, Dapeng Tun, Panguwang, Quanda Village, Xiapeng Tun.

Distribution in China: Anhui, Guangdong, Guangxi, Guizhou, Hainan, Hongkong, Shaanxi, Taiwan, Yunnan.

▶ A. 种群；B. 植物体一段带雌苞；C. 植物体一段；D. 侧叶；E. 茎一段，示腹面局部植物体；F. 腹瓣；G. 腹叶；H. 茎横切面；I. 背瓣中部细胞，示角质层细疣；J. 背瓣中部细胞，示油体。（凭证标本：韦玉梅等 201110-25）

A. Population; B. Portion of plant with a gynoecium; C. Portion of plant; D. Leaves; E. Portion of stem showing ventral merophyte; F. Leaf lobule; G. Underleaves; H. Transverse section of stem; I. Median cells of leaf lobe showing dorsal cuticle; J. Median cells of leaf lobe showing oil bodies. (All from *Wei et al.* 201110-25)

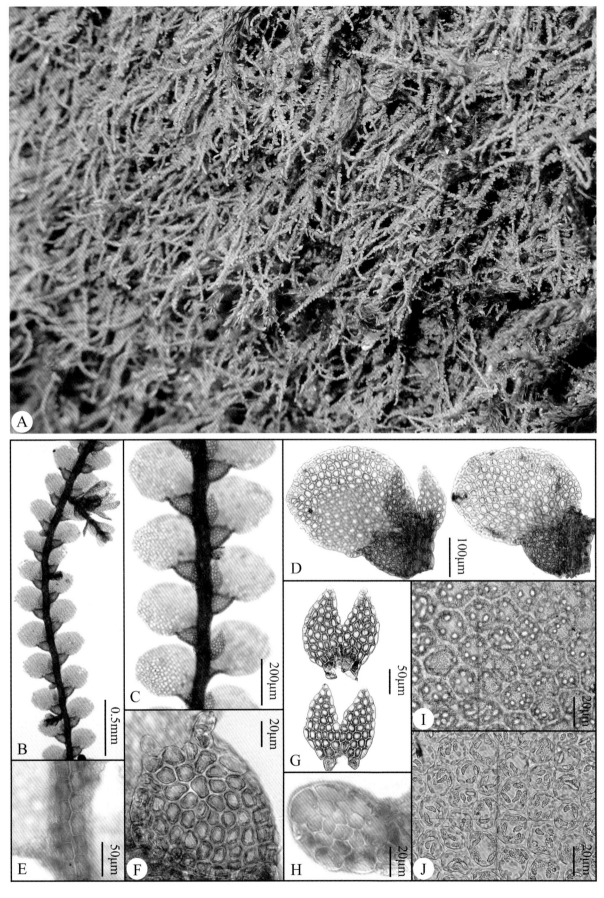

图 163 疣萼细鳞苔

Fig. 163 *Lejeunea tuberculosa* Steph.

巴氏薄鳞苔

Leptolejeunea balansae Steph., Hedwigia 35(3): 105. 1896.

植物体黄绿色，带叶宽 0.51—0.83 mm，不规则分枝，分枝细鳞苔型。茎横切面 7 个表皮细胞和 3 个内部细胞，腹面局部植物体 2 个细胞宽。侧叶覆瓦状排列至远生，背瓣卵形，顶端钝，边缘全缘；腹瓣卵形或椭圆形，长为背瓣的 1/3—2/5，中齿单细胞，角齿退化，透明疣位于中齿基部的近轴侧。侧叶细胞薄壁，角质层平滑，三角体小至大，简单三角形，中部球状加厚常见。油体未见；油胞散生，每个腹瓣具 2—6（—10）个，常在中间成不连续的一列排列，近基部具有一个大油胞，其余分散在叶中部。腹叶远生，宽为茎的 1.5—2.0 倍，深 2 裂，基部边缘由 6 个加长的细胞构成，裂瓣 2—4 个细胞长，基部 1—2 个细胞宽。雄苞和雌苞未见。

生境：枯枝，树干，树基，叶附生。海拔 1048—1136 m。

雅长分布：黄猄洞天坑。

中国分布：福建，广西，贵州，江西，西藏，云南。

Plants yellowish green when fresh, 0.51–0.83 mm wide, irregularly branched, branching of the *Lejeunea*-type. Stem in transverse section with 7 epidermal cells and 3 medullary cells; ventral merophyte 2 cells wide. Leaves imbricate to distant, lobe ovate, apex obtuse, margin entire; lobule ovate or oblong, 1/3–2/5 as long as the lobe, first tooth unicellular, second tooth obsolete, hyaline papilla at the proximal side of the first tooth. Lobe cell thin-walled, trigones small to large, simple-triangulate, intermediate thickenings frequent. Cuticle smooth. Oil bodies not seen. Ocelli 2–6(–10), scattered in the leaf lobe, usually arranged in a non-continuous longitudinal series, suprabasal ocellus 1, larger than other ocelli. Underleaves distant, 1.5–2.0 times as wide as the stem, deeply bilobed, margin of basal portion consisting of 6 rectangular or rectangular-linear cells, lobes 2–4 cells long, 1–2 cells wide at base. Androecia and gynoecia not seen.

Habitat: On dead branches, tree trunks, tree bases and leaves. Elev. 1048–1136 m.

Distribution in Yachang: Huangjingdong Tiankeng, .

Distribution in China: Fujian, Guangxi, Guizhou, Jiangxi, Tibet, Yunnan.

▶ A. 种群；B. 植物体；C. 植物体一段；D. 侧叶；E. 植物体一段（背面观）；F. 腹瓣；G. 腹叶；H. 茎横切面。（凭证标本：*韦玉梅等 201104-23*）

A. Population; B. Plant; C. Portion of plant; D. Leaves; E. Portion of plant (dorsal view); F. Leaf lobule; G. Underleaf; H. Transverse section of stem. (All from *Wei et al. 201104-23*)

图 164 巴氏薄鳞苔

Fig. 164 *Leptolejeunea balansae* Steph.

斑叶纤鳞苔（斑叶细鳞苔）

Microlejeunea punctiformis (Taylor) Steph., Hedwigia 29(2): 90. 1890.

Lejeunea punctiformis Taylor, London J. Bot. 5: 398. 1846.

植物体浅绿色至黄绿色，带叶宽 0.20—0.30 mm，不规则分枝，分枝细鳞苔型。茎横切面 7 个表皮细胞和 3 个内部细胞，腹面局部植物体 2 个细胞宽。侧叶毗邻至远生，背瓣卵形，顶端钝或圆钝，边缘全缘；腹瓣卵形，长为背瓣的 1/2—2/3，中齿单细胞，角齿退化，透明疣位于中齿基部的近轴侧。侧叶细胞薄壁，角质层具细疣，三角体小，简单三角形，中部球状加厚有时可见。油体聚合型，每个细胞 2—4 个；油胞 1—2 个，位于背瓣基部。腹叶远生，宽为茎的 1—2 倍，顶端 2 裂达 1/2—3/4 深。雌雄同株。雄苞顶生或间生，雄苞叶 2—4 对，雄苞腹叶生于整个雄苞。雌苞具 1 个新生枝，新生枝叶发生顺序细鳞苔型。蒴萼未见。

生境：倒木，腐木，枯枝，树干，树基，叶附生。海拔 1100—1759 m。

雅长分布：草王山，黄猄洞天坑，逻家田屯，盘古王，旁墙屯，下棚屯，中井屯竖井，中井屯。

中国分布：安徽，重庆，福建，广东，广西，贵州，海南，湖北，湖南，江西，四川，台湾，西藏，香港，云南，浙江。

Plants light green to yellowish green when fresh, 0.20−0.30 mm wide, irregularly branched, branching of the *Lejeunea*-type. Stem in transverse section with 7 epidermal cells and 3 medullary cells; ventral merophyte 2 cells wide. Leaves contiguous to distant, lobe ovate, apex obtuse to rounded-obtuse, margin entire; lobule ovate, 1/2−2/3 as long as the lobe, first tooth unicellular, second tooth obsolete, hyaline papilla at the proximal side of the first tooth. Lobe cell thin-walled, trigones small, simple-triangulate, intermediate thickenings frequent. Cuticle finely punctate. Oil bodies segmented, 2−4 per median cell. Ocelli 1−2 in the base of leaf lobe. Underleaves distant, 1−2 times as wide as the stem, bilobed to 1/2−3/4 underleaf length. Monoicous. Androecia terminal or intercalary, bracts in 2−4 pairs, bracteoles present throughout the androecium. Gynoecia with 1 innovation, innovation leaf sequence lejeuneoid. Perianths not seen.

Habitat: On fallen trees, rotten logs, dead branches, tree trunks, tree bases and leaves. Elev. 1100−1759 m.

Distribution in Yachang: Caowangshan, Huangjingdong Tiankeng, Luojiatian Tun, Panguwang, Pangqiang Tun, Xiapeng Tun, Zhongjing Tun.

Distribution in China: Anhui, Chongqing, Fujian, Guangdong, Guangxi, Guizhou, Hainan, Hongkong, Hubei, Hunan, Jiangxi, Sichuan, Taiwan, Tibet, Yunnan, Zhejiang.

A. 种群；B. 植物体；C. 植物体一段，示腹瓣；D. 植物体一段，示雌苞；E. 植物体一段；F. 植物体一段（背面观）；G. 腹叶；H. 茎横切面；I. 叶基部细胞，示油胞和油体；J. 背瓣中部细胞，示角质层细疣。（凭证标本：韦玉梅等 201105-33）

A. Population; B. Plant; C. Portion of plant showing lobules; D. Portion of plant showing a gynoecium; E. Portion of plant; F. Portion of plant (dorsal view); G. Underleaf; H. Transverse section of stem; I. Basal cells of leaf lobe showing ocelli and oil bodies; J. Median cells of leaf lobe showing dorsal cuticle. (All from *Wei et al. 201105-33*)

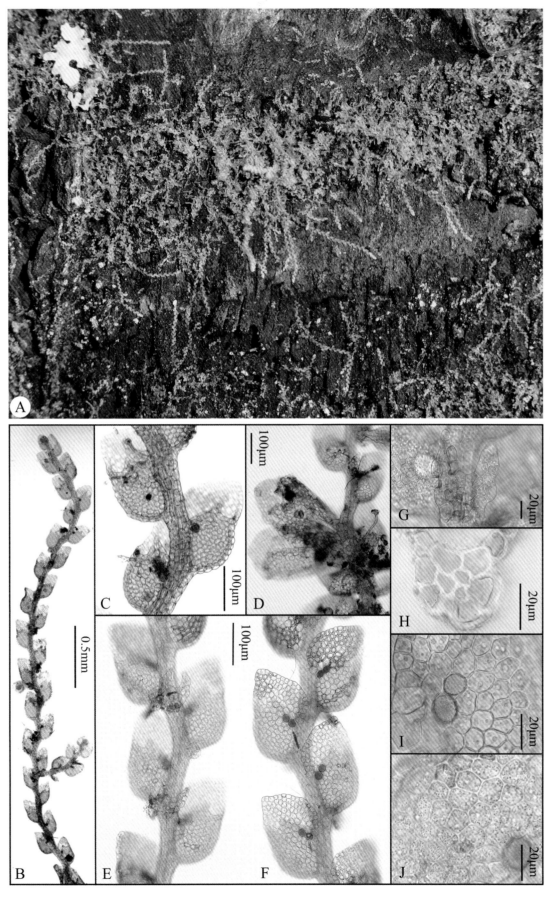

图 165 斑叶纤鳞苔

Fig. 165 *Microlejeunea punctiformis* (Taylor) Steph.

疏叶纤鳞苔（疏叶细鳞苔）

Microlejeunea ulicina (Taylor) Steph., Hedwigia 29(2): 88. 1890.

Lejeunea ulicina (Taylor) Gottsche, Lindenb. et Nees, Syn. Hepat.: 387. 1845.

植物体浅绿色，带叶宽 0.20—0.26 mm，不规则分枝，分枝细鳞苔型。茎横切面 7 个表皮细胞和 3 个内部细胞，腹面局部植物体 2 个细胞宽。侧叶远生，背瓣卵形，顶端圆，常内弯，边缘全缘；腹瓣卵形，与背瓣近于等长，中齿单细胞，角齿退化，透明疣位于中齿基部的近轴侧。侧叶细胞薄壁，角质层近于平滑，三角体小，简单三角形，中部球状加厚少见。油体未见；油胞 1 (—2) 个，位于背瓣基部。腹叶远生，宽为茎的 1.0—1.5 倍，顶端 2 裂达 1/2—2/3 深。雌雄异株。雄苞顶生或间生，雄苞叶 2—4 对，雄苞腹叶仅生于雄苞基部。雌苞和蒴萼未见。

生境：石生，树干。海拔 1099—1215 m。

雅长分布：香当天坑，旁墙屯。

中国分布：安徽，福建，广东，广西，贵州，海南，湖南，江西，上海，四川，台湾，西藏，香港，云南，浙江。

Plants light green when fresh, 0.20−0.26 mm wide, irregularly branched, branching of the *Lejeunea*-type. Stem in transverse section with 7 epidermal cells and 3 medullary cells; ventral merophyte 2 cells wide. Leaves distant, lobe ovate, apex rounded, margin entire; lobule ovate, nearly as long as the lobe, first tooth unicellular, second tooth obsolete, hyaline papilla at the proximal side of the first tooth. Lobe cell thin-walled, trigones small, simple-triangulate, intermediate thickenings scarce. Cuticle nearly smooth. Oil bodies not seen. Ocelli usually 1(−2) in the base of each leaf lobe. Underleaves distant, 1.0−1.5 times as wide as the stem, bilobed to 1/2−2/3 underleaf length. Dioicous. Androecia terminal or intercalary, bracts in 2−4 pairs, bracteoles present only at the androecium. Gynoecia and perianths not seen.

Habitat: On rocks and tree trunks. Elev. 1099−1215 m.

Distribution in Yachang: Xiangdang Tiankeng, Pangqiang Tun.

Distribution in China: Anhui, Fujian, Guangdong, Guangxi, Guizhou, Hainan, Hongkong, Hunan, Jiangxi, Shanghai, Sichuan, Taiwan, Tibet, Yunnan, Zhejiang.

A. 种群；B. 植物体；C. 植物体一段（背面观）；D. 植物体一段；E. 茎一段，示腹叶；F. 雄苞；G. 侧叶。（凭证标本：*韦玉梅等 191019-350*）

A. Population; B. Plants; C. Portion of plant (dorsal view); D. Portion of plant; E. Portion of stem showing underleaves; F. Androecium; G. Leaf. (All from *Wei et al. 191019-350*)

图 166 疏叶纤鳞苔

Fig. 166 *Microlejeunea ulicina* (Taylor) Steph.

圆叶拟多果苔（圆叶疣鳞苔）

Myriocoleopsis minutissima (Sm.) R. L. Zhu, Y. Yu et Pócs, Phytotaxa 183(4): 293. 2014.

Cololejeunea minutissima (Sm.) Steph., Bot. Gaz. 17(6): 171. 1892.

植物体浅绿色，带叶宽 0.26—0.35 mm，不规则分枝，分枝细鳞苔型。茎横切面 5 个表皮细胞和 1 个内部细胞；腹面局部植物体 1 个细胞宽。侧叶远生至毗邻，背瓣圆形，顶端圆，边缘具圆齿；腹瓣阔卵形，与背瓣近于等长，中齿 2 个细胞长，基部 1 个细胞宽，角齿 1—2 个细胞长，基部 1—2 个细胞宽，有时退化，透明疣位于中齿基部内表面；副体 1 个细胞。侧叶细胞薄壁，角质层平滑，三角体小，简单三角形，中部球状加厚缺。油体聚合型，每个细胞具 2—5 个油体。腹叶缺。雌雄同株。雄苞常顶生，雄苞叶 4—7 对，雄苞腹叶缺。雌苞具 1 个新生枝，雌苞腹叶缺。蒴萼梨形，具 5 个短脊，喙 1—2 个细胞长。芽胞圆盘状，由 18—24 个细胞构成，生于背瓣腹面。

生境： 树干环境常见，石生、树基上也有分布。海拔 1136—1361 m。

雅长分布： 大宴坪天坑漏斗，吊井天坑，黄猄洞天坑，拉洞天坑，李家坨屯，旁墙屯，下棚屯，香当天坑，中井屯。

中国分布： 安徽，重庆，海南，台湾，云南。首次记录于广西。

Plants light green when fresh, 0.26−0.35 mm wide, irregularly branched, branching of the *Lejeunea*-type. Stem in transverse section with 5 epidermal cells and 1 medullary cell; ventral merophyte 1 cell wide. Leaves distant to contiguous, lobe orbicular, apex rounded, margin crenulate; lobule broad ovate, as long as the lobe, free margin strongly involute, first tooth 2 cells long, 1 cell wide at base, second tooth 1−2 cells long, 1−2 cell wide at base, sometimes obsolete, hyaline papilla on the inner surface of lobule at the base of the first tooth. Stylus unicellular. Lobe cell thin-walled, trigones small, simple-triangulate, intermediate thickenings absent. Oil bodies segmented, 2−5 per median cell. Underleaves absent. Monoicous. Androecia usually terminal, bracts in 4−7 pairs, bracteoles absent. Gynoecia with 1 innovation, bracteole absent. Perianths pyriform, with 5 short keels, beak 1−2 cells long. Gemmae discoid, 18−24-celled, on ventral surface of leaf lobe.

Habitat: Often on tree trunks, sometimes on rocks and tree bases. Elev. 1136−1361 m.

Distribution in Yachang: Dayanping Tiankeng, Diaojing Tiankeng, Huangjingdong Tiankeng, Ladong Tiankeng, Lijiatuo Tun, Pangqiang Tun, Xiapeng Tun, Xiangdang Tiankeng, Zhongjing Tun.

Distribution in China: Anhui, Chongqing, Hainan, Taiwan, Yunnan. New to Guangxi.

▶ A. 种群；B. 植物体；C. 植物体一段带蒴萼；D. 植物体一段；E. 茎横切面；F. 雌苞叶；G. 芽胞；H. 侧叶；I. 背瓣中部细胞，示油体。（凭证标本：韦玉梅等 201116-9）

A. Population; B. Plant; C. Portion of Plant with a perianth; D. Portion of plant; E. Transverse section of stem; F. Female bracts; G. Gemmae; H. Leaf; I. Median cells of leaf lobe showing oil bodies. (All from *Wei et al. 201116-9*)

图 167　圆叶拟多果苔

Fig. 167　*Myriocoleopsis minutissima* (Sm.) R. L. Zhu, Y. Yu & Pócs

大绿片苔

Aneura maxima (Schiffn.) Steph., Bull. Herb. Boissier 7 (10): 760 (270). 1899.

叶状体背面鲜绿色，宽带状，边缘常呈褶皱状。无气孔和气室分化。中肋不明显。叶状体横切面呈线形，中央可达 12 层细胞厚，向边缘渐变薄，单层细胞翼部（3—）5—10 个细胞宽。表皮细胞五边形或六边形，薄壁，小于内层细胞。油体小，聚合型，在表皮细胞和内层细胞中均存在。假根无色透明，仅生于叶状体腹面中部。雌雄异株。雄株未见。雌生殖枝堆状，着生于叶状体边缘缺刻下。蒴被圆柱形，表面光滑。孢子体未成熟时包裹在蒴被内。孢蒴椭球形，成熟时 4 瓣开裂。孢子球形，表面具颗粒状纹饰。弹丝单螺旋加厚，成束聚生于孢蒴裂瓣顶端。

生境： 土生，钙华基质，腐木。海拔 1238—1822 m。
雅长分布： 草王山，下岩洞屯。
中国分布： 广西，河南，陕西，台湾。

Thalli dorsally light green when fresh, widely ribbon-like, weakly to strongly undulate along margin. Air chambers and epidermal pores absent. Midribs indistinct. Transverse sections of thallus linear, up to 12 cells thick in middle, gradually becomes thinner towards margin, Unistratose wings (3−)5−10 cells wide. Epidermal cells 5−6 angled, thin-walled, smaller than inner cells. Oil bodies small, segmented, present in all epidermal and inner cells. Rhizoids hyaline, restricted to the median part of ventral thallus. Dioicous. Male plants not seen. Female branches mound-like, situating beneath a lateral notch of thallus. Calyptrae cylindrical, smooth. Sporophyte enclosed by a calyptra before maturity. Capsules ellipsoidal, longitudinally dehiscing by 4 regular valves at maturity. Spores globose, surface granulate. Elaters with 1 spiral thickening, elaterophores adhere to the tips of the capsule valves.

Habitat: On soil, calcareous substrates and rotten logs. Elev. 1238−1822 m.
Distribution in Yachang: Caowangshan, Xiayandong Tun.
Distribution in China: Guangxi, Henan, Shaanxi, Taiwan.

▶ A. 种群；B. 叶状体带孢子体；C. 表皮细胞，示油体；D. 叶状体横切面；E. 叶状体横切面中部分；F. 叶状体近中部横切面；G. 叶状体横切面边缘部分；H. 开裂的孢蒴和弹丝束；I. 弹丝；J. 孢子。（凭证标本：*唐启明等 20201108-219*）

A. Population; B. Thalli with sporophytes; C. Epidermal cells showing oil bodies; D. Transverse section of thallus; E. Transverse section of thallus at median part; F. Transverse section of thallus near median part; G. Transvertse section of thallus at marginal part; H. Dehisced capsule with elaterophores; I. Elaters; J. Spores. (All from *Tang et al. 20201108-219*)

图 168 大绿片苔

Fig. 168 *Aneura maxima* (Schiffn.) Steph.

波叶片叶苔

Riccardia chamedryfolia (With.) Grolle, Trans. Brit. Bryol. Soc. 5(4): 772. 1969.

叶状体新鲜时黄绿色至深绿色，干燥后灰棕色，2—3回羽状分枝，匍匐生长，无向地枝。叶状体横切面表皮细胞略小于内层细胞，大小为内层细胞的1/2—2/3。主轴横切面平凸形，中央5—9层细胞厚，向边缘渐变薄，边缘锐尖至翼状，单层细胞翼部1—2个细胞宽。末端羽枝横切面平凸形，边缘翼状，单层细胞翼部2—3个细胞宽。油体棕色，聚合型，圆形、卵圆形或蠕虫形，仅见于内层细胞，内层细胞每个细胞含有1—5个油体。假根生于主轴腹面。雌雄同株。雄生殖枝侧生在主轴或主分枝上，藏精器2列，3—6对。雌生殖枝侧生在主轴或主分枝上，藏卵器2列，基部有纤毛状或鳞片状侧丝。

生境：腐木。海拔1084 m。

雅长分布：黄猄洞天坑。

中国分布：全国范围广布。首次记录于广西。

备注：该种油体通常在表皮细胞、翼部细胞和内层细胞均存在，但在雅长的种群中表皮细胞、翼部细胞均未见油体的存在。

Thalli yellowish green to dark green when fresh, pale brown in herbarium, 2–3-pinnately branched, prostrate, without geotropic stolons. Epidermal cells slightly smaller than inner cells in transverse section, 1/2–2/3 the inner cells in size. Transverse sections of main axis plano-convex, 5–9 cells thick in middle, become gradually thinner towards margin, acute to winged at margin, unistratose wings 1–2 cells wide. Transverse sections of ultimate pinnules plano-convex, winged at margin, unistratose wings 2–3 cells wide. Oil bodies brown, segmented, globose, ovoid or vermiform; oil bodies totaly lacking in epidermal and wing cells; oil bodies in inner cells 1–5 per cell. Rhizoids often on ventral surface of thallus. Monoicous. Male branches lateral on main axes or primary pinnae, antheridial chambers in 2 regular rows, with 3–6 pairs. Female branches lateral on main axes or primary pinnae, archegonia in 2 regular rows, with hair- to scale-like paraphyses at base.

Habitat: On rotten logs. Elev. 1084 m.

Distribution in Yachang: Huangjingdong Tiankeng.

Distribution in China: Widely distributed in China. New to Guangxi.

Note: Oil bodies of Riccardia chamedryfolia usually present in epidermal, wing, and inner cells. But epidermal and wing cells of thallus in the population from Yachang lack oil bodies.

▶ A. 种群；B. 叶状体；C. 末羽片横切面；D. 主轴横切面；E. 叶状体一段；F. 雄生殖枝；G. 表皮细胞，油体缺失；H. 内层细胞，示油体。（凭证标本：唐启明等20210717-26）

A. Population; B. Thallus; C. Transverse section of ultimate pinnule; D. Transverse section of main axis; E. Portion of thallus; F. Male branch; G. epidermal cells without oil bodies; H. Inner cells showing oil bodies. (All from *Tang et al.* 20210717-26)

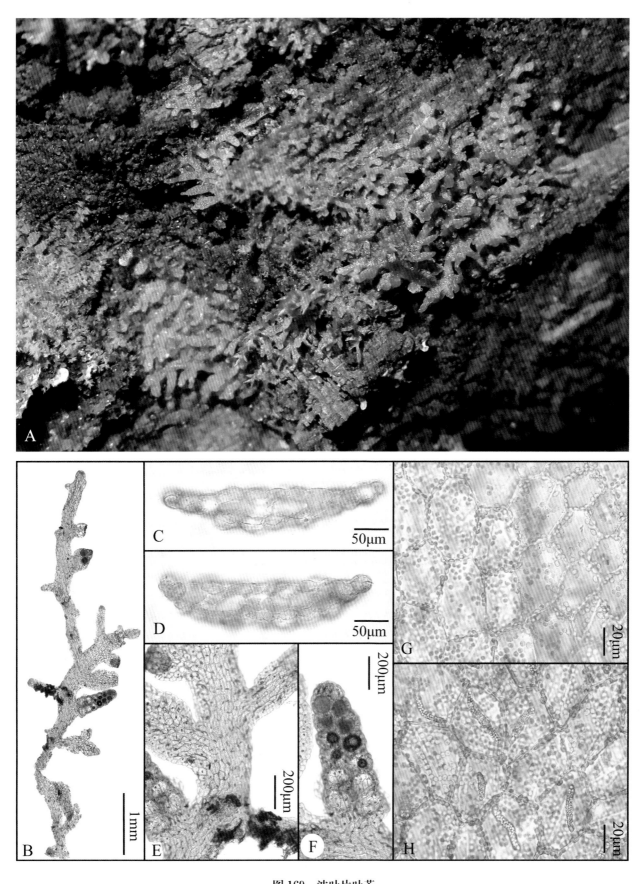

图 169 波叶片叶苔

Fig. 169 *Riccardia chamedryfolia* (With.) Grolle

黄片叶苔

Riccardia flavovirens Furuki, J. Hattori Bot. Lab. 70: 333. 1991.

叶状体新鲜时黄绿色至深绿色，干燥后灰棕色，2—3回规则的羽状分枝，匍匐生长，具有向地枝。叶状体横切面表皮细胞小于内层细胞，大小为内层细胞的1/4—1/2。主轴横切面平凸形，中央5—8层细胞厚，向边缘渐变薄，边缘锐尖至翼状，单层细胞翼部1—2个细胞宽。末端羽枝横切面平凸形至双凸型，边缘翼状，单层细胞翼部2—4个细胞宽。油体棕褐色，聚合型，圆形至卵圆形，表皮细胞、翼部细胞和内层细胞均存在，表皮细胞和翼部细胞中每个细胞含有1个油体，内层细胞中每个细胞含有1—2个油体。假根主要生于主轴贴基质生长的部分以及向地枝上。雌雄生殖枝未见。

生境：石生。海拔1606 m。

雅长分布：盘古王。

中国分布：台湾。首次记录于广西。

Thalli yellowish green to dark green when fresh, pale brown in herbarium, 2–3-pinnately branched, prostrate, with geotropic stolons. Epidermal cells smaller than inner cells in transverse section, 1/4–1/2 the inner cells in size. Transverse sections of main axis plano-convex, 5–8 cells thick in middle, become gradually thinner towards margin, acute to winged at margin, unistratose wings 1–2 cells wide. Transverse sections of ultimate pinnules plano-convex to concave-convex, winged at margin, unistratose wings 2–4 cells wide. Oil bodies dark-brown, segmented, globose to ovoid; oil bodies in epidermal and wing cells 1 per cell; oil bodies in inner cells 1(–2) per cell. Rhizoids mainly on prostrate thalli and geotropic stolons. Male and female branches not seen.

Habitat: On rocks. Elev. 1606 m.

Distribution in Yachang: Panguwang.

Distribution in China: Taiwan. New to Guangxi.

▶ A. 种群；B. 叶状体；C. 叶状体一段；D. 末羽片横切面；E. 主轴横切面；F. 内层细胞，示油体；G. 表皮细胞，示油体。（凭证标本：*韦玉梅等 201110-49B*）

A. Population; B. Thallus; C. Portion of thallus; D. Transverse section of ultimate pinnule; E. Transverse section of main axis; F. inner cells showing oil bodies; G. Epidermal cells showing oil bodies. (All from *Wei et al. 201110-49B*)

图 170　黄片叶苔

Fig. 170　*Riccardia flavovirens* Furuki

长崎片叶苔

Riccardia nagasakiensis (Steph.) S. Hatt., Bull. Tokyo Sci. Mus. 11: 164. 1944.

叶状体新鲜时深绿色至黑绿色，干燥后黑褐色，2—3 回规则的羽状分枝，匍匐至上倾生长，有时具有向地枝。叶状体横切面表皮细胞明显小于内层细胞，大小为内层细胞的 1/5—1/3。主轴横切面平凸形，中央 4—12 层细胞厚，向边缘渐变薄，边缘锐尖，有时呈翼状，单层细胞翼部 0—1 个细胞宽。末端羽枝横切面平凸形至双凸型，边缘翼状，单层细胞翼部 0—2 个细胞宽。油体棕褐色，聚合型，圆形至卵圆形，仅内层细胞中存在，内层细胞中每个细胞含有 1—3 个油体。假根主要生于主轴贴基质生长的部分以及向地枝上。雌雄生殖枝未见。

生境： 石生。海拔 1606 m。
雅长分布： 盘古王。
中国分布： 海南，台湾。首次记录于广西。

Thalli yellowish dark green to blackish green when fresh, blackish brown in herbarium, 2–3-pinnately branched, prostrate to ascending, sometimes with geotropic stolons. Epidermal cells distinctly smaller than inner cells in transverse section, 1/5–1/3 the inner cells in size. Transverse sections of main axis plano-convex, 4–12 cells thick in middle, become gradually thinner towards margin, acute to winged at margin, unistratose wings 0–1 cells wide. Transverse sections of ultimate pinnules plano-convex to concave-convex, winged at margin, unistratose wings 0–2 cells wide. Oil bodies dark-brown, segmented, globose to ovoid; oil bodies totaly lacking in epidermal and wing cells; oil bodies in inner cells 1–3 per cell. Rhizoids mainly on prostrate thalli and geotropic stolons. Male and female branches not seen.

Habitat: On rock. Elev. 1606 m.
Distribution in Yachang: Panguwang.
Distribution in China: Hainan, Taiwan. New to Guangxi.

A. 种群；B. 叶状体；C. 叶状体一段；D. 末羽片横切面；E. 主轴横切面；F. 内部细胞，示油体；G. 表皮细胞，示油体缺失。（凭证标本：*韦玉梅等 201110-49A*）

A. Population; B. Plant; C. Portion of plant; D. Transverse section of ultimate pinnule; E. Transverse sections of main axis; F. inner cells showing oil bodies; G. Epidermal cells without oil bodies. (All from *Wei et al. 201110-49A*)

图 171　长崎片叶苔

Fig. 171　*Riccardia nagasakiensis* (Steph.) S. Hatt.

掌状片叶苔

Riccardia palmata (Hedw.) Carruth., J. Bot. 3(10): 302. 1865.

叶状体新鲜时黄绿色至深绿色，干燥后黑棕色，1—3回不规则的羽状分枝，匍匐生长，具有向地枝。叶状体横切面表皮细胞略小于内层细胞，大小为内层细胞的1/3—1/2。主轴横切面平凸形，中央5—11层细胞厚，向边缘渐变薄，边缘锐尖。末端羽枝横切面线形至平凸形，边缘锐尖至翼状，单层细胞翼部0—1个细胞宽。油体棕褐色，聚合型，圆形、卵圆形至椭圆形，表皮细胞、翼部细胞和内层细胞均存在，表皮细胞中并非所有细胞均含有油体，具油体的细胞每个含有1—2个油体，内层细胞中每个细胞含有1—5个油体。假根主要生于主轴贴基质生长的部分以及向地枝上。雌雄异株。雄生殖枝侧生在主轴上或主分枝基部，藏精器2列，5—15对。雌生殖枝侧生在主轴上或主分枝基部，藏卵器2列，基部有流苏鳞片状侧丝。蒴被圆柱形，表面被有2—3个细胞组成的瘤状细胞团。孢子体未成熟时包裹在蒴被内。孢蒴椭球形，成熟时4瓣开裂。孢子球形，表面具颗粒状纹饰。弹丝单螺旋加厚，成束聚生于孢蒴裂瓣顶端。

生境：石生，土生。海拔1823—1829 m。

雅长分布：草王山。

中国分布：全国大部分省区均有分布。

Thalli yellowish green to dark green when fresh, dark brown in herbarium, irregularly 1–3-pinnately branched, prostrate, with geotropic stolons. Epidermal cells slightly smaller than inner cells in transverse section, 1/3–1/2 the inner cells in size. Transverse sections of main axis plano-convex, 5–11 cells thick in middle, become gradually thinner towards margin, acute at margin. Transverse sections of ultimate pinnules linear to plano-convex, acute to winged at margin, unistratose wings 0–1 cell wide. Oil bodies dark-brown, segmented, globose, ovoid to elliptical; oil bodies in epidermal cells scattered, 1–2 per cell; oil bodies in inner cells 1–5 per cell. Rhizoids mainly on prostrate thalli and geotropic stolons. Dioicous. Male branches lateral on main axes or base of primary pinnae, antheridial chambers in 2 regular rows, with 5–15 pairs. Female branches lateral on main axes or base of primary pinnae, archegonia in 2 regular rows, with fringed scale-like paraphyses at base. Calyptrae cylindrical, covered with enlarged, 2–3-celled mass. Sporophyte enclosed by a calyptra before maturity. Capsules ellipsoidal, longitudinally dehiscing by 4 regular valves at maturity. Spores globose, surface granulate. Elaters with 1 spiral thickening, elaterophores adhere to the tips of the capsule valves.

Habitat: On rocks and soil. Elev. 1823–1829 m.

Distribution in Yachang: Caowangshan.

Distribution in China: Widely distributed in most provinces of China.

▶ A. 种群；B. 叶状体；C. 叶状体一段；D. 末羽片横切面；E. 主轴横切面；F. 内部细胞，示油体；G. 表皮细胞及油体；H. 蒴被；I. 裂开的孢蒴；J. 孢子和弹丝。（凭证标本：*韦玉梅等 201108-25*）

A. Population; B. Thallus; C. Portion of thallus; D. Transverse section of ultimate pinnule; E. Transverse section of main axis; F. inner cells with oil bodies; G. Epidermal cells showing oil bodies; H. Calyptrae; I. Dehisced capsule; J. Spores and elaters. (All from *Wei et al. 201108-25*)

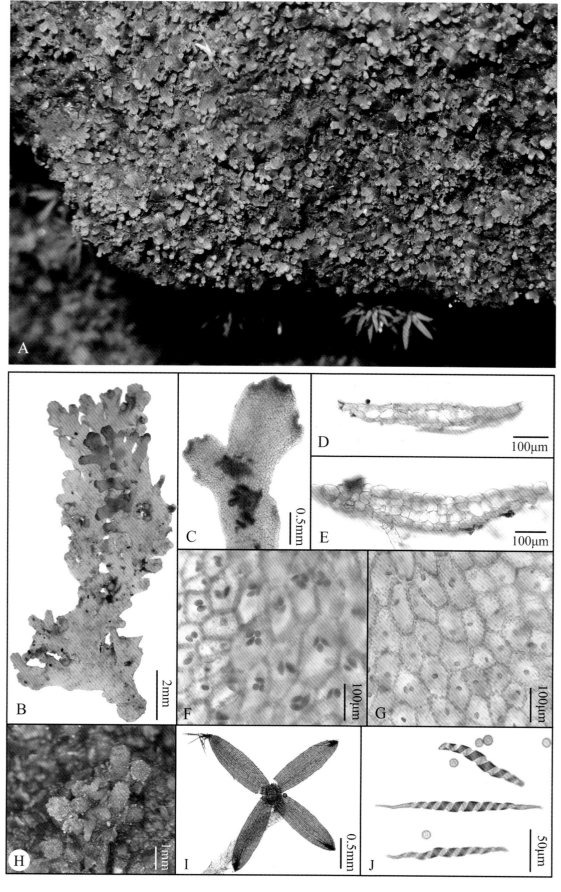

图 172 掌状片叶苔

Fig. 172 *Riccardia palmata* (Hedw.) Carr.

纤细片叶苔

Riccardia pusilla Grolle, J. Jap. Bot. 41(8): 231. 1966.

叶状体新鲜时浅绿色至黄绿色，干燥后灰棕色，2—3回不规则的羽状分枝，匍匐生长，具有向地枝。叶状体横切面表皮细胞小于内层细胞，大小为内层细胞的1/3—1/2。主轴横切面线形至平凸形，中央3—8层细胞厚，向边缘渐变薄，边缘翼状，单层细胞翼部1—3个细胞宽。末端羽枝横切面线形至平凸形，边缘翼状，单层细胞翼部2—5个细胞宽。油体棕褐色，聚合型，圆形至椭圆形，仅内层细胞存在，表皮细胞几乎不含油体，内层细胞中每个细胞含有1—2个油体。假根主要生于主轴贴基质生长的部分以及向地枝上。雌雄生殖枝未见。

生境： 石生，腐木。海拔931—1773 m。

雅长分布： 草王山，深洞。

中国分布： 浙江。首次记录于广西。

Thalli light green to yellowish green when fresh, pale brown in herbarium, irregularly 2–3-pinnately branched, prostrate, with geotropic stolons. Epidermal cells smaller than inner cells in transverse section, 1/3–1/2 the inner cells in size. Transverse sections of main axis linear to plano-convex, 3–8 cells thick in middle, become gradually thinner towards margin, winged at margin, unistratose wings 1–3 cells wide.. Transverse sections of ultimate pinnules linear to plano-convex, winged at margin, unistratose wings 2–5 cells wide. Oil bodies dark-brown, segmented, globose to elliptical; oil bodies rarely scattered in epidermal cells, 1 per cell; oil bodies in inner cells 1–2 per cell. Rhizoids mainly on prostrate thalli and geotropic stolons. Male and female branches not seen.

Habitat: On rocks and rotten logs. Elev. 931–1773 m.

Distribution in Yachang: Caowangshan, Shendong.

Distribution in China: Zhejiang. New to Guangxi.

▶ A. 种群；B. 叶状体；C–D. 叶状体一段；E. 末羽片横切面；F. 主轴横切面；G. 内部细胞，示油体；H. 表皮细胞，示油体缺失。（凭证标本：*韦玉梅等 191017-245*）

A. Population; B. Thallus; C–D. Portions of thallus; E. Transverse section of ultimate pinnule; F. Transverse section of main axis; G. inner cells showing oil bodies; H. Epidermal cells without oil bodies. (All from *Wei et al. 191017-245*)

图 173 纤细片叶苔

Fig. 173 *Riccardia pusilla* Grolle

狭尖叉苔

Metzgeria consanguinea Schiffn., Nova Acta Acad. Caes. Leop.-Carol. German. Nat. Cur. 60(2): 271. 1893.

叶状体淡绿色至黄绿色，叉状分枝，顶端圆钝或形成具芽胞的长狭尖。中肋明显，背腹面均为2个细胞宽，腹面具疏松长毛。翼部单层细胞，10—18个细胞宽，两侧边缘平展或向腹面卷曲，具疏松的单一直长毛。叶翼细胞五边形或六边形，薄壁，三角体不明显。油体未见。雌雄生殖器官未见。

生境： 腐木，树干，树枝。海拔 1643—1823 m。

雅长分布： 草王山，大棚屯。

中国分布： 重庆，甘肃，广西，贵州，湖北，陕西，四川，台湾，香港，云南，浙江。

Thalli light green to yellowish green, furcate, apex obtuse or attenuated forming gemmiferous tip. Midrib sharply defined, 2 cells wide on both dorsal and ventral sides, with sparse hairs on ventral side. Thallus wings unistratose, 10–18 cells wide, margins plane to slightly recurved, ciliated with single-celled straight hairs. Wing cells 5–6 angled, thin-walled, trigones indistinct. Oil bodies not seen. Male and female branches not seen.

Habitat: On rotten logs, tree trunks and tree branches. Elev. 1643–1823 m.

Distribution in Yachang: Caowangshan, Dapeng Tun.

Distribution in China: Chongqing, Gansu, Guangxi, Guizhou, Hongkong, Hubei, Shaanxi, Sichuan, Taiwan, Yunnan, Zhejiang.

▶ A. 种群；B. 叶状体一段，示具长狭尖和芽胞的分枝；C. 叶状体一段（腹面观）；D. 翼部边缘的长毛；E. 叶状体一段（背面观）；F. 芽胞；G. 叶状体横切面；H. 叶状体横切面边缘部分；I. 叶状体横切面中部部分。（凭证标本：*韦玉梅等 201108-17*）

A. Population; B. Portion of thallus showing attenuated branches with gemmae at apex; C. Portion of thallus (ventral view); D. Maginal hairs; E. Portion of thallus (dorsal view); F. Gemmae; G. Transverse section of thallus; H. Transverse section of thallus at marginal part; I. Transverse section of thallus at median part. (All from *Wei et al. 201108-17*)

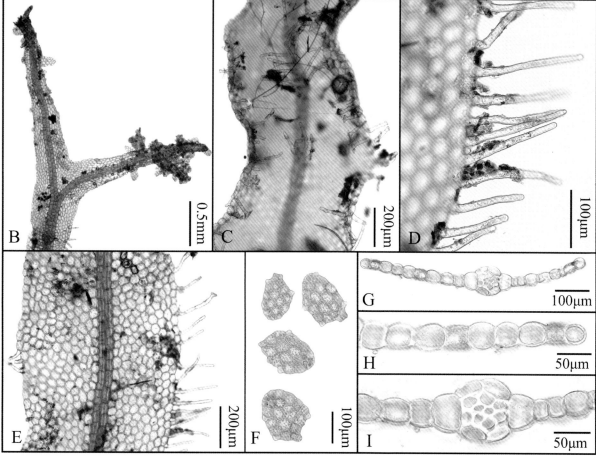

图 174 狭尖叉苔

Fig. 174 *Metzgeria consanguinea* Schiffn.

林氏叉苔（喜马拉雅叉苔）

Metzgeria lindbergii Schiffn., Denkschr. Kaiserl. Akad. Wiss., Math.-Naturwiss. Kl. 67: 182. 1898.

Metzgeria himalayensis Kashyap, J. Bombay Nat. Hist. Soc. 26: 280. 1917.

叶状体淡绿色至黄绿色，叉状分枝，顶端圆钝或心脏形。中肋明显，背面2个细胞宽，腹面2—3个细胞宽，腹面具疏松长毛。翼部单层细胞，12—18个细胞宽，两侧边缘常向腹面卷曲，具单一或成对的直长毛，长毛有时也见于叶状体背面。叶翼细胞五边形或六边形，薄壁，三角体不明显。油体未见。雌雄同株。生殖枝均生于中肋腹面。雄枝内卷球形，外表面平滑，精子器着生于中肋两侧。蒴被梨形，外表面密生长毛，基部有苞膜包围。孢蒴球形，成熟时4瓣纵裂。孢子球形，具颗粒状纹饰。弹丝单螺旋加厚。

生境：潮湿石生，石生，土生，枯枝，树干，树基。海拔1176—1753 m。

雅长分布：草王山，大宴坪天坑漏斗，逻家田屯，盘古王，下棚屯。

中国分布：安徽，福建，广西，贵州，海南，湖南，江西，四川，台湾，云南，浙江。

Thalli light green to yellowish green, furcate, apex obtuse to emarginate. Midrib sharply defined, 2–3 cells wide on both dorsal and ventral sides, with sparse hairs on ventral side. Thallus wings unistratose, 12–18 cells wide, margins usually recurved, with single or paired hairs, hairs straight. Wing cells 5–6 angled, thin-walled, trigones indistinct. Oil bodies not seen. Monoicous. Gametoecia scattered along the ventral side of the midrib. Male branches spherical, midrib, without hair. Calyptrae pyriform, with hairs, surrounded by involucres at base. Capsules spherical, longitudinally dehiscing by 4 regular valves at maturity. Spores globose, surface granulate. Elaters with 1 spiral thickening.

Habitat: On (wet) rocks, soil, dead branches, tree trunks and tree bases. Elev. 1176–1753 m.

Distribution in Yachang: Caowangshan, Dayanping Tiankeng, Luojiatian Tun, Panguwang, Xiapeng Tun.

Distribution in China: Anhui, Fujian, Guangxi, Guizhou, Hainan, Hunan, Jiangxi, Sichuan, Taiwan, Yunnan, Zhejiang.

▶ A. 种群；B. 叶状体一段带雌雄枝（腹面观）；C. 叶状体一段；D. 叶状体横切面；E. 叶状体横切面中部部分；F. 翼部边缘的长毛；G. 弹丝；H. 翼部细胞；I. 孢子。（凭证标本：*韦玉梅等 201105-18*）

A. Population; B. Portion of thallus with male branches and female branch (ventral view); C. Portion of thallus; D. Transverse section of thallus; E. Transverse section of thallus at median part; F. Marginal hairs; G. Elaters; H. Wing cells; I. Spore. (All from *Wei et al. 201105-18*)

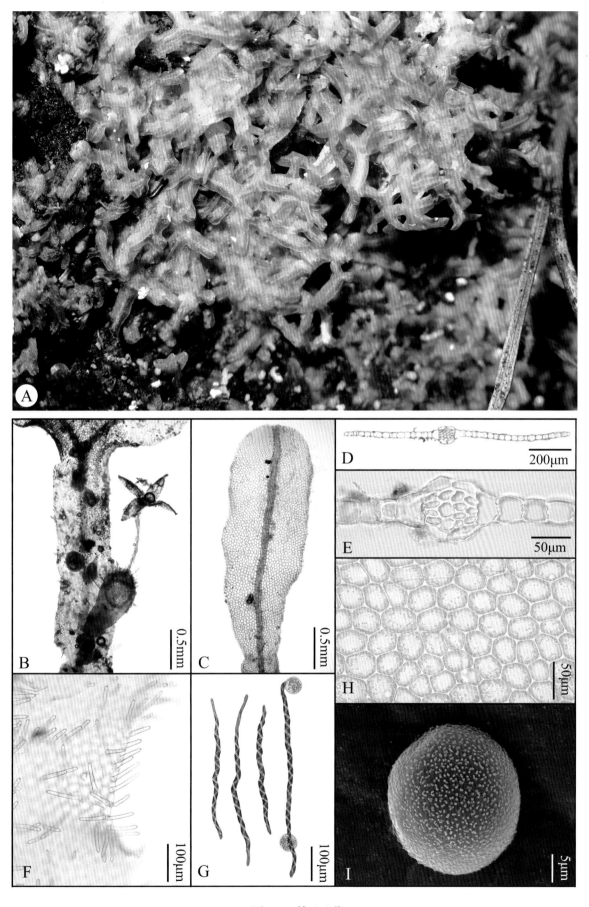

图 175 林氏叉苔

Fig. 175 *Metzgeria lindbergii* Schiffn.

角苔 HORNWORTS

▶ 东亚短角苔　*Notothylas japonica* Horik.

塔拉加角苔（新拟中文名）

Anthoceros telaganus Steph., Sp. Hepat. (Stephani) 5: 1005. 1916.

叶状体丛生成莲座状，背面绿色，边缘全缘或有时具钝齿，内部具黏液腔。叶状体每个表皮细胞具1个不规则形状的叶绿体，叶绿体内含1个蛋白核。念珠藻生于叶状体腹面的黏液腔中。雌雄同株。精子器腔生于叶状体背面表皮下，不规则排列，每个精子器腔中具5—12个精子器，精子器具柄，精子器壁由4层排列规则的细胞构成。苞膜直立，圆筒形。孢子体直立，长角状，明显突出苞膜。孢蒴成熟后纵向两瓣开裂，蒴轴发育良好，孢蒴外壁具分散的气孔。孢子棕色至黑棕色，远极面具刺突状的片层突起，突起的顶部通常具一个圆疣，近极面具刺状的凸起，且这些凸起通常在基部相连，并具明显的三射线，沿着三射线具有一条光滑区域。假弹丝2—3个细胞长。

生境：土生。海拔1216—1235 m。

雅长分布：李家坨屯。

中国分布：广西。

Thalli forming rosettes, dorsally green when fresh, margins entire or somewhat crenulate, with mucilage chambers. Epidermal cells with 1 chloroplast per cell, pyrenoid present. Nostoc colonies scattered within mucilage chambers. Monoicous. Antheridial chambers scattered under the epidermal layer of the dorsal thallus, with 5–12 antheridia per cavity, each antheridium with four-tiered jacket cell arrangement and a long stalk. Involucres erect, cylindrical. Sporophytes erect, long horn-shaped, projecting from involucres. Capsules bivalved, longitudinally dehiscing from apex to base at maturity, with well-developed columella, stomata scattered in epidermal layer of capsule wall. Spores brown to dark brown, distal surface with numerous spinose-lamellate projections which often topped by a papilla, proximal surface with spine-like projections which often confluent at the base, triradiate mark distinct, bordered on both sides by unsculptured strip. Pseudoelaters 2–3 cells long.

Habitat: On soil. Elev. 1216–1235 m.

Distribution in Yachang: Lijiatuo Tun.

Distribution in China: Guangxi.

▶ A. 种群；B. 孢蒴壁横切面；C. 叶状体横切面，示精子器腔；D. 表皮细胞，示叶绿体；E. 精子器；F. 假弹丝；G. 孢蒴外壁细胞，示气孔；H. 孢子近极面观；I. 孢子远极面观。（凭证标本：*韦玉梅等 201116-1*）

A. Population; B. Transverse section of capsule wall; C. Transverse section of thallus showing antheridial chambers; D. Epidermal cells of thallus showing chloroplasts; E. Antheridia; F. Epidermal cells of capsule showing stoma; G. Pseudoelaters; H. Proximal view of spore; I. Distal view of spore. (All from *Wei et al. 201116-1*)

图 176　塔拉加角苔

Fig. 176　*Anthoceros telaganus* Steph.

东亚短角苔

Notothylas javanica (Sande Lac.) Gottsche, Bot. Zeitung (Berlin) Beil. 16: 20. 1858.

叶状体丛生成莲座状，有时叶状体重叠生长，背面绿色，边缘具不规则裂瓣或钝齿，内部不具黏液腔。叶状体每个表皮细胞具1个不规则形状的叶绿体，叶绿体内含一个蛋白核。念珠藻生于叶状体腹面内。雌雄同株。精子器腔生于叶状体背面表皮下，不规则排列，每个精子器腔中具2—4个精子器，精子器具短柄，精子器壁细胞排列不规则。苞膜水平生长或稍微上倾，椭圆形至圆锥形。孢子体水平生长或稍微上倾，隐藏于苞膜或顶端略微突出苞膜。孢蒴成熟后无任何特殊加厚的细胞形成的开裂线，不规则开裂，蒴轴发育，孢蒴外壁不具气孔。孢子黄色，远极面蠕虫状纹饰，近极面蠕虫状纹饰，并具明显的三射线。假弹丝缺。

生境：土生。海拔1153—1185 m。

雅长分布：逻家田屯，悬崖天坑。

中国分布：澳门，广东，广西，台湾，西藏，云南。

Thalli forming rosettes, in well-developed conditions densely overlapped, dorsally green when fresh, margins irregularly lobulated or crenulate, without mucilage chamber. Epidermal cells with 1 chloroplast per cell, pyrenoid present. Nostoc colonies scattered within ventral thallus. Monoicous. Antheridial chambers scattered under the epidermal layer of the dorsal thallus, with 2−4 antheridia per cavity, each antheridium with non-tiered jacket cell arrangement and a short stalk. Involucres horizontal or subhorizontal, oval or conical. Sporophytes horizontal or subhorizontal, embedded in or protruding slightly from involucres. Capsules dehiscing irregularly at maturity, without special dehiscence line, with developed columella, stoma absent. Spores yellow, distal surface vermiculate, proximal surface with similar ornamentation, triradiate mark distinct. Pseudoelater absent.

Habitat: On soil. Elev. 1153−1185 m.

Distribution in Yachang: Luojiatian Tun, Xuanya Tiankeng.

Distribution in China: Guangdong, Guangxi, Macao, Taiwan, Tibet, Yunnan.

▶ A. 种群；B. 孢子体；C. 孢蒴外壁；D. 孢子近极面观；E. 孢子远极面观。（凭证标本：*韦玉梅等 201107-16*）

A. Population; B. Sporophyte; C. Epidermal layer of capsule wall; D. Proximal view of spore; E. Distal view of spore. (All from *Wei et al. 201107-16*)

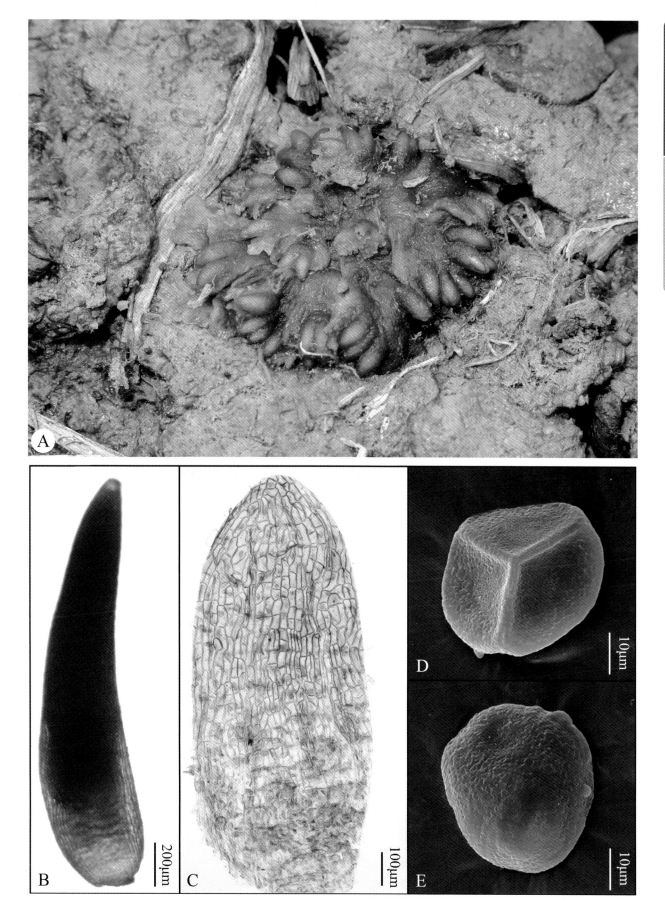

图 177　东亚短角苔

Fig. 177　*Notothylas japonica* (Sande Lac.) Gottsche

短角苔

Notothylas orbicularis (Schwein.) Sull., Amer. J. Sci. Arts (ser. 2) 1(1): 75. 1846.

叶状体丛生成莲座状，背面绿色，边缘具不规则裂瓣，内部不具黏液腔。叶状体每个表皮细胞具1个不规则形状的叶绿体，叶绿体内含一个蛋白核。念珠藻生于叶状体腹面内。雌雄同株。精子器腔着生于叶状体背面表皮下，不规则排列，每个精子器腔中具1—2个精子器，精子器具短柄，精子器壁细胞排列不规则。苞膜水平生长或稍微上倾，椭圆形至圆锥形。孢子体水平生长或稍微上倾，隐藏于苞膜或顶端略微突出苞膜。孢蒴成熟后沿着特殊加厚的细胞形成的开裂线纵裂成2瓣，蒴轴发育，孢蒴外壁不具气孔。孢子黄色至棕色，远极面密集蠕虫状纹饰，近极面密集蠕虫状状纹饰，并具明显的三射线。假弹丝1—4个细胞，细胞壁具不规则螺旋加厚。

生境：潮湿土生。海拔1288 m。

雅长分布：中井屯。

中国分布：广西，黑龙江，湖南，吉林，江苏，辽宁，四川，台湾，云南，浙江。

Thalli forming rosettes, dorsally green when fresh, margins irregularly lobulated, without mucilage chamber. Epidermal cells with 1 chloroplast per cell, pyrenoid present. Nostoc colonies scattered within ventral thallus. Monoicous. Antheridial chambers scattered under the epidermal layer of the dorsal thallus, with 1−2 antheridia per cavity, each antheridium with non-tiered jacket cell arrangement and a short stalk. Involucres horizontal or subhorizontal, oval or conical. Sporophytes horizontal or subhorizontal, embedded in or protruding slightly from involucres. Capsules dehiscing along special dehiscence line consisting of thick-walled cells at maturity, with developed columella, stoma absent. Spores yellow to brown, distal surface densely vermiculate, proximal surface with similar ornamentation, triradiate mark distinct. Pseudoelaters 1−4 cells long, with irregularly spiral thickenings..

Habitat: On wet soil. Elev. 1288 m.

Distribution in Yachang: Zhongjing Tun.

Distribution in China: Guangxi, Heilongjiang, Hunan, Jilin, Jiangsu, Liaoning, Sichuan, Taiwan, Yunnan, Zhejiang.

A. 种群；B. 孢蒴；C. 孢蒴外壁，示开裂线；D. 假弹丝；E. 孢子近极面观；F. 孢子远极面观。（凭证标本：*唐启明等 20201105-79*）

A. Population; B. Capsule; C. Epidermal layer of capsule wall showing the special dehiscence line; D. Pseudoelaters; E. Proximal view of spore; F. Distal view of spore. (All from *Tang et al. 20201105-79*)

图 178 短角苔
Fig. 178 *Notothylas orbicularis* (Schwein.) Sull.

高领黄角苔

Phaeoceros carolinianus (Michx.) Prosk., Bull. Torrey Bot. Club 78(4): 347. 1951.

叶状体成片生长或形成莲座状，背面绿色至深绿色，边缘具不规则裂瓣，内部不具黏液腔。叶状体每个表皮细胞具1个的叶绿体，叶绿体内含一个蛋白核。念珠藻生于叶状体腹面内。雌雄同株。精子器腔着生于叶状体背面表皮下，不规则排列，每个精子器腔中具2—4个精子器，精子器具柄，精子器壁细胞不规则排列。苞膜直立，圆筒形。孢子体直立，长角状，明显突出苞膜。孢蒴长可达5 cm，成熟后顶端呈黄色，纵向两瓣开裂，干时扭曲，蒴轴发育良好，孢蒴外壁具分散的气孔。孢子黄色，远极面刺状或乳头状纹饰，近极面具明显的三射线，每个面中央具颗粒状纹饰。假弹丝2—4个细胞，常分叉，有时具狭窄的不规则带状加厚。

生境： 土生环境常见，石生、潮湿土生、岩面薄土生环境也有。海拔620—1553 m。

雅长分布： 草王山，大宴坪天坑漏斗，二沟，九十九堡，拉雅沟，李家坨屯，落花生屯，逻家田屯，旁墙屯，全达村，山干屯，塘英村，下棚屯，下岩洞屯，香当天坑。

中国分布： 福建，贵州，湖北，湖南，陕西，台湾，云南。首次记录于广西。

Thalli forming patches or rosettes, dorsally green to dark green when fresh, margins irregularly lobulated, without mucilage chamber. Epidermal cells with 1 chloroplast per cell, pyrenoid present. Nostoc colonies scattered within ventral thallus. Monoicous. Antheridial chambers scattered under the epidermal layer of the dorsal thallus, with 2–4 antheridia per cavity, each antheridium with non-tiered jacket cell arrangement and a stalk. Involucres erect, cylindrical. Sporophytes erect, long horn-shaped, projecting from involucres. Capsules long, up to 5 cm in length, yellow at the tip when mature, bivalved, longitudinally dehiscing from apex to base at maturity, becoming twisted when dry, with well-developed columella, stomata scattered in epidermal layer of capsule wall. Spores yellow, distal surface spinulate to papillate, proximal surface granulate confined to the center of each triradiate face, triradiate mark distinct. Pseudoelaters 2–4 cells long, frequently branched, sometimes with irregular bands of thickenings.

Habitat: Often on soil, sometimes on rocks, wet soil and on rocks with a thin layer of soil. Elev. 620–1553 m.

Distribution in Yachang: Caowangshan, Dayanping Tiankeng, Ergou, Jiushijiubao, Layagou, Lijiatuo Tun, Luohuasheng Tun, Luojiatian Tun, Pangqiang Tun, Quanda Village, Shangan Tun, Tangying Village, Xiapeng Tun, Xiayandong Tun, Xiangdang Tiankeng.

Distribution in China: Fujian, Guizhou, Hubei, Hunan, Shaanxi, Taiwan, Yunnan. New to Guangxi.

A. 种群；B. 叶状体横切面；C. 叶状体表皮细胞，示叶绿体；D. 孢蒴外壁细胞，示气孔；E. 精子器腔；F. 精子器；G. 假弹丝；H. 孢子近极面观；I. 孢子远极面观。（凭证标本：*韦玉梅等 201105-11*）

A. Population; B. Transverse section of thallus; C. Epidermal cells of thallus showing chloroplasts; D. Epidermal cells of capsule showing stoma; E. Antheridial chambers with Antheridia; F. Antheridia; G. Pseudoelaters; H. Proximal view of spore; I. Distal view of spore. (All from *Wei et al. 201105-11*)

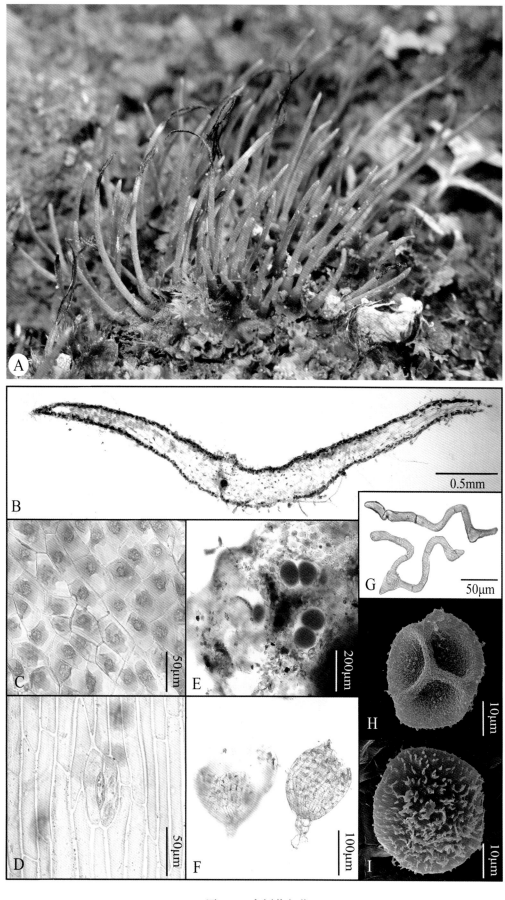

图 179 高领黄角苔

Fig. 179 *Phaeoceros carolinianus* (Michx.) Prosk

小黄角苔

Phaeoceros exiguus (Steph.) J. Haseg., J. Hattori Bot. Lab. 60: 387. 1986.

叶状体丛生成莲座状，背面绿色至深绿色，边缘平滑，内部不具黏液腔。叶状体每个表皮细胞具1个的叶绿体，叶绿体内含一个蛋白核。念珠藻生于叶状体腹面的黏液细胞内。雌雄同株。精子器腔着生于叶状体背面表皮下，不规则排列，每个精子器腔中具2—5个精子器，精子器具柄，精子器壁细胞不规则排列。苞膜直立，圆筒形。孢子体直立，长角状，明显突出苞膜。孢蒴短，一般不超过1 cm，成熟后顶端呈黄色，纵向两瓣开裂，蒴轴发育良好，孢蒴外壁具分散的气孔。孢子黄色，远极面分散的乳头状纹饰，还具有一些粗糙的、较大的圆形隆起物分散在其中，近极面密集的乳头状纹饰，并具明显的三射线。假弹丝2—4个细胞，有时具狭窄的不规则带状加厚。

生境：土生。海拔1185—1376 m。

雅长分布：盘古王，悬崖天坑，悬崖屯。

中国分布：广西，台湾。

Thalli forming rosettes, dorsally green to dark green when fresh, margins entire, without mucilage chamber. Epidermal cells with 1 chloroplast per cell, pyrenoid present. Nostoc colonies scattered within ventral thallus. Monoicous. Antheridial chambers scattered under the epidermal layer of the dorsal thallus, with 2−5 antheridia per cavity, each antheridium with non-tiered jacket cell arrangement and a stalk. Involucres erect, cylindrical. Sporophytes erect, long horn-shaped, projecting from involucres. Capsules short, not exceeding to 1 cm in length, yellow at the tip when mature, bivalved, longitudinally dehiscing from apex to base at maturity, not twisted when dry, with well-developed columella, stomata scattered in epidermal layer of capsule wall. Spores yellow, distal surface distal surface loosely verrucate, with coarse, larger protuberances scattered in between smaller verrucae, proximal surface densely verrucate, triradiate mark distinct. Pseudoelaters 2−4 cells long, sometimes with irregular bands of thickenings.

Habitat: On soil. Elev. 1185−1376 m.

Distribution in Yachang: Panguwang, Xuanya Tiankeng, Xuanya Tun.

Distribution in China: Guangxi, Taiwan.

▶ A. 种群；B. 叶状体横切面中部部分；C. 孢蒴外壁细胞，示气孔；D. 精子器；E. 叶状体表皮细胞，示叶绿体；F. 精子器腔；G. 假弹丝；H. 孢子近极面观；I. 孢子远极面观。（凭证标本：韦玉梅等 201110-7）

A. Population; B. Transverse section of thallus at median part; C. Epidermal cells of capsule showing stoma; D. Antheridium; E. Epidermal cells of thallus showing chloroplasts; F. Antheridial chamber; G. Pseudoelaters; H. Proximal view of spore; I. Distal view of spore. (All from *Wei et al. 201110-7*)

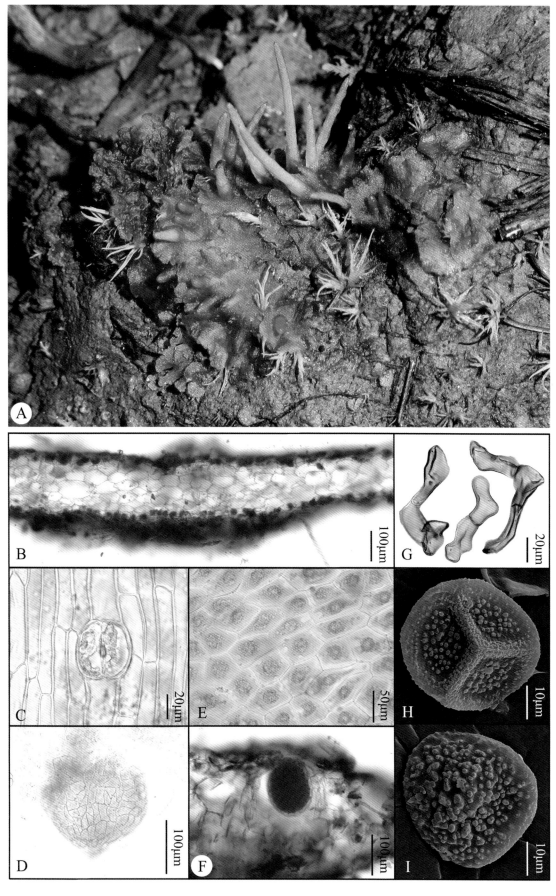

图 180 小黄角苔

Fig. 180 *Phaeoceros exiguus* (Steph.) J. Haseg.

东亚大角苔

Megaceros flagellaris (Mitt.) Steph., Sp. Hepat. 5: 951. 1916.

叶状体成片生长，背面绿色至深绿色，边缘平滑或具鹿角状小裂片，内部不具黏液腔。叶状体每个表皮细胞具1—5个的叶绿体，叶绿体内不含蛋白核。念珠藻生于叶状体腹面内。雌雄同株。精子器腔着生于叶状体背面表皮下，不规则排列，每个精子器腔中具1—3个精子器，精子器具短柄，精子器壁细胞不规则排列。苞膜直立，圆筒形。孢子体直立，长角状，明显突出苞膜。孢蒴成熟后纵向两瓣开裂，裂瓣干时扭曲，蒴轴发育良好，孢蒴外壁不具气孔。孢子新鲜时绿色，干时无色至浅棕色，远极面密集乳头状凸起纹饰，近极面疏松的乳头状凸起纹饰，三射线发育不好，但明显。假弹丝单螺旋加厚。

生境：石生。海拔1635 m。

雅长分布：盘古王。

中国分布：安徽，福建，广东，广西，贵州，海南，湖南，江西，四川，台湾，香港，云南。

Thalli forming patches, dorsally green to dark green when fresh, margins entire or irregularly lobulated, without mucilage chamber. Epidermal cells with 1–5 chloroplasts per cell, pyrenoid absent. Nostoc colonies scattered within ventral thallus. Monoicous. Antheridial chambers scattered under the epidermal layer of the dorsal thallus, with 1–3 antheridia per cavity, each antheridium with non-tiered jacket cell arrangement and a short stalk. Involucres erect, cylindrical. Sporophytes erect, long horn-shaped, projecting from involucres. Capsules bivalved, longitudinally dehiscing from apex to base at maturity, becoming twisted when dry, with well-developed columella, without stoma. Spores green when fresh, hyaline to pale brown when dry, distal surface densely verrucate, proximal surface loosely verrucate, triradiate mark poorly developed, but distinct. Pseudoelaters with 1 spiral thickening.

Habitat: On rocks. Elev. 1635 m.

Distribution in Yachang: Panguwang.

Distribution in China: Anhui, Fujian, Guangdong, Guangxi, Guizhou, Hainan, Hongkong, Hunan, Jiangxi, Sichuan, Taiwan, Yunnan.

▶ A. 种群；B. 叶状体；C. 弹丝；D. 叶状体一段，示分散的精子器腔；E. 叶状体横切面中部部分；F. 孢蒴外壁细胞；G. 叶状体表皮细胞，示叶绿体；H. 孢子近极面观；I. 孢子远极面观。（凭证标本：*专玉梅等 201110-40*）

A. Population; B. Thallus; C. Elaters; D. Portion of thallus showing scattered antheridial chambers; E. Transverse section of thallus at median part; F. Epidermal cells of capsule; G. Epidermal cells of thallus showing chloroplasts; H. Proximal view of spore; I. Distal view of spore. (All from *Wei et al. 201110-40*)

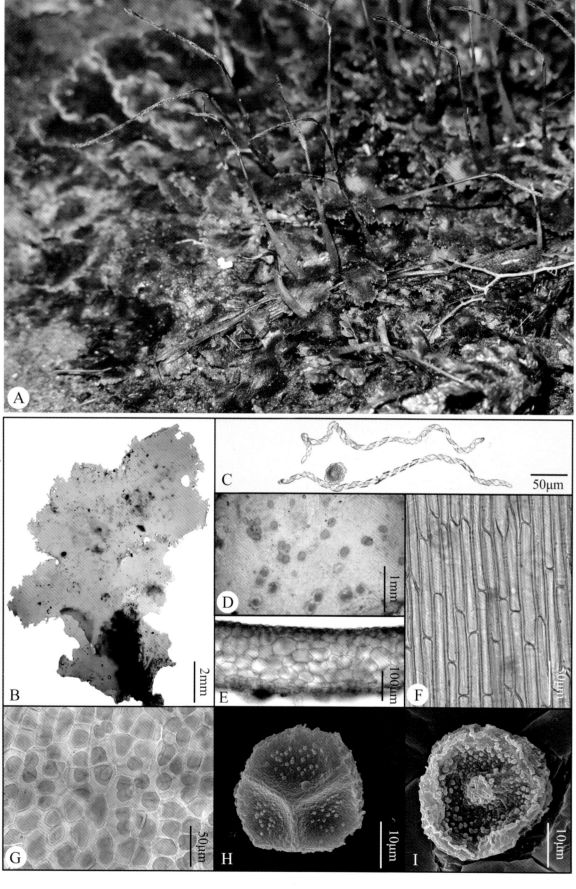

图 181 东亚大角苔

Fig. 181 *Megaceros flagellaris* (Mitt.) Steph.

致 谢
ACKNOWLEDGMENTS

感谢华东师范大学朱瑞良教授在物种鉴定和文稿修改上提供的宝贵意见，感谢深圳市仙湖植物园张力研究员和贵阳生产力促进中心韩国营副研究员在文稿的编排上提供的宝贵意见，感谢新加坡植物园 Lily M. J. Chen 陈美君老师在文稿的设计上提供的宝贵意见。感谢广西植物所刘演研究员、许为斌研究员以及陈海玲助理研究员为本书提供的精美图片素材以及在本书开展野外考察和编研过程中给予的各种帮助。

We thank Professor Rui-Liang Zhu from the East China Normal University for his valuable advices on species identification and manuscript revision. Thanks are extended to researcher Li Zhang of the Fairy Lake Botanical Garden, Shenzhen, and associate researcher Guo-Ying Han of the Productivity Promotion Center, Guiyang, for their recomendations on the layout of the manuscript. We are grateful to Lily M. J. Chen from the Singapore Botanic Gardens for her helpful suggestions to improve on the design of this book. Appreciation is given to researcher Yan Liu, researcher Wei-Bin Xu and assistant researcher Hai-Ling Chen, all from the Guangxi Institute of Botany, for providing exquisite photographs and giving their full support to the field investigations and administrative facilitation of this project.

主要参考文献
THE MAIN REFERENCES

白学良, 赵东平. 2018. 五大连池火山岩溶地貌苔藓植物彩图志[M]. 呼和浩特: 内蒙古大学出版社.
BAI X L, ZHAO D P. 2018. Bryophytes of Wudalianchi Volcano Region in Colour[M]. Huhehaote: Inner Mongolia University Press.

陈凤彬, 吴淑玉, 雷平. 2020. 江西苔藓植物名录[M]. 南昌: 红星电子音像出版社.
CHEN F B, WU S Y, LEI P. 2020. Bryophyte Checklist of Jiangxi Province, China[M]. Nanchang: Hongxing Electronic Audio Video Publishing House.

高谦. 2003. 中国苔藓志: 第九卷[M]. 北京: 科学出版社.
GAO C. 2003. Flora Bryophytarum Sinicorum: Vol. 9[M]. Beijing: Science Press.

高谦, 曹同. 2000. 云南植物志: 第十七卷[M]. 北京: 科学出版社.
GAO C, CAO T. 2000. Flora Yunnanica: Vol. 17[M]. Beijing: Science Press.

高谦, 吴玉环. 2010. 中国苔纲和角苔纲植物属志[M]. 北京: 科学出版社.
GAO C, WU Y H. 2010. Genera Hepaticopsida et Anthocerotopsida Sinicorum [M]. Beijing: Science Press.

广西林业勘测设计院. 2007. 广西雅长兰科植物自治区级自然保护区综合科学考察报告（油印稿）[R]. 南宁: 广西林业勘测设计院.
GUANGXI ZHUANG AUTONOMOUS REGION FORESTRY RECONNAISSANCE DESIGN INSTITUTE. 2007. The Comprehensive Investigation Report of Guangxi Yachang Orchids Natural Reserve (Mimeograph)[R]. Nanning: Guangxi Zhuang Autonomous Region Forestry Reconnaissance Design Institute.

贾渝, 何思. 2013. 中国生物物种名录: 第一卷植物 苔藓植物[M]. 北京: 科学出版社.
JIA Y, HE S. 2013. Species Catalogue of China: Vol. 1 Plants Bryophytes[M]. Beijing: Science Press.

马文章, 喻智勇. 2020. 云南金平分水岭苔藓植物野外识别手册[M]. 昆明: 云南美术出版社.
MA W Z, YU Z Y. 2020. Field Guide to the Bryophytes of Jinping Fenshuiling, Yunnan[M]. Kunming: Yunnan Fine Arts Publishing House.

师雪芹, 王健. 2021. 安徽省苔藓植物名录[J]. 生物多样性, 29(6): 798-804.

SHI X Q, WANG J. 2021. Bryophyte checklist of Anhui Province, China[J]. Biodiversity Science, 29(6): 798−804.

王立松, 贾渝, 张宪春, 等. 2018. 中国生物物种名录: 第一卷植物 总名录（上册）[M]. 北京: 科学出版社.
WANG L S, JIA Y, ZHANG X C, et al. 2018. Species Catalogue of China: Vol. 1 Plants A Synoptic Checklist (I)[M]. Beijing: Science Press.

吴德邻, 张力. 2013. 广东苔藓志[M]. 广州: 广东科技出版社.
WU D L, ZHANG L. 2013. Bryophyte Flora of Guangdong[M]. Guangzhou: Guangdong Science and Technology Press.

熊源新, 曹威. 2018. 贵州苔藓志: 第三卷[M]. 贵阳: 贵州科技出版社.
XIONG Y X, CAO W. 2018. Bryophyte Flora of Guizhou: Vol. 3[M]. Guiyang: Guizhou Science and Technology Publishing House.

AKIYAMA H, ODRZYKOSKI I J. 2020. Phylogenetic re-examination of the genus *Conocephalum* Hill. (Marchantiales: Conocephalaceae)[J]. Bryophyte Diversity and Evolution, 42(1): 1−18.

ASTHANA A K, SRIVASTAVA S C. 1991. Indian hornworts (a taxonomic study)[J]. Bryophytorum Bibliotheca, 42: 1−158.

BAKALIN V A, KLIMOVA K G, NGUYEN V S. 2020. A review of *Calypogeia* (Marchantiophyta) in the eastern Sino-Himalaya and Meta-Himalaya based mostly on types[J]. PhytoKeys, 153: 111−154.

BISCHLER H. Marchantia L. 1989. the Asiatic and Oceanic taxa[J]. Bryophytorum Bibliotheca, 38: 1−317.

CAO W, XIONG Y X, ZHAO D G, et al. 2019. *Porella perrottetiana* var. *angustifolia* (Porellaceae, Marchantiophyta), new to China, and taxonomic revision of *Porella perrottetiana* in the East Yunnan-Guizhou Plateau[J]. Herzogia, 32(2): 344−356.

CHANTANAORRAPINT S, SRIDITH K. 2014. The genus *Plagiochasma* (Aytoniaceae, Marchantiopsida) in Thailand[J]. Cryptogamie, Bryologie, 35(2): 127−132.

FREY W, STECH M, FISCHER E, 2009. Bryophytes and seedless vascular plants. 3: I−IX + 419 pp. In FREY W, Syllabus of Plant Families. Adolf Engler's Syllabus der Pflanzenfamilien, 13th edition[M]. Berlin, Stuttgart: Gebr. Borntraeger Verlagsbuchhandlung.

FURUKI T. 1991. A taxonomical revision of the Aneuraceae (Hepaticae) of Japan[J]. Journal of the

Hattori Botanical Laboratory, 70: 293−397.

GAO C, WU Y H. 2005. *Radula stellatogemmipara* (Radulaceae, Hepaticae), a new species from Fujian and Guangxi, China[J]. Nova Hedwigia, 80(1−2): 237−240.

HAN G Y, ZHAO Z T. 2020. The liverworts and hornworts of Gansu Province, Northwest China[J]. Chenia, 14: 169−179.

HASEGAWA J. 1998. Taxonomic status of Anthoceros subbrevis Haseg. and taxonomy of its allied taxa[J]. Bryological Research, 7: 173−177.

HATTORI S. 1981. Notes on the Asiatic species of the genus *Frullania*, Hepaticae. XⅢ[J]. Journal of the Hattori Botanical Laboratory, 49: 147−168.

JIA Y, HE Q, LI F X, et al. 2016. A newly updated and annotated checklist of the Anthocerotae and Hepaticae of Qinling Mts., China[J]. Journal of Bryology, 38(4): 312−326.

KITAGAWA N. 1966. A revision of the family Lophoziaceae of Japan and its adjacent regions. II.[J]. Journal of the Hattori Botanical Laboratory, 29: 101−149.

KRAYESKY D M, CRANDALL-STOTLER B J, STOTLER R E. 2005. A revision of the genus *Fossombronia* Raddi in East Asia and Oceania[J]. Journal of the Hattori Botanical Laboratory, 98: 1−45.

LEE G E, GRADSTEIN S R. 2021. Guide to the Genera of Liverworts and Hornworts of Malaysia[M]. Tokyo: Hattori Botanical Laboratory.

LONG D G. 2006. Revision of the genus *Asterella* P. Beauv. in Eurasia[J]. Bryophytorum Bibliotheca, 63: 1−299.

PENG T, ZHU R L. 2014. A revision of the genus *Notothylas* (Notothyladaceae, Anthocerotophyta) in China[J]. Phytotaxa, 156(3): 156−164.

SO M L. 2001. *Plagiochila* (Hepaticae, Plagiochilaceae) in China[J]. Systematic Botany, 60: 1−214.

SÖDERSTRÖM L, HAGBORG A, VAN KONRAT M, et al. 2016. World checklist of hornworts and liverworts[J]. PhytoKeys, 59: 1−828.

SRIVASTAVA S C, DIXIT R. 1996. The genus *Cyathodium* Kunze[J]. Journal of the Hattori Botanical Laboratory, 80: 149−215.

SUKKHARAK P. 2018. A revision of the genus *Frullania* (Marchantiophyta: Frullaniaceae) in Thailand[J]. Nova Hedwigia, 106(1–2): 115–207.

SUKKHARAK P. 2022. Liverwort Genera of Thailand[M]. Chiang Mai: Chiang Mai University Press.

SZWEYKOWSKI J, BUCZKOWSKA K, ODRZYKOSKI I J. 2005. *Conocephalum salebrosum* (Marchantiopsida, Conocephalaceae) – a new Holarctic liverwort species[J]. Plant Systematics and Evolution, 253: 133-158.

WANG J, ZHU R L, GRADSTEIN S R. 2016. Taxonomic revision of Lejeuneaceae subfamily Ptychanthoideae (Marchantiophyta) in China[J]. Bryophytorum Bibliotheca, 65: 1-141.

WEI Y M, TANG Q M, HO B C, et al. 2018. An annotated checklist of the bryophytes of Guangxi, China[J]. Chenia, 13(Special Issue No.1): 1-132.

WIGGINTON M J E W. 2004. Jone's Liverowrt and Hornwort Flora of West Africa[M]. Belgium: Natonal Botanic Garden of Belgium.

XIANG Y L, ZHANG Z X, CHEN S W, et al. 2022. Morphological and molecular evidence confirms a new species, *Riccia subcrinita* YouL.Xiang & R.L.Zhu and *Riccia junghuhniana* Nees & Lindenb. (Ricciaceae, Marchantiophyta) new to China[J]. Phytotaxa, 531(1): 41-53.

YAMADA K. 1979. A revision of Asian taxa of *Radula*, Hepaticae[J]. Journal of the Hattori Botanical Laboratory, 45: 201-322.

YANG J D. 2009. Liverworts and Hornworts of Taiwan I. Lejeuneaceae[M]. Nantou: Endemic Species Research Istitute.

YANG J D. 2011. Liverworts and Hornworts of Taiwan II.[M]. Nantou: Endemic Species Research Istitute.

YANG J D, YAO K Y, LIN S H. 2013. Two species of *Colura* (family Lejeuneaceae) newly recorded to Taiwan[J]. Taiwan Journal of Biodiversity, 15(4): 331-341.

YU X J, LIU X F, HONG L, et al. 2020. A new checklist of bryophytes in Hubei Province, China[J]. Chenia, 14: 180-224.

YUZAWA Y. 2001. Species of the genus *Frullania* (Family Frullaniaceae, Hepaticae) in Japan. II.[J]. Natural and Environmental Science Research, 14: 1-47.

ZHU R L, GRADSTEIN S R. 2005. Monograph of *Lopholejeunea* (Lejeuneaceae, Hepaticae) in Asia[J]. Systematic Botany, 74: 1-98.

ZHU R L, SO M L. 2001. Epiphyllous liverworts of China[J]. Nova Hedwigia Beihefte, 121: 1-418.

凭证标本
VOUCHER SPECIMENS

1. 十字花萼苔 Asterella cruciata (Steph.) Horik.

凭证标本：拉雅沟，*唐启明 & 张仕艳 20180930-35*；盘古王，*唐启明等 20201110-261*；二沟，*唐启明等 20201109-255*；一沟，*韦玉梅等 191014-143*。

Representative specimens examined: Layagou, *Tang & Zhang 20180930-35*; Panguwang, *Tang et al. 20201110-261*; Ergou, *Tang et al. 20201109-255*; Yigou, *Wei et al. 191014-143*.

2. 加萨花萼苔 Asterella khasyana (Griff.) Grolle

凭证标本：草王山，*唐启明 & 张仕艳 20181002-269B*，*唐启明 & 张仕艳 20181002-261*，*唐启明 & 韦玉梅 20191013-96*。

Representative specimens examined: Caowangshan, *Tang & Zhang 20181002-269B*, *Tang & Zhang 20181002-261*, *Tang & Wei 20191013-96*.

3. 瓦氏花萼苔 Asterella wallichiana (Lehm. et Lindenb.) Grolle

凭证标本：草王山，*唐启明 & 张仕艳 20181002-271*；李家坨屯，*唐启明 & 韦玉梅 20191018-303*；里郎天坑，*唐启明等 20210723-12*；逻家田屯，*韦玉梅等 191020-403*；盘古王，*唐启明等 20201110-264*；深洞，*韦玉梅等 191017-237*；下岩洞屯，*唐启明 & 张仕艳 20181001-197*；中井天坑，*唐启明 & 张仕艳 20181007-590*；中井屯，*唐启明 & 张仕艳 20181007-594B*。

Representative specimens examined: Caowangshan, *Tang & Zhang 20181002-271*; Lijiatuo Tun, *Tang & Wei 20191018-303*; Lilang Tiankeng, *Tang et al. 20210723-12*; Luojiatian Tun, *Wei et al. 191020-403*; Panguwang, *Tang et al. 20201110-264*; Shendong, *Wei et al. 191017-237*; Xiayandong Tun, *Tang & Zhang 20181001-197*; Zhongjing Tiankeng, *Tang & Zhang 20181007-590*; Zhongjing Tun, *Tang & Zhang 20181007-594B*.

4. 钝鳞紫背苔 Plagiochasma appendiculatum Lehm. et Lindenb.

凭证标本：白岩坨屯，*唐启明 & 韦玉梅 20191020-480*；草王山，*唐启明 & 张仕艳 20181002-267*；大宴坪竖井，*唐启明 & 韦玉梅 20191017-267*；黄猄洞天坑，*唐启明 20180612-325A1*；拉雅沟，*唐启明 & 张仕艳 20180930-93A*；蓝家湾天坑，*唐启明 & 韦玉梅 20191019-330*；盘古王，*唐启明等 20201110-295*；下岩洞屯，*唐启明 & 张仕艳 20181001-195*；中井天坑，*唐启明 & 张仕艳 20181007-547*。

Representative specimens examined: Baiyantuo Tun, *Tang & Wei 20191020-480*; Caowangshan, *Tang & Zhang 20181002-267*; Dayanping Tun, *Tang & Wei 20191017-267*; Huangjingdong Tiankeng, *Tang 20180612-325A1*; Layagou, *Tang & Zhang 20180930-93A*; Lanjiawan Tiankeng, *Tang & Wei 20191019-330*; Panguwang, *Tang et al. 20201110-295*; Xiayandong Tun, *Tang & Zhang 20181001-195*; Zhongjing Tiankeng, *Tang & Zhang 20181007-547*.

5. 无纹紫背苔 Plagiochasma intermedium Lindenb. et Gottsche

凭证标本：草王山，韦玉梅等 191013-99；大宴坪天坑漏斗，唐启明等 20210718-117；吊井天坑，唐启明等 20210722-65；黄猄洞天坑，唐启明等 20210717-47；蓝家湾天坑，韦玉梅等 191018-297；老屋基天坑，唐启明等 20210721-108；里郎天坑，唐启明等 20210723-2；盘古王，唐启明等 20210716-11；悬崖天坑，唐启明等 20210721-54；中井屯，韦玉梅等 191016-219。

Representative specimens examined: Caowangshan, *Wei et al. 191013-99*; Dayanping Tiankeng, *Tang et al. 20210718-117*; Diaojing Tiankeng, *Tang et al. 20210722-65*; Huangjingdong Tiankeng, *Tang et al. 20210717-47*; Lanjiawan Tiankeng, *Wei et al. 191018-297*; Laowuji Tiankeng, *Tang et al. 20210721-108*; Lilang Tiankeng, *Tang et al. 20210723-2*; Panguwang, *Tang et al. 20210716-11*; Xuanya Tiankeng, *Tang et al. 20210721-54*; Zhongtuo Tun, *Wei et al. 191016-219*.

6. 石地钱 Reboulia hemisphaerica (L.) Raddi

凭证标本：吊井天坑，唐启明等 20210722-21；黄猄洞天坑，唐启明等 20201104-11。

Representative specimens examined: Diaojing Tiankeng, *Tang et al. 20210722-21*; Huangjingdong Tiankeng, *Tang et al. 20201104-11*.

7. 小蛇苔 Conocephalum japonicum (Thunb.) Grolle

凭证标本：吊井天坑，唐启明等 20210722-20；李家坨屯，唐启明 & 韦玉梅 20191018-308；逻家田屯，唐启明 & 张仕艳 20181003-348；旁墙屯，韦玉梅等 191018-276；全达村，唐启明等 20201108-230；山干屯，唐启明 20190522-443；中井屯，唐启明 & 韦玉梅 20191016-218。

Representative specimens examined: Diaojing Tiankeng, *Tang et al. 20210722-20*; Lijiatuo Tun, *Tang & Wei 20191018-308*; Luojiatian Tun, *Tang & Zhang 20181003-348*; Pangqiang Tun, *Wei et al. 191018-276*; Quanda Village, *Tang et al. 20201108-230*; Shangan Tun, *Tang 20190522-443*; Zhongjing Tun, *Tang & Wei 20191016-218*.

8. 暗色蛇苔 Conocephalum salebrosum Szweyk.

凭证标本：草王山，唐启明等 20210719-29；里郎天坑，唐启明等 20210723-42；盘古王，唐启明等 20201110-259；中井屯霄罗湾洞穴，唐启明 20190515-33。

Representative specimens examined: Caowangshan, *Tang et al. 20210719-29*; Lilang Tiankeng, *Tang et al. 20210723-42*; Panguwang, *Tang et al. 20201110-259*; Xiaoluowan Cave, *Tang 20190515-33*.

9. 楔瓣地钱东亚亚种 Marchantia emarginata subsp. tosana (Steph.) Bischl.

凭证标本：白岩坨屯，唐启明 & 韦玉梅 20191020-486；草王山，唐启明 & 张仕艳 20181002-252A；吊井天坑，唐启明等 20210722-34；拉雅沟，唐启明 & 韦玉梅 20191014-161；逻家田屯，唐启明 20190521-420；二沟，唐启明等 20210715-9。

Representative specimens examined: Baiyantuo Tun, *Tang & Wei 20191020-486*; Caowangshan, *Tang & Zhang 20181002-252A*; Diaojing Tiankeng, *Tang et al. 20210722-34*; Layagou, *Tang & Wei 20191014-161*;

Luojiatian Tun, *Tang 20190521-420*; Ergou, *Tang et al. 20210715-9*.

10. 粗裂地钱风兜亚种 Marchantia paleacea subsp. diptera (Nees et Mont.) Inoue

凭证标本：吊井天坑，*唐启明等 20210722-24*；拉雅沟，*唐启明 & 张仕艳 20180930-103*；逻家田屯，*唐启明 & 韦玉梅 20191020-477*；中井屯，*唐启明 & 韦玉梅 20191016-216*。

Representative specimens examined: Diaojing Tiankeng, *Tang et al. 20210722-24*; Layagou, *Tang & Zhang 20180930-103*; Luojiatian Tun, *Tang & Wei 20191020-477*; Zhongjing Tun, *Tang & Wei 20191016-216*.

11. 地钱土生亚种 Marchantia polymorpha subsp. ruderalis Bischl. et Boissel.-Dub.

凭证标本：草王山，*韦玉梅等 221112-7*；全达村，*唐启明等 20201108-229*。

Representative specimens examined: Caowangshan, *Wei et al. 221112-7*; Quanda Village, *Tang et al. 20201108-229*.

12. 毛地钱 Dumortiera hirsuta (Sw.) Nees

凭证标本：吊井天坑，*唐启明等 20210722-53*；拉雅沟，*唐启明 & 张仕艳 20180930-27*；李家坨屯，*唐启明 & 韦玉梅 20191018-291A*；里郎天坑，*唐启明等 20210723-30*；逻家田屯，*唐启明 & 韦玉梅 20191020-461*；二沟，*唐启明等 20201109-253*；中井天坑，*唐启明 & 张仕艳 20181007-567*。

Representative specimens examined: Diaojing Tiankeng, *Tang et al. 20210722-53*; Layagou, *Tang & Zhang 20180930-27*; Lijiatuo Tun, *Tang & Wei 20191018-291A*; Lilang Tiankeng, *Tang et al. 20210723-30*; Luojiatian Tun, *Tang & Wei 20191020-461*; Ergou, *Tang et al. 20201109-253*; Zhongjing Tiankeng, *Tang & Zhang 20181007-567*.

13. 单月苔 Monosolenium tenerum Griff.

凭证标本：拉雅沟，*唐启明 & 韦玉梅 20191014-139*；二沟，*唐启明等 20210715-3*；一沟，*韦玉梅等 191014-145*；中井天坑，*唐启明 & 张仕艳 20181007-574*。

Representative specimens examined: Layagou, *Tang & Wei 20191014-139*; Ergou, *Tang et al. 20210715-3*; Yigou, *Wei et al. 191014-145*; Zhongjing Tiankeng, *Tang & Zhang 20181007-574*.

14. 光苔 Cyathodium cavernarum Kunze ex Lehm.

凭证标本：霄罗湾洞穴，*唐启明 & 张仕艳 20181001-190*。

Specimens examined: Xiaoluowan Cave, *Tang & Zhang 20181001-190*.

15. 芽胞光苔 Cyathodium tuberosum Kashyap

凭证标本：草王山，*唐启明等 20210719-35*；黄猄洞天坑，*唐启明等 20210717-48*；拉雅沟，*唐启明 & 张仕艳 20180930-70*；李家坨屯，*唐启明 & 韦玉梅 20191018-286*；盘古王，*唐启明等 20210716-18*；下岩洞屯，*唐启明 & 张仕艳 20181001-223*；悬崖天坑，*唐启明等 20210721-52*；中井屯竖井，*唐启明 20190517-223*；中井屯霄罗湾洞穴，*唐启明 20190515-34*。

Representative specimens examined: Caowangshan, *Tang et al. 20210719-35*; Huangjingdong Tiankeng, *Tang et al. 20210717-48*; Layagou, *Tang & Zhang 20180930-70*; Lijiatuo Tun, *Tang & Wei 20191018-286*; Panguwang, *Tang et al. 20210716-18*; Xiayandong Tun, *Tang & Zhang 20181001-223*; Xuanya Tiankeng, *Tang et al. 20210721-52*; Zhongjing Tun, *Tang 20190517-223*; Xiaoluowan Cave, *Tang 20190515-34*.

16. 无翼钱苔 **Riccia billardierei** Mont. et Nees

凭证标本：隆合朝屯，*黄萍等 210724-5A*。

Specimens examined: Longhechao Tun, *Huang et al. 210724-5A*.

17. 稀枝钱苔 **Riccia huebeneriana** Lindenb.

凭证标本：逻家田屯，*唐启明 & 张仕艳 20181003-347*。

Specimens examined: Luojiatian Tun, *Tang & Zhang 20181003-347*.

18. 印尼钱苔 **Riccia junghuhniana** Nees et Lindenb.

凭证标本：隆合朝屯，*黄萍等 210724-7B*。

Specimens examined: Longhechao Tun, *Huang et al. 210724-7B*.

19. 厚壁钱苔 **Riccia oryzicola** T. Tominaga et Furuki

凭证标本：隆合朝屯，*黄萍等 210724-5B*；悬崖屯，*唐启明等 20210721-64*。

Specimens examined: Longhechao Tun, *Huang et al. 210724-5B*; Xuanya Tun, *Tang et al. 20210721-64*.

20. 花边钱苔 **Riccia rhenana** Lorb. ex Müll.Frib.

凭证标本：大宴坪屯，*唐启明等 20201105-127*。

Specimens examined: Dayanping Tun, *Tang et al. 20201105-127*.

21. 日本小叶苔 **Fossombronia japonica** Schiffn.

凭证标本：草王山，*唐启明 & 韦玉梅 20191013-93*；李家坨屯，*韦玉梅等 191017-273*；逻家田屯，*韦玉梅等 191020-405*；盘古王，*唐启明 & 张仕艳 20181004-413B*；旁墙屯，*唐启明 & 韦玉梅 20191019-417*。

Representative specimens examined: Caowangshan, *Tang & Wei 20191013-93*; Lijiatuo Tun, *Wei et al. 191017-273*; Luojiatian Tun, *Wei et al. 191020-405*; Panguwang, *Tang & Zhang 20181004-413B*; Pangqiang Tun, *Tang & Wei 20191019-417*.

22. 小叶苔 **Fossombronia pusilla** (L.) Nees

凭证标本：草王山，*韦玉梅等 191013-114*，*韦玉梅等 201108-27*。

Specimens examined: Caowangshan, *Wei et al. 191013-114*, *Wei et al. 201108-27*.

23. 带叶苔 **Pallavicinia lyellii** (Hook.) Gray

凭证标本：拉雅沟，*唐启明 & 韦玉梅 20191014-151*；旁墙屯，*韦玉梅等 201115-5*；二沟，*韦玉梅等 201109-17*。

Representative specimens examined: Layagou, *Tang & Wei 20191014-151*; Pangqiang Tun, *Wei et al. 201115-5*; Ergou, *Wei et al. 201109-17*.

24. 异溪苔 **Apopellia endiviifolia** (Dicks.) Nebel et D.Quandt

凭证标本：草王山，*唐启明 & 韦玉梅 20191013-125*；吊井天坑，*唐启明等 20210722-15*；里郎天坑，*唐启明等 20210723-46*；逻家田屯，*唐启明 & 张仕艳 20181003-349A*；盘古王，*唐启明等 20201110-266*；旁墙屯，*韦玉梅等 191018-274*；全达村，*唐启明等 20201108-228*；下岩洞屯，*唐启明 & 张仕艳 20181001-220*；中井天坑，*唐启明 & 张仕艳 20181007-566B1*。

Representative specimens examined: Caowangshan, *Tang & Wei 20191013-125*; Diaojing Tiankeng, *Tang et al. 20210722-15*; Lilang Tiankeng, *Tang et al. 20210723-46*; Luojiatian Tun, *Tang & Zhang 20181003-349A*; Panguwang, *Tang et al. 20201110-266*; Pangqiang Tun, *Wei et al. 191018-274*; Quanda Village, *Tang et al. 20201108-228*; Xiayandong Tun, *Tang & Zhang 20181001-220*; Zhongjing Tiankeng, *Tang & Zhang 20181007-566B1*.

25. 短萼狭叶苔 **Liochlaena subulata** (A.Evans) Schljakov

凭证标本：草王山，*韦玉梅等 191013-83*；黄猄洞天坑，*韦玉梅等 201104-12*；蓝家湾天坑，*唐启明 20190521-431*；盘古王，*韦玉梅等 201110-37A*；旁墙屯，*韦玉梅等 191019-371A*；下岩洞屯，*唐启明 & 张仕艳 20181001-133*；中井屯，*黄萍等 210722-3*。

Representative specimens examined: Caowangshan, *Wei et al. 191013-83*; Huangjingdong Tiankeng, *Wei et al. 201104-12*; Lanjiawan Tiankeng, *Tang 20190521-431*; Panguwang, *Wei et al. 201110-37A*; Pangqiang Tun, *Wei et al. 191019-371A*; Xiayandong Tun, *Tang & Zhang 20181001-133*; Zhongjing Tun, *Huang et al. 210722-3*.

26. 秩父无褶苔 **Mesoptychia chichibuensis** (Inoue) L.Söderstr. et Váňa

凭证标本：蓝家湾天坑，*韦玉梅等 191019-391*；深洞，*韦玉梅等 191017-263*。

Representative specimens examined: Lanjiawan Tiankeng, *Wei et al. 191019-391*; Shendong, *Wei et al. 191017-263*.

27. 中华无褶苔 **Mesoptychia chinensis** Bakalin

凭证标本：黄猄洞天坑，*韦玉梅等 201104-50*；盘古王，*唐启明等 20210716-16*；深洞，*韦玉梅等 191017-259*。

Specimens examined: Huangjingdong Tiankeng, *Wei et al. 201104-50*; Panguwang, *Tang et al. 20210716-16*; Shendong, *Wei et al. 191017-259*.

28. 玉山无褶苔 Mesoptychia morrisoncola (Horik.) L.Söderstr. et Váňa

凭证标本：大宴坪竖井，*唐启明 & 韦玉梅 20191017-268*；拉洞天坑，*韦玉梅等 191016-232*；蓝家湾天坑，*唐启明 & 韦玉梅 20191019-379*；李家坨屯，*唐启明 & 韦玉梅 20191018-316*；深洞，*韦玉梅等 191017-262*；中井天坑，*唐启明 & 张仕艳 20181007-580*；中井屯竖井，*唐启明 20190517-240*。

Representative specimens examined: Dayanping Shaft, *Tang & Wei 20191017-268*; Ladong Tiankeng, *Wei et al. 191016-232*; Lanjiawan Tiankeng, *Tang & Wei 20191019-379*; Lijiatuo Tun, *Tang & Wei 20191018-316*; Shendong, *Wei et al. 191017-262*; Zhongjing Tiankeng, *Tang & Zhang 20181007-580*; Zhongjing Tun, *Tang 20190517-240*.

29. 南亚被蒴苔 Nardia assamica (Mitt.) Amakawa

凭证标本：草王山，*韦玉梅等 201108-30*；九十九堡，*韦玉梅等 201111-12*。

Representative specimens examined: Caowangshan, *Wei et al. 201108-30*; Jiushijiubao, *Wei et al. 201111-12*.

30. 假苞苔 Notoscyphus lutescens (Lehm. et Lindenb.) Mitt.

凭证标本：黄猄洞天坑，*韦玉梅等 201104-27*；旁墙屯，*韦玉梅等 201115-7*；二沟，*韦玉梅等 201109-5*。

Representative specimens examined: Huangjingdong Tiankeng, *Wei et al. 201104-27*; Pangqiang Tun, *Wei et al. 201115-7*; Ergou, *Wei et al. 201109-5*.

31. 偏叶管口苔 Solenostoma comatum (Nees) C.Gao

凭证标本：拉雅沟，*唐启明 & 韦玉梅 20191014-147*；二沟，*韦玉梅等 201109-20*。

Representative specimens examined: Layagou, *Tang & Wei 20191014-147*; Ergou, *Wei et al. 201109-20*.

32. 截叶管口苔 Solenostoma truncatum (Nees) R.M.Schust. ex Váňa et D.G.Long

凭证标本：草王山，*唐启明 & 张仕艳 20181002-246*；吊井天坑，*唐启明等 20210722-22*；黄猄洞天坑，*韦玉梅等 191012-30*；九十九堡，*韦玉梅等 201111-5*；拉雅沟，*唐启明 & 张仕艳 20180930-105*；瞭望台，*唐启明 & 张仕艳 20181006-515A*；逻家田屯，*唐启明 & 张仕艳 20181003-405*；盘古王，*韦玉梅等 201110-12*；山干屯，*唐启明 20190522-446*；塘英村，*韦玉梅等 191013-135A*；下棚屯，*韦玉梅等 191015-180*；下岩洞屯，*唐启明 & 张仕艳 20181001-126*；一沟，*韦玉梅等 191014-155*。

Representative specimens examined: Caowangshan, *Tang & Zhang 20181002-246*; Diaojing Tiankeng, *Tang et al. 20210722-22*; Huangjingdong Tiankeng, *Wei et al. 191012-30*; Jiushijiubao, *Wei et al. 201111-5*; Layagou, *Tang & Zhang 20180930-105*; Liaowangtai, *Tang & Zhang 20181006-515A*; Luojiatian Tun, *Tang & Zhang 20181003-405*; Panguwang, *Wei et al. 201110-12*; Shangan Tun, *Tang 20190522-446*; Tangying Village, *Wei et al. 191013-135A*; Xiapeng Tun, *Wei et al. 191015-180*; Xiayandong Tun, *Tang & Zhang 20181001-126*; Yigou, *Wei et al. 191014-155*.

33. 刺叶护蒴苔 Calypogeia arguta Nees et Mont.

凭证标本：草王山，*韦玉梅等 201108-33*；一沟，*黄萍等 210715-27*。

Representative specimens examined: Caowangshan, *Wei et al. 201108-33*; Yigou, *Huang et al. 210715-27*.

34. 全缘护蒴苔 Calypogeia japonica Steph.

凭证标本：草王山，*韦玉梅等 201108-5*；九十九堡，*韦玉梅等 201111-7*；下棚屯，*唐启明等 20201111-323*。

Representative specimens examined: Caowangshan, *Wei et al. 201108-5*; Jiushijiubao, *Wei et al. 201111-7*; Xiapeng Tun, *Tang et al. 20201111-323*.

35. 双齿护蒴苔 Calypogeia tosana (Steph.) Steph.

凭证标本：草王山，*唐启明 20190518-298*；黄猄洞天坑，*韦玉梅等 191012-69*；旁墙屯，*韦玉梅等 191019-370A*。

Representative specimens examined: Caowangshan, *Tang 20190518-298*; Huangjingdong Tiankeng, *Wei et al. 191012-69*; Pangqiang Tun, *Wei et al. 191019-370A*.

36. 东亚对耳苔（雌株）Syzygiella nipponica (S. Hatt.) K. Feldberg

凭证标本：草王山，*韦玉梅等 191013-131*。

Specimens examined: Caowangshan, *Wei et al. 191013-131*.

37. 东亚对耳苔（雄株）Syzygiella nipponica (S. Hatt.) K. Feldberg

凭证标本：草王山，*韦玉梅等 221112-51，221112-55*。

Specimens examined: Caowangshan, *Wei et al. 221112 -51,221112-55*.

38. 弯叶大萼苔 Cephalozia hamatiloba Steph.

凭证标本：草王山，*唐启明 & 张仕艳 20181002-300B*；黄猄洞天坑，*唐启明 20180612-271*；盘古王，*韦玉梅等 201110-19*；旁墙屯，*韦玉梅等 201115-6B*。

Representative specimens examined: Caowangshan, *Tang & Zhang 20181002-300B*; Huangjingdong Tiankeng, *Tang 20180612-271*; Panguwang, *Wei et al. 201110-19*; Pangqiang Tun, *Wei et al. 201115-6B*.

39. 拳叶苔 Nowellia curvifolia (Dicks.) Mitt.

凭证标本：盘古王，*韦玉梅等 201110-18*；旁墙屯，*韦玉梅等 191019-356*；下棚屯，*唐启明等 20201111-324*。

Representative specimens examined: Panguwang, *Wei et al. 201110-18*; Pangqiang Tun, *Wei et al. 191019-356*; Xiapeng Tun, *Tang et al. 20201111-324*.

40. 合叶裂齿苔 Odontoschisma denudatum (Mart.) Dumort.

凭证标本：盘古王，*唐启明 & 张仕艳 20181005-504*；旁墙屯，*韦玉梅等 191019-371B*；下岩洞屯，*唐启明 & 张仕艳 20181001-157*。

Representative specimens examined: Panguwang, *Tang & Zhang 20181005-504*; Pangqiang Tun, *Wei et al. 191019-371B*; Xiayandong Tun, *Tang & Zhang 20181001-157*.

41. 粗齿拟大萼苔 Cephaloziella dentata (Raddi) Steph.

凭证标本：草王山，*唐启明 & 张仕艳 20181002-300B*；瞭望台，*唐启明 & 张仕艳 20181006-514C*；下岩洞屯，*唐启明 & 张仕艳 20181001-116B*。

Representative specimens examined: Caowangshan, *Tang & Zhang 20181002-300B*; Liaowangtai, *Tang & Zhang 20181006-514C*; Xiayandong Tun, *Tang & Zhang 20181001-116B*.

42. 小叶拟大萼苔 Cephaloziella microphylla (Steph.) Douin

凭证标本：黄猄洞天坑，*韦玉梅等 201104-15, 唐启明等 20210717-5*；里郎天坑，*唐启明等 20210723-49*；瞭望台，*唐启明 & 张仕艳 20181006-515C*；下岩洞屯，*唐启明 & 张仕艳 20181001-152*；中井天坑，*唐启明 & 张仕艳 20181007-585A*。

Representative specimens examined: Huangjingdong Tiankeng, *Wei et al. 201104-15, Tang et al. 20210717-5*; Lilang Tiankeng, *Tang et al. 20210723-49*; Liaowangtai, *Tang & Zhang 20181006-515C*; Xiayandong Tun, *Tang & Zhang 20181001-152*; Zhongjing Tiankeng, *Tang & Zhang 20181007-585A*.

43. 鳞叶筒萼苔 Cylindrocolea kiaeri (Austin) Váňa

凭证标本：下岩洞屯，*唐启明 & 张仕艳 20181001-190*；中井天坑，*唐启明 & 张仕艳 20181007-594A-1*；盘古王，*唐启明等 20210716-18*。

Representative specimens examined: Xiayandong Tun, *Tang & Zhang 20181001-190*; Zhongjing Tiankeng, *Tang & Zhang 20181007-594A-1*; Panguwang, *Tang et al. 20210716-18*.

44. 甲克苔 Jackiella javanica Schiffn.

凭证标本：草王山，*唐启明等 20210719-27*；九十九堡，*韦玉梅等 201111-10*；盘古王，*韦玉梅等 201110-1*；旁墙屯，*韦玉梅等 201115-6A*；下棚屯，*韦玉梅等 191015-199*。

Representative specimens examined: Caowangshan, *Tang et al. 20210719-27*; Jiushijiubao, *Wei et al. 201111-10*; Panguwang, *Wei et al. 201110-1*; Pangqiang Tun, *Wei et al. 201115-6A*; Xiapeng Tun, *Wei et al. 191015-199*.

45. 柯氏合叶苔 Scapania koponenii Potemkin

凭证标本：九十九堡，*韦玉梅等 201111-6*。

Specimens examined: Jiushijiubao, *Wei et al. 201111-6*.

46. 小睫毛苔 Blepharostoma minus Horik.

凭证标本：盘古王，*韦玉梅等 201110-37B*；塘英村，*韦玉梅等 191013-135B*。

Representative specimens examined: Panguwang, *Wei et al. 201110-37B*; Tangying Village, *Wei et al. 191013-135B*.

47. 卵叶鞭苔 Bazzania angustistipula N.Kitag.

凭证标本：草王山，*韦玉梅等 221112-68*，*韦玉梅等 221112-70*。

Representative specimens examined: Caowangshan, *Wei et al. 221112-68*, *Wei et al. 221112-70*.

48. 三裂鞭苔 Bazzania tridens (Reinw., Blume et Nees) Trevis.

凭证标本：旁墙屯，*韦玉梅等 191019-369*；塘英村，*韦玉梅等 191013-133*。

Representative specimens examined: Pangqiang Tun, *Wei et al. 191019-369*; Tangying Village, *Wei et al. 191013-133*.

49. 指叶苔 Lepidozia reptans (L.) Dumort.

凭证标本：草王山，*韦玉梅等 221112-40*。

Representative specimens examined: Caowangshan, *Wei et al. 221112-40*.

50. 长角剪叶苔 Herbertus dicranus (Taylor) Trevis.

凭证标本：草王山，*韦玉梅等 221112-52*，*韦玉梅等 221112-60*。

Representative specimens examined: Caowangshan, *Wei et al. 221112-52*, *Wei et al. 221112-60*.

51. 树羽苔 Chiastocaulon dendroides (Nees) Carl

凭证标本：草王山，*韦玉梅等 221112-61*。

Representative specimens examined: Caowangshan, *Wei et al. 221112-61*.

52. 埃氏羽苔 Plagiochila akiyamae Inoue

凭证标本：大宴坪天坑漏斗，*唐启明 & 韦玉梅 20191017-256*。

Specimens examined: Dayanping Tiankeng, *Tang & Wei 20191017-256*.

53. 阿萨羽苔 Plagiochila assamica Steph.

凭证标本：草王山，*唐启明等 20210719-15*。

Specimens examined: Caowangshan, *Tang et al. 20210719-15*.

54. 中华羽苔 **Plagiochila chinensis** Steph.

凭证标本：草王山，*唐启明 & 韦玉梅 20191013-119*。
Specimens examined: Caowangshan, *Tang & Wei 20191013-119*.

55. 树生羽苔 **Plagiochila corticola** Steph.

凭证标本：蓝家湾天坑，*韦玉梅等 191019-374*。
Representative specimens examined: Lanjiawan Tiankeng, *Wei et al. 191019-374*.

56. 德式羽苔 **Plagiochila delavayi** Steph.

凭证标本：逻家田屯，*唐启明 20190521-424*。
Specimens examined: Luojiatian Tun, *Tang 20190521-424*.

57. 裂叶羽苔 **Plagiochila furcifolia** Mitt.

凭证标本：中井天坑，*唐启明 & 张仕艳 20181007-568*。
Specimens examined: Zhongjing Tiankeng, *Tang & Zhang 20181007-568*.

58. 裸茎羽苔 **Plagiochila gymnoclada** Sande Lac.

凭证标本：草王山，*唐启明 20190518-269A*；蓝家湾天坑，*唐启明 & 韦玉梅 20191019-384*。
Representative specimens examined: Caowangshan, *Tang 20190518-269A*; Lanjiawan Tiankeng, *Tang & Wei 20191019-384*.

59. 容氏羽苔 **Plagiochila junghuhniana** Sande Lac.

凭证标本：黄猄洞天坑，*唐启明 20180612-305A*；逻家田屯，*韦玉梅等 191020-414*。
Representative specimens examined: Huangjingdong Tiankeng, *Tang 20180612-305A*; Luojiatian Tun, *Wei et al. 191020-414*.

60. 加萨羽苔 **Plagiochila khasiana** Mitt.

凭证标本：草王山，*韦玉梅等 221112-69*。
Representative specimens examined: Caowangshan, *Wei et al. 221112-69*.

61. 昆明羽苔 **Plagiochila kunmingensis** Piippo

凭证标本：草王山，*唐启明 & 韦玉梅 20191013-99*；大棚屯，*唐启明 20190520-396B*；吊井天坑，*唐启明等 20210722-14*；黄猄洞天坑，*韦玉梅等 191012-74*；逻家田屯，*唐启明 20190521-422B*；盘古王，*韦玉梅等 201110-46*；下棚屯，*唐启明 & 韦玉梅 20191015-186*；香当天坑，*唐启明 20190516-173*。

Representative specimens examined: Caowangshan, *Tang & Wei 20191013-99*; Dapeng Tun, *Tang 20190520-396B*; Diaojing Tiankeng, *Tang et al. 20210722-14*; Huangjingdong Tiankeng, *Wei et al. 191012-74*; Luojiatian Tun, *Tang 20190521-422B*; Panguwang, *Wei et al. 201110-46*; Xiapeng Tun, *Tang & Wei 20191015-186*; Xiangdang Tiankeng, *Tang 20190516-173*.

62. 尼泊尔羽苔 **Plagiochila nepalensis** Lindenb.

凭证标本：大宴坪天坑漏斗，*韦玉梅等 201105-45A*；黄猄洞天坑，*韦玉梅等 191012-77*；蓝家湾天坑，*韦玉梅等 191018-283A*；老屋基天坑，*唐启明等 20210721-146*；盘古王，*唐启明 & 张仕艳 20181005-486A*；香当天坑，*韦玉梅等 201107-13*；中井屯霄罗湾洞穴，*唐启明 20190515-18*。

Representative specimens examined: Dayanping Tiankeng, *Wei et al. 191012-77*; Huangjingdong Tiankeng, *Wei et al. 201104-24*; Lanjiawan Tiankeng, *Wei et al. 191018-283A*; Laowuji Tiankeng, *Tang et al. 20210721-146*; Panguwang, *Tang & Zhang 20181005-486A*; Xiangdang Tiankeng, *Wei et al. 201107-13*; Xiaoluowan Cave, *Tang 20190515-18*.

63. 卵叶羽苔 **Plagiochila ovalifolia** Mitt.

凭证标本：蓝家湾天坑，*唐启明 20190520-382*；盘古王，*韦玉梅等 201110-42*。

Representative specimens examined: Lanjiawan Tiankeng, *Tang 20190520-382*; Panguwang, *Wei et al. 201110-42*.

64. 圆头羽苔 **Plagiochila parvifolia** Lindenb.

凭证标本：黄猄洞天坑，*韦玉梅等 191012-42*；蓝家湾天坑，*唐启明 & 韦玉梅 20191019-427*；盘古王，*唐启明等 20201110-271*；下棚屯，*唐启明 & 韦玉梅 20191015-200*；悬崖天坑，*韦玉梅等 201107-23*。

Representative specimens examined: Huangjingdong Tiankeng, *Wei et al. 191012-42*; Lanjiawan Tiankeng, *Tang & Wei 20191019-427*; Panguwang, *Tang et al. 20201110-271*; Xiapeng Tun, *Tang & Wei 20191015-200*; Xuanya Tiankeng, *Wei et al. 201107-23*.

65. 刺叶羽苔 **Plagiochila sciophila** Nees ex Lindenb.

凭证标本：草王山，*韦玉梅等 201108-10*；大宴坪竖井，*唐启明 & 韦玉梅 20191017-262*；大宴坪天坑漏斗，*韦玉梅等 201105-46*；吊井天坑，*唐启明等 20210722-25*；黄猄洞天坑，*韦玉梅等 201104-49*；拉洞天坑，*韦玉梅等 191016-229*；拉雅沟，*唐启明 & 张仕艳 20180930-90*；蓝家湾天坑，*韦玉梅等 201115-3*；里郎天坑，*唐启明等 20210723-24*；逻家田屯，*唐启明 20190521-411*；盘古王，*唐启明 & 张仕艳 20181004-460*；深洞，*韦玉梅等 191017-251*；下棚屯，*唐启明 & 韦玉梅 20191015-192A*；香当天坑，*唐启明 20190516-129*；一沟，*韦玉梅等 191014-147*；中井屯，*唐启明等 20210718-49*；中井屯竖井，*唐启明 20190517-210*。

Representative specimens examined: Caowangshan, *Wei et al. 201108-10*; Dayanping Shaft, *Tang & Wei 20191017-262*; Dayanping Tiankeng, *Wei et al. 201105-46*; Diaojing Tiankeng, *Tang et al. 20210722-25*; Huangjingdong Tiankeng, *Wei et al. 201104-49*; Ladong Tiankeng, *Wei et al. 191016-229*; Layagou, *Tang & Zhang*

20180930-90; Lanjiawan Tiankeng, *Wei et al. 201115-3*; Lilang Tiankeng, *Tang et al. 20210723-24*; Luojiatian Tun, *Tang 20190521-411*; Panguwang, *Tang & Zhang 20181004-460*; Shendong, *Wei et al. 191017-251*; Xiapeng Tun, *Tang & Wei 20191015-192A*; Xiangdang Tiankeng, *Tang 20190516-129*; Yigou, *Wei et al. 191014-147*; Zhongjing Tun, *Tang et al. 20210718-49*; Zhongjing Tun, *Tang 20190517-210*.

66. 大耳羽苔 Plagiochila subtropica Steph.

凭证标本：草王山，*韦玉梅等 191013-108*，*唐启明 20181002-286*，*唐启明等 20210719-7*。

Representative specimens examined: Caowangshan, *Wei et al. 191013-108*, *Tang 20181002-286*, *Tang et al. 20210719-7*.

67. 短齿羽苔 Plagiochila vexans Schiffn. ex Steph.

凭证标本：草王山，*唐启明等 20210719-14*。

Specimens examined: Caowangshan, *Tang et al. 20210719-14*.

68. 韦氏羽苔 Plagiochila wightii Nees ex Lindenb.

凭证标本：大宴坪天坑漏斗，*韦玉梅等 201105-52*；黄猄洞天坑，*韦玉梅等 191012-11*；蓝家湾天坑，*韦玉梅等 191019-377*；逻家田屯，*唐启明 & 张仕艳 20181003-368*；下棚屯，*唐启明 & 韦玉梅 20191015-171B*；中井屯，*唐启明等 20210718-48*。

Representative specimens examined: Dayanping Tiankeng, *Wei et al. 201105-52*; Huangjingdong Tiankeng, *Wei et al. 191012-11*; Lanjiawan Tiankeng, *Wei et al. 191019-377*; Luojiatian Tun, *Tang & Zhang 20181003-368*; Xiapeng Tun, *Tang & Wei 20191015-171B*; Zhongjing Tun, *Tang et al. 20210718-48*.

69. 裂萼苔 Chiloscyphus polyanthos (L.) Corda

凭证标本：逻家田屯，*唐启明 20190521-428*。

Specimens examined: Luojiatian Tun, *Tang 20190521-428*.

70. 四齿异萼苔 Heteroscyphus argutus (Reinw., Blume et Nees) Schiffn.

凭证标本：吊井天坑，*唐启明等 20210722-100*；黄猄洞天坑，*韦玉梅等 201104-54*；拉雅沟，*唐启明 & 韦玉梅 20191014-138*；蓝家湾天坑，*韦玉梅等 201114-8*；盘古王，*唐启明 & 张仕艳 20181004-443*；深洞，*韦玉梅等 191017-243A*；二沟，*韦玉梅等 201109-21*；一沟，*韦玉梅等 191014-139*；中井天坑，*唐启明 & 张仕艳 20181007-570*；中井屯竖井，*唐启明 20190517-220A*。

Representative specimens examined: Diaojing Tiankeng, *Tang et al. 20210722-100*; Huangjingdong Tiankeng, *Wei et al. 201104-54*; Layagou, *Tang & Wei 20191014-138*; Lanjiawan Tiankeng, *Wei et al. 201114-8*; Panguwang, *Tang & Zhang 20181004-443*; Shendong, *Wei et al. 191017-243A*; Ergou, *Wei et al. 201109-21*; Yigou, *Wei et al. 191014-139*; Zhongjing Tiankeng, *Tang & Zhang 20181007-570*; Zhongjing Tun, *Tang 20190517-220A*.

71. 双齿异萼苔 Heteroscyphus coalitus (Hook.) Schiffn.

凭证标本：拉洞天坑，*韦玉梅等 191016-233*；拉雅沟，*唐启明 & 张仕艳 20180930-31A*；蓝家湾天坑，*韦玉梅等 191019-395-1*；李家坨屯，*唐启明 & 韦玉梅 20191018-291B*；逻家田屯，*韦玉梅等 191020-410*；盘古王，*韦玉梅等 201110-14*；深洞，*韦玉梅等 191017-250*；中井屯竖井，*唐启明 20190517-220B*；中井屯霄罗湾洞穴，*唐启明 20190515-35A*。

Representative specimens examined: Ladong Tiankeng, *Wei et al. 191016-233*; Layagou, *Tang & Zhang 20180930-31A*; Lanjiawan Tiankeng, *Wei et al. 191019-395-1*; Lijiatuo Tun, *Tang & Wei 20191018-291B*; Luojiatian Tun, *Wei et al. 191020-410*; Panguwang, *Wei et al. 201110-14*; Shendong, *Wei et al. 191017-250*; Zhongjing Tun, *Tang 20190517-220B*; Xiaoluowan Cave, *Tang 20190515-35A*.

72. 平叶异萼苔 Heteroscyphus planus (Mitt.) Schiffn.

凭证标本：黄猄洞天坑，*韦玉梅等 191012-61*；中井屯，*韦玉梅等 191016-200*。

Representative specimens examined: Huangjingdong Tiankeng, *Wei et al. 191012-61*; Zhongtuo Tun, *Wei et al. 191016-200*.

73. 南亚异萼苔 Heteroscyphus zollingeri (Gottsche) Schiffn.

凭证标本：逻家田屯，*唐启明 & 张仕艳 20181003-384*。

Specimens examined: Luojiatian Tun, *Tang & Zhang 20181003-384*.

74. 尖叶齿萼苔 Lophocolea bidentata (L.) Dumort.

凭证标本：草王山，*韦玉梅等 201108-12*；吊井天坑，*唐启明等 20210722-17*；黄猄洞天坑，*唐启明等 20210717-7*；拉洞天坑，*唐启明 20190517-187*；老屋基天坑，*唐启明等 20210721-140*；盘古王，*唐启明 & 张仕艳 20181005-506*；旁墙屯，*韦玉梅等 191019-348*；全达村，*韦玉梅等 201108-45*；深洞，*韦玉梅等 191017-252*；悬崖天坑，*唐启明等 20210721-17*；二沟，*唐启明等 20201109-245*；一沟，*韦玉梅等 191014-138*。

Representative specimens examined: Caowangshan, *Wei et al. 201108-12*; Diaojing Tiankeng, *Tang et al. 20210722-17*; Huangjingdong Tiankeng, *Tang et al. 20210717-7*; Ladong Tiankeng, *Tang 20190517-187*; Laowuji Tiankeng, *Tang et al. 20210721-140*; Panguwang, *Tang & Zhang 20181005-506*; Pangqiang Tun, *Wei et al. 191019-348*; Quanda Village, *Wei et al. 201108-45*; Shendong, *Wei et al. 191017-252*; Xuanya Tiankeng, *Tang et al. 20210721-17*; Ergou, *Tang et al. 20201109-245*; Yigou, *Wei et al. 191014-138*.

75. 拟异叶齿萼苔 Lophocolea concreta Mont.

凭证标本：白岩坨屯，*韦玉梅等 191020-435*；大宴坪天坑漏斗，*韦玉梅等 201105-21*；吊井天坑，*唐启明等 20210722-18*；黄猄洞天坑，*韦玉梅等 201104-37*；拉雅沟，*唐启明 & 张仕艳 20180930-100*；旁墙屯，*唐启明 & 韦玉梅 20191019-407B*；悬崖天坑，*韦玉梅等 201107-14*。

Representative specimens examined: Baiyantuo Tun, *Wei et al. 191020-435*; Dayanping Tiankeng, *Wei et al. 201105-21*; Diaojing Tiankeng, *Tang et al. 20210722-18*; Huangjingdong Tiankeng, *Wei et al. 201104-37*; Layagou,

Tang & Zhang 20180930-100; Pangqiang Tun, *Tang & Wei 20191019-407B*; Xuanya Tiankeng, *Wei et al. 201107-14*.

76. 疏叶齿萼苔 Lophocolea itoana Inoue

凭证标本：草王山，*韦玉梅等 191013-128*；蓝家湾天坑，*韦玉梅等 191018-328*；里郎天坑，*唐启明等 20210723-44*；逻家田屯，*唐启明 20190521-402B*；中井屯竖井，*唐启明 20190517-211A*。

Representative specimens examined: Caowangshan, *Wei et al. 191013-128*; Lanjiawan Tiankeng, *Wei et al. 191018-328*; Lilang Tiankeng, *Tang et al. 20210723-44*; Luojiatian Tun, *Tang 20190521-402B*; Zhongjing Tun, *Tang 20190517-211A*.

77. 芽胞齿萼苔 Lophocolea minor Nees

凭证标本：草王山，*唐启明 & 张仕艳 20181002-332*；黄猄洞天坑，*韦玉梅等 201104-53*；九十九堡，*韦玉梅等 201111-2*；拉雅沟，*唐启明 & 张仕艳 20180930-46*；逻家田屯，*韦玉梅等 191020-404*；盘古王，*韦玉梅等 201110-2*；全达村，*唐启明等 20201108-231*；下棚屯，*韦玉梅等 191015-170A*；悬崖天坑，*韦玉梅等 201107-1*；二沟，*韦玉梅等 201109-2*；一沟，*韦玉梅等 191014-156*；中井天坑，*唐启明 & 张仕艳 20181007-573*；中井屯，*唐启明 20190515-49*。

Representative specimens examined: Caowangshan, *Tang & Zhang 20181002-332*; Huangjingdong Tiankeng, *Wei et al. 201104-53*; Jiushijiubao, *Wei et al. 201111-2*; Layagou, *Tang & Zhang 20180930-46*; Luojiatian Tun, *Wei et al. 191020-404*; Panguwang, *Wei et al. 201110-2*; Quanda Village, *Tang et al. 20201108-231*; Xiapeng Tun, *Wei et al. 191015-170A*; Xuanya Tiankeng, *Wei et al. 201107-1*; Ergou, *Wei et al. 201109-2*; Yigou, *Wei et al. 191014-156*; Zhongjing Tiankeng, *Tang & Zhang 20181007-573*; Zhongtuo Tun, *Tang 20190515-49*.

78a. 尖瓣光萼苔原亚种 Porella acutifolia (Lehm. et Lindenb.) Trevis. subsp. acutifolia

凭证标本：下棚屯，*韦玉梅等 191015-166*。

Specimens examined: Xiapeng Tun, *Wei et al. 191015-166*.

78b. 尖瓣光萼苔东亚亚种 Porella acutifolia subsp. tosana (Steph.) S.Hatt.

凭证标本：下棚屯，*韦玉梅等 191015-172*。

Specimens examined: Xiapeng Tun, *Wei et al. 191015-172*.

79a. 丛生光萼苔原变种 Porella caespitans (Steph.) S.Hatt. var. caespitans

凭证标本：草王山，*韦玉梅等 191013-85*；大棚屯，*唐启明 20190520-368A*；黄猄洞天坑，*唐启明 20180612-305C*；拉洞天坑，*唐启明 20190517-191*。

Specimens examined: Caowangshan, *Wei et al. 191013-85*; Dapeng Tun, *Tang 20190520-368A*; Huangjingdong Tiankeng, *Tang 20180612-305C*; Ladong Tiankeng, *Tang 20190517-191*.

79b. 丛生光萼苔心叶变种 Porella caespitans var. cordifolia (Steph.) S.Hatt. ex T.Katag. et T.Yamag.

凭证标本：白岩坨屯，*韦玉梅等 191020-437*；草王山，*韦玉梅等 201108-15*；大棚屯，*唐启明 20190520-360*；大宴坪天坑漏斗，*韦玉梅等 201105-15*；逻家田屯，*韦玉梅等 191020-412*；盘古王，*唐启明等 20201110-263*；下棚屯，*韦玉梅等 191015-162*；下岩洞屯，*唐启明 & 张仕艳 20181001-217B*；悬崖天坑，*韦玉梅等 201107-35*。

Representative specimens examined: Baiyantuo Tun, *Wei et al. 191020-437*; Caowangshan, *Wei et al. 201108-15*; Dapeng Tun, *Tang 20190520-360*; Dayanping Tiankeng, *Wei et al. 201105-15*; Luojiatian Tun, *Wei et al. 191020-412*; Panguwang, *Tang et al. 20201110-263*; Xiapeng Tun, *Wei et al. 191015-162*; Xiayandong Tun, *Tang & Zhang 20181001-217B*; Xuanya Tiankeng, *Wei et al. 201107-35*.

80a. 密叶光萼苔原亚种 Porella densifolia (Steph.) S.Hatt. subsp. **densifolia**

凭证标本：大棚屯，*唐启明 20190520-390*；大宴坪天坑漏斗，*韦玉梅等 201105-12*；黄猄洞天坑，*韦玉梅等 191012-40*；盘古王，*唐启明 & 张仕艳 20181004-430B*；旁墙屯，*唐启明 & 韦玉梅 20191019-398*；下岩洞屯，*唐启明 & 张仕艳 20181001-199*；悬崖天坑，*韦玉梅等 201107-34*；中井屯，*唐启明 20190515-50*。

Representative specimens examined: Dapeng Tun, *Tang 20190520-390*; Dayanping Tiankeng, *Wei et al. 201105-12*; Huangjingdong Tiankeng, *Wei et al. 191012-40*; Panguwang, *Tang & Zhang 20181004-430B*; Pangqiang Tun, *Tang & Wei 20191019-398*; Xiayandong Tun, *Tang & Zhang 20181001-199*; Xuanya Tiankeng, *Wei et al. 201107-34*; Zhongtuo Tun, *Tang 20190515-50*.

80b. 密叶光萼苔长叶亚种 Porella densifolia subsp. **appendiculata** (Steph.) S.Hatt.

凭证标本：黄猄洞天坑，*韦玉梅等 191012-28*，*唐启明 20180612-316*；下岩洞屯，*唐启明 & 张仕艳 20181001-216C*。

Representative specimens examined: Huangjingdong Tiankeng, *Wei et al. 191012-28*, *Tang 20180612-316*; Xiayandong Tun, *Tang & Zhang 20181001-216C*.

81. 大叶光萼苔 Porella grandifolia (Steph.) S.Hatt.

凭证标本：大宴坪天坑漏斗，*唐启明等 20210718-119*；黄猄洞天坑，*唐启明等 20210717-16*；蓝家湾天坑，*韦玉梅等 191018-313*；悬崖天坑，*唐启明等 20210721-42A*；霄罗湾洞穴，*唐启明 20190515-24*。

Representative specimens examined: Dayanping Tiankeng, *Tang et al. 20210718-119*; Huangjingdong Tiankeng, *Tang et al. 20210717-16*; Lanjiawan Tiankeng, *Wei et al. 191018-313*; Xuanya Tiankeng, *Tang et al. 20210721-42A*; Xiaoluowan Cave, *Tang 20190515-24*.

82. 尾尖光萼苔 Porella handelii S.Hatt.

凭证标本：草王山，*唐启明 & 张仕艳 20181002-274*；吊井天坑，*唐启明等 20210722-75*；黄猄洞天坑，*韦玉梅等 191012-20*；兰花园，*唐启明 & 张仕艳 20181008-611*；老屋基天坑，*唐启明等 20210721-138*；里郎天坑，*唐启明等 20210723-48*；隆合朝屯，*黄萍等 210724-1*；香当天坑，*韦玉梅等 201117-7*；悬崖天坑，韦玉梅等 201107-15；一沟，韦玉梅等 201109-32；中井天坑，唐启明 & 张仕艳 20181007-566A。

Representative specimens examined: Caowangshan, *Tang & Zhang 20181002-274*; Diaojing Tiankeng, *Tang et al. 20210722-75*; Huangjingdong Tiankeng, *Wei et al. 191012-20*; Orchid Garden, *Tang & Zhang 20181008-611*; Laowuji Tiankeng, *Tang et al. 20210721-138*; Lilang Tiankeng, *Tang et al. 20210723-48*; Longhechao Tun, *Huang et al. 210724-1*; Xiangdang Tiankeng, *Wei et al. 201117-7*; Xuanya Tiankeng, *Wei et al. 201107-15*; Yigou, *Wei et al. 201109-32*; Zhongjing Tiankeng, *Tang & Zhang 20181007-566A*.

83. 日本光萼苔 Porella japonica (Sande Lac.) Mitt.

凭证标本：盘古王，韦玉梅等 201110-31；下棚屯，韦玉梅等 191015-168。

Representative specimens examined: Panguwang, *Wei et al. 201110-31*; Xiapeng Tun, *Wei et al. 191015-168*.

84. 基齿光萼苔 Porella madagascariensis (Nees et Mont.) Trevis.

凭证标本：蓝家湾天坑，唐启明 & 韦玉梅 20191019-363；盘古王，唐启明 & 张仕艳 20181004-421A。

Representative specimens examined: Lanjiawan Tiankeng, *Tang & Wei 20191019-363*; Panguwang, *Tang & Zhang 20181004-421A*.

85. 亮叶光萼苔 Porella nitens (Steph.) S.Hatt.

凭证标本：下棚屯，韦玉梅等 191015-169。

Specimens examined: Xiapeng Tun, *Wei et al. 191015-169*.

86. 钝叶光萼苔鳞叶变种 Porella obtusata var. macroloba (Steph.) S.Hatt. et M.X.Zhang

凭证标本：草王山，韦玉梅等 191013-87；达陇坪屯，韦玉梅等 191016-223；大棚屯，唐启明 20190520-359；大宴坪天坑漏斗，韦玉梅等 201105-54；黄猄洞天坑，韦玉梅等 191012-32；逻家田屯，韦玉梅等 191020-418；盘古王，唐启明等 20201110-265；下棚屯，唐启明 & 韦玉梅 20191015-189；香当天坑，韦玉梅等 201107-9；悬崖天坑，韦玉梅等 201107-43；中井屯，韦玉梅等 191016-214；里郎天坑，唐启明等 20210723-34。

Representative specimens examined: Caowangshan, *Wei et al. 191013-87*; Dalongping Tun, *Wei et al. 191016-223*; Dapeng Tun, *Tang 20190520-359*; Dayanping Tiankeng, *Wei et al. 201105-54*; Huangjingdong Tiankeng, *Wei et al. 191012-32*; Luojiatian Tun, *Wei et al. 191020-418*; Panguwang, *Tang et al. 20201110-265*; Xiapeng Tun, *Tang & Wei 20191015-189*; Xiangdang Tiankeng, *Wei et al. 201107-43*; Xuanya Tiankeng, *Wei et al. 201107-9*; Zhongtuo Tun, *Wei et al. 191016-214*; Lilang Tiankeng, *Tang et al. 20210723-34*.

87a. 毛边光萼苔原变种 Porella perrottetiana (Mont.) Trevis. var. perrottetiana

凭证标本：大宴坪天坑漏斗，唐启明等 20210718-69；吊井天坑，唐启明等 20210722-60；黄猄洞天坑，韦玉梅等 201104-22；拉洞天坑，唐启明 20190517-191；里郎天坑，唐启明等 20210723-26；香当天坑，韦玉梅等 201107-8。

Representative specimens examined: Dayanping Tiankeng, *Tang et al. 20210718-69*; Diaojing Tiankeng, *Tang et al. 20210722-60*; Huangjingdong Tiankeng, *Wei et al. 201104-22*; Ladong Tiankeng, *Tang 20190517-191*; Lilang Tiankeng, *Tang et al. 20210723-26*; Xiangdang Tiankeng, *Wei et al. 201107-8*.

87b. 毛边光萼苔狭叶变种 Porella perrottetiana var. angustifolia Pócs

凭证标本：大宴坪天坑漏斗，*唐启明等 20210718-79*；蓝家湾天坑，*韦玉梅等 191018-288*。

Representative specimens examined: Dayanping Tiankeng, *Tang et al. 20210718-79*; Lanjiawan Tiankeng, *Wei et al. 191018-288*.

87c. 毛边光萼苔齿叶变种 Porella perrottetiana var. ciliatodentata (P.C.Chen et P.C.Wu) S.Hatt.

凭证标本：九十九堡，*韦玉梅等 201111-20*；老屋基天坑，*唐启明等 20210721-99*；逻家田屯，*唐启明 & 张仕艳 20181003-408*；悬崖天坑，*韦玉梅等 201107-27*。

Representative specimens examined: Jiushijiubao, *Wei et al. 201111-20*; Laowuji Tiankeng; *Tang et al. 20210721-99*; Luojiatian Tun, *Tang & Zhang 20181003-408*; Xuanya Tiankeng, *Wei et al. 201107-27*.

88. 小瓣光萼苔 Porella plumosa (Mitt.) Parihar

凭证标本：草王山，*韦玉梅等 191013-96*；大宴坪天坑漏斗，*唐启明等 20210718-52*；黄猄洞天坑，*唐启明等 20210717-35*；拉洞屯，*韦玉梅等 191016-230*；蓝家湾天坑，*韦玉梅等 191018-326*；盘古王，*唐启明 & 张仕艳 20181004-462*；下棚屯，*唐启明 & 韦玉梅 20191015-194*；中井屯霄罗湾洞穴，*唐启明 20190515-28*。

Representative specimens examined: Caowangshan, *Wei et al. 191013-96*; Dayanping Tiankeng, *Tang et al. 20210718-52*; Huangjingdong Tiankeng, *Tang et al. 20210717-35*; Ladong tun, *Wei et al. 191016-230*; Lanjiawan Tiankeng, *Wei et al. 191018-326*; Panguwang, *Tang & Zhang 20181004-462*; Xiapeng Tun, *Tang & Wei 20191015-194*; Xiaoluowan Cave, *Tang 20190515-28*. Distribution in China: Gansu, Guangxi, Guizhou, Jiangxi, Shaanxi, Sichuan, Taiwan, Yunnan, Zhejiang.

89. 齿边光萼苔 Porella stephaniana (C.Massal.) S.Hatt.

凭证标本：蓝家湾天坑，*韦玉梅等 191018-331*，*韦玉梅等 191018-336*。

Specimens examined: Lanjiawan Tiankeng, *Wei et al. 191018-331*, *Wei et al. 191018-336*.

90. 多瓣光萼苔 Porella ulophylla (Steph.) S.Hatt.

凭证标本：香当天坑，*唐启明等 20201107-190*。

Representative specimens examined: Xiangdang Tiankeng, *Tang et al. 20201107-190*.

91. 尖舌扁萼苔 Radula acuminata Steph.

凭证标本：深洞，*韦玉梅等 191017-241B*；二沟，*韦玉梅等 201109-29*。

Representative specimens examined: Shendong, *Wei et al. 191017-241B*; Ergou, *Wei et al. 201109-29*.

92. 大瓣扁萼苔 Radula cavifolia Hampe ex Gottsche

凭证标本： 草王山，*韦玉梅等 221112-33*，*韦玉梅等 221112-43*，*韦玉梅等 221112-86*。

Representative specimens examined: Caowangshan, *Wei et al. 221112-33*, *Wei et al. 221112-43*, *Wei et al. 221112-86*.

93. 扁萼苔 Radula complanata (L.) Dumort.

凭证标本： 草王山，*韦玉梅等 191013-95*；黄猄洞天坑，*唐启明等 20210717-18*；蓝家湾天坑，*韦玉梅等 191018-315*。

Representative specimens examined: Caowangshan, *Wei et al. 191013-95*; Huangjingdong Tiankeng, *Tang et al. 20210717-18*; Lanjiawan Tiankeng, *Wei et al. 191018-315*.

94. 爪哇扁萼苔 Radula javanica Gottsche

凭证标本： 盘古王，*唐启明 & 张仕艳 20181004-428B*。

Specimens examined: Panguwang, *Tang & Zhang 20181004-428B*.

95. 尖叶扁萼苔 Radula kojana Steph.

凭证标本： 蓝家湾天坑，*韦玉梅等 191018-289*；里郎天坑，*唐启明等 20210723-49*；盘古王，*韦玉梅等 201110-21*；深洞，*韦玉梅等 191017-253*。

Representative specimens examined: Lanjiawan Tiankeng, *Wei et al. 191018-289*; Lilang Tiankeng, *Tang et al. 20210723-49*; Panguwang, *Wei et al. 201110-21*; Shendong, *Wei et al. 191017-253*.

96. 刺边扁萼苔 Radula lacerata Steph.

凭证标本： 蓝家湾天坑，*韦玉梅等 191018-280A*。

Specimens examined: Lanjiawan Tiankeng, *Wei et al. 191018-280A*.

97. 芽胞扁萼苔 Radula lindenbergiana Gottsche ex C.Hartm.

凭证标本： 白岩坨屯，*韦玉梅等 191020-421*；吊井天坑，*唐启明等 20210722-101*；黄猄洞天坑，*韦玉梅等 191012-12*；蓝家湾天坑，*韦玉梅等 191018-296*；逻家田屯，*韦玉梅等 191020-407*；下岩洞屯，*唐启明 & 张仕艳 20181001-208*；悬崖天坑，*韦玉梅等 201107-5*；中井屯竖井，*唐启明 20190517-228B*；中井屯霄罗湾洞穴，*唐启明 20190515-16*。

Representative specimens examined: Baiyantuo Tun, *Wei et al. 191020-421*; Diaojing Tiankeng, *Tang et al. 20210722-101*; Huangjingdong Tiankeng, *Wei et al. 191012-12*; Lanjiawan Tiankeng, *Wei et al. 191018-296*;

Luojiatian Tun, *Wei et al. 191020-407*; Xiayandong Tun, *Tang & Zhang 20181001-208*; Xuanya Tiankeng, *Wei et al. 201107-5*; Zhongjing Tun, *Tang 20190517-228B*; Xiaoluowan Cave, *Tang 20190515-16*.

98. 星苞扁萼苔 **Radula stellatogemmipara** C.Gao et Y.H.Wu.

凭证标本：白岩坨屯，*韦玉梅等 191020-436*；黄猄洞天坑，*韦玉梅等 191012-18*。

Representative specimens examined: Baiyantuo Tun, *Wei et al. 191020-436*; Huangjingdong Tiankeng, *Wei et al. 191012-18*.

99. 黑耳叶苔 **Frullania amplicrania** Steph.

凭证标本：黄猄洞天坑，*唐启明 20180612-301B*；悬崖天坑，*韦玉梅等 201107-37*；下岩洞屯，*唐启明 & 张仕艳 20181001-125*。

Representative specimens examined: Huangjingdong Tiankeng, *Tang 20180612-301B*; Laowuji Tiankeng, *Wei et al. 201107-37*; Xiayandong Tun, *Tang & Zhang 20181001-125*.

100. 细茎耳叶苔 **Frullania bolanderi** Austin

凭证标本：下棚屯，*韦玉梅等 191015-188*；二沟，*韦玉梅等 201109-7B*；中井天坑，*唐启明 & 张仕艳 20181007-543*。

Representative specimens examined: Xiapeng Tun, *Wei et al. 191015-188*; Ergou, *Wei et al. 201109-7B*; Zhongjing Tiankeng, *Tang & Zhang 20181007-543*.

101. 达乌里耳叶苔 **Frullania davurica** Hampe ex Gottsche, Lindenb. et Nees

凭证标本：盘古王，*唐启明 & 张仕艳 20181004-441A*；悬崖天坑，*韦玉梅等 201107-40*。

Representative specimens examined: Panguwang, *Tang & Zhang 20181004-441A*; Xuanya Tiankeng, *Wei et al. 201107-40*.

102. 皱叶耳叶苔 **Frullania ericoides** (Nees) Mont.

凭证标本：白岩坨屯，*韦玉梅等 191020-423*；草王山，*唐启明 & 张仕艳 20181002-342*；大棚屯，*唐启明 20190520-357*；大宴坪天坑漏斗，*韦玉梅等 201105-17*；吊井天坑，*唐启明等 20210722-52*；黄猄洞天坑，*韦玉梅等 201104-7*；九十九堡，*韦玉梅等 201111-17*；拉雅沟，*唐启明 & 张仕艳 20180930-67*；里郎天坑，*唐启明等 20210723-5*；盘古王，*韦玉梅等 201110-5*；下棚屯，*韦玉梅等 191015-171*；下岩洞屯，*唐启明 & 张仕艳 20181001-202*；悬崖天坑，*韦玉梅等 201107-33*；二沟，*韦玉梅等 201109-11*；中井天坑，*唐启明 & 张仕艳 20181007-527*。

Representative specimens examined: Baiyantuo Tun, *Wei et al. 191020-423*; Caowangshan, *Tang & Zhang 20181002-342*; Dapeng Tun, *Tang20190520-357*; Dayanping Tiankeng, *Wei et al. 201105-17*; Diaojing Tiankeng, *Tang et al. 20210722-52*; Huangjingdong Tiankeng, *Wei et al. 201104-7*; Jiushijiubao, *Wei et al. 201111-17*; Layagou, *Tang & Zhang 20180930-67*; Lilang Tiankeng, *Tang et al. 20210723-5*; Panguwang, *Wei et al. 201110-5*; Xiapeng

Tun, *Wei et al. 191015-171*; Xiayandong Tun, *Tang & Zhang 20181001-202*; Xuanya Tiankeng, *Wei et al. 201107-33*; Ergou, *Wei et al. 201109-11*; Zhongjing Tiankeng, *Tang & Zhang 20181007-527*.

103. 细瓣耳叶苔 Frullania hypoleuca Nees

凭证标本：大棚屯，*唐启明 20190520-392*。

Specimens examined: Dapeng Tun, *Tang 20190520-392*.

104. 石生耳叶苔 Frullania inflata Gottsche

凭证标本：大宴坪天坑漏斗，*唐启明等 20210718-65*；吊井天坑，*唐启明等 20210722-3*；黄猄洞天坑，*唐启明 20180612-293B*；拉洞天坑，*唐启明 20190517-200*；蓝家湾天坑，*韦玉梅等 191018-319*；里郎天坑，*黄泙等 210723-6*；悬崖天坑，*唐启明等 20210721-10*；中井天坑，*唐启明 & 张仕艳 20181007-537B*；中井屯，*韦玉梅等 191016-208*。

Representative specimens examined: Dayanping Tiankeng, *Tang et al. 20210718-65*; Diaojing Tiankeng, *Tang et al. 20210722-3*; Huangjingdong Tiankeng, *Tang 20180612-293B*; Ladong Tiankeng, *Tang 20190517-200*; Lanjiawan Tiankeng, *Wei et al. 191018-319*; Lilang Tiankeng, *Huang et al. 210723-6*; Xuanya Tiankeng, *Tang et al. 20210721-10*; Zhongjing Tiankeng, *Tang & Zhang 20181007-537B*; Zhongtuo Tun, *Wei et al. 191016-208*.

105. 列胞耳叶苔 Frullania moniliata (Reinw., Blume et Nees) Mont.

凭证标本：草王山，*韦玉梅等 191013-109A*；大棚屯，*唐启明 20190520-369*；逻家田屯，*唐启明 & 张仕艳 20181003-407*；下棚屯，*唐启明等 20201111-322*。

Representative specimens examined: Caowangshan, *Wei et al. 191013-109A*; Dapeng Tun, *Tang 20190520-369*; Luojiatian Tun, *Tang & Zhang 20181003-407*; Xiapeng Tun, *Tang et al. 20201111-322*.

106. 羊角耳叶苔喙尖变种 Frullania monocera var. acutiloba (Mitt.) Hentschel et von Konrat

凭证标本：草王山，*唐启明 20190518-260*；中井屯，*唐启明 & 张仕艳 20181007-596*。

Representative specimens examined: Dapeng Tun, *Tang 20190520-392*; Zhongjingtun, *Tang & Zhang 20181007-596*.

107. 盔瓣耳叶苔 Frullania muscicola Steph.

凭证标本：草王山，*唐启明 20190518-261*；拉雅沟，*唐启明 & 张仕艳 20180930-99*；罗家田，*唐启明 & 张仕艳 20181003-376*；中井屯，*韦玉梅等 191016-206*；香当天坑，*韦玉梅等 201107-25*。

Representative specimens examined: Caowangshan, *Tang 20190518-261*; Layagou, *Tang & Zhang 20180930-99*; Luojiatian, *Tang & Zhang 20181003-376*; Zhongjing Tun, *Wei et al. 191016-206*; Xiangdang Tiankeng, *Wei et al. 201107-25*.

108. 尼泊尔耳叶苔 Frullania nepalensis (Spreng.) Lehm. et Lindenb.

凭证标本：黄猄洞天坑，*韦玉梅等 191012-37*；李家坨屯，*唐启明 & 韦玉梅 20191018-314*；盘古王，*韦玉梅等 201110-3*；旁墙屯，*韦玉梅等 191018-346*；二沟，*韦玉梅等 201109-8*。

Representative specimens examined: Huangjingdong Tiankeng, *Wei et al. 191012-37*; Lijiatuo Tun, *Tang & Wei 20191018-314*; Panguwang, *Wei et al. 201110-3*; Pangqiang Tun, *Wei et al. 191018-346*; Ergou, *Wei et al. 201109-8*.

109. 大隅耳叶苔 Frullania osumiensis (S.Hatt.) S.Hatt.

凭证标本：草王山，*唐启明 & 张仕艳 20181002-343D*；九十九堡，*韦玉梅等 201108-18*，*韦玉梅等 201111-1*。

Representative specimens examined: Caowangshan, *Tang & Zhang 20181002-343D*; Jiushijiubao, *Wei et al. 201108-18*, *Wei et al. 201111-1*.

110. 钟瓣耳叶苔 Frullania parvistipula Steph.

凭证标本：大棚屯，*唐启明 20190520-379B*；大宴坪天坑漏斗，*韦玉梅等 201105-1*；吊井天坑，*唐启明等 20210722-99*；黄猄洞天坑，*韦玉梅等 201104-35*；九十九堡，*韦玉梅等 201111-3*；蓝家湾天坑，*韦玉梅等 201113-3*；老屋基天坑，*唐启明等 20210721-133*；里郎天坑，*唐启明等 20210723-4*；瞭望台，*唐启明 & 张仕艳 20181006-515D*；逻家田屯，*唐启明 & 张仕艳 20181003-360*；旁墙屯，*韦玉梅等 191019-349*；下棚屯，*韦玉梅等 191015-189*；下岩洞屯，*唐启明 & 张仕艳 20181001-143A*；悬崖天坑，*唐启明等 20210721-6*。

Representative specimens examined: Dapeng Tun, *Tang 20190520-379B*; Dayanping Tiankeng, *Wei et al. 201105-1*; Diaojing Tiankeng, *Tang et al. 20210722-99*; Huangjingdong Tiankeng, *Wei et al. 201104-35*; Jiushijiubao, *Wei et al. 201111-3*; Lanjiawan Tiankeng, *Wei et al. 201113-3*; Laowuji Tiankeng, *Tang et al. 20210721-133*; Lilang Tiankeng, *Tang et al. 20210723-4*; Liaowangtai, *Tang & Zhang 20181006-515D*; Luojiatian Tun, *Tang & Zhang 20181003-360*; Pangqiang Tun, *Wei et al. 191019-349*; Xiapeng Tun, *Wei et al. 191015-189*; Xiayandong Tun, *Tang & Zhang 20181001-143A*; Xuanya Tiankeng, *Tang et al. 20210721-6*.

111. 喙瓣耳叶苔 Frullania pedicellata Steph.

凭证标本：白岩坨屯，*韦玉梅等 191020-433*；蓝家湾天坑，*韦玉梅等 191018-304*。

Representative specimens examined: Baiyantuo Tun, *Wei et al. 191020-433*; Lanjiawan Tiankeng, *Wei et al. 191018-304*.

112. 大萼耳叶苔 Frullania physantha Mitt.

凭证标本：大宴坪天坑漏斗，*韦玉梅等 201105-3*；黄猄洞天坑，*韦玉梅等 201104-32*；蓝家湾天坑，*韦玉梅等 201113-2*；悬崖天坑，*韦玉梅等 201107-39*；旁墙屯，*韦玉梅等 191019-354*；中井天坑，*唐启明 & 张仕艳 20181007-534*；中井屯，*唐启明等 20210718-7*；中井屯，*韦玉梅等 191016-202*。

Representative specimens examined: Dayanping Tiankeng, *Wei et al. 201105-3*; Huangjingdong Tiankeng,

Wei et al. 201104-32; Lanjiawan Tiankeng, Wei et al. 201113-2; Xuanya Tiankeng, Wei et al. 201107-24; Pangqiang Tun, Wei et al. 191019-354; Zhongjing Tiankeng, Tang & Zhang 20181007-534; Zhongjing Tun, Tang et al. 20210718-7; Zhongtuo Tun, Wei et al. 191016-202.

113. 微齿耳叶苔 Frullania rhytidantha S.Hatt.

凭证标本：白岩垱漏斗，*唐启明等 20210718-58*；九十九堡，*韦玉梅等 201111-19*；老屋基天坑，*唐启明等 20210721-72*；龙坪村，*韦玉梅等 201116-11*；盘古王，*唐启明 & 张仕艳 20181004-432A*；下棚屯，*韦玉梅等 191015-190*；香当天坑，*韦玉梅等 201107-3*；中井屯，*韦玉梅等 191016-215*。

Representative specimens examined: Baiyandangloudou, *Tang et al. 20210718-58*; Jiushijiubao, *Wei et al. 201111-19*; Laowuji Tiankeng, *Tang et al. 20210721-72*; Longping Village, *Wei et al. 201116-11*; Panguwang, *Tang & Zhang 20181004-432A*; Xiapeng Tun, *Wei et al. 191015-190*; Xiangdang Tiankeng, *Wei et al. 201107-3*; Zhongtuo Tun, *Wei et al. 191016-215*.

114. 陕西耳叶苔 Frullania schensiana C.Massal.

凭证标本：草王山，*韦玉梅等 191013-109B，唐启明 & 韦玉梅 20191013-115*；香当天坑，*唐启明 20190516-166*。

Representative specimens examined: Caowangshan, *Wei et al. 191013-109B, Tang & Wei 20191013-115*; Xiangdang Tiankeng, *Tang 20190516-166*.

115. 欧耳叶苔长叶变种 Frullania tamarisci var. elongatistipula (Vard.) S.Hatt.

凭证标本：草王山，*韦玉梅等 221112-67，韦玉梅等 221114-21*。

Representative specimens examined: Caowangshan, *Wei et al. 221112-67, Wei et al. 221114-21*.

116. 云南耳叶苔密叶变种 Frullania yuennanensis var. siamensis (N.Kitag.,Thaithong et S.Hatt.) S.Hatt. et P.J.Lin

凭证标本：大棚屯，*唐启明 20190520-396A*。

Specimens examined: Dapeng Tun, *Tang 20190520-396A*.

117. 汤泽耳叶苔 Frullania yuzawana S.Hatt.

凭证标本：蓝家湾天坑，*韦玉梅等 191018-285，韦玉梅等 191018-281*。

Representative specimens examined: Lanjiawan Tiankeng, *Wei et al. 191018-285, Wei et al. 191018-281*.

118. 南亚顶鳞苔 Acrolejeunea sandvicensis (Gottsche) Steph.

凭证标本：草王山，*唐启明 & 张仕艳 20181002-343A*；大棚屯，*唐启明 20190520-383*；大宴坪天坑漏斗，*唐启明等 20210718-84*；吊井天坑，*唐启明等 20210720-18*；黄猄洞天坑，*韦玉梅等 201104-42*；拉雅沟，

唐启明 & 张仕艳 20180930-101；里郎天坑，唐启明等 20210723-6；逻家田屯，唐启明 & 张仕艳 20181003-375A；盘古王，唐启明 & 张仕艳 20181004-431B；旁墙屯，韦玉梅等 191019-351-1；下棚屯，唐启明 & 韦玉梅 20191015-181；下岩洞屯，唐启明 & 张仕艳 20181001-150A；悬崖天坑，韦玉梅等 201107-4；中井屯，唐启明等 20210718-31；中井屯，韦玉梅等 191016-212。

Representative specimens examined: Caowangshan, *Tang & Zhang 20181002-343A*; Dapeng Tun, *Tang 20190520-383*; Dayanping Tiankeng, *Tang et al. 20210718-84*; Diaojing Tiankeng, *Tang et al. 20210720-18*; Huangjingdong Tiankeng, *Wei et al. 201104-42*; Layagou, *Tang & Zhang 20180930-101*; Lilang Tiankeng, *Tang et al. 20210723-6*; Luojiatian Tun, *Tang & Zhang 20181003-375A*; Panguwang, *Tang & Zhang 20181004-431B*; Pangqiang Tun, *Wei et al. 191019-351-1*; Xiapeng Tun, *Tang & Wei 20191015-181*; Xiayandong Tun, *Tang & Zhang 20181001-150A*; Xuanya Tiankeng, *Wei et al. 201107-4*; Zhongjing Tun, *Tang et al. 20210718-31*; Zhongtuo Tun, *Wei et al. 191016-212*.

119. 中华顶鳞苔 Acrolejeunea sinensis (Jian Wang bis, R.L.Zhu et Gradst.) Jian Wang bis et Gradst.

凭证标本：蓝家湾天坑，韦玉梅等 191018-287。

Representative specimens examined: Lanjiawan Tiankeng, *Wei et al. 191018-287*.

120. 大叶冠鳞苔 Lopholejeunea eulopha (Taylor) Schiffn.

凭证标本：盘古王，唐启明 & 张仕艳 20181004-449A。

Specimens examined: Panguwang, *Tang & Zhang 20181004-449A*.

121. 黑冠鳞苔 Lopholejeunea nigricans (Lindenb.) Steph. ex Schiffn.

凭证标本：黄猄洞天坑，唐启明 20180612-320C；拉雅沟，唐启明 & 张仕艳 20180930-47A。

Representative specimens examined: Huangjingdong Tiankeng, *Tang 20180612-320C*; Layagou, *Tang & Zhang 20180930-47A*.

122. 皱萼苔 Ptychanthus striatus (Lehm. et Lindenb.) Nees

凭证标本：老屋基天坑，唐启明等 20210721-115；逻家田屯，唐启明 & 张仕艳 20181003-373B；深洞，韦玉梅等 191017-238；下棚屯，韦玉梅等 191015-179；二沟，韦玉梅等 201109-32。

Representative specimens examined: Laowuji Tiankeng, *Tang et al. 20210721-115*; Luojiatian Tun, *Tang & Zhang 20181003-373B*; Shendong, *Wei et al. 191017-238*; Xiapeng Tun, *Wei et al. 191015-179*; Ergou, *Wei et al. 201109-32*.

123. 东亚多褶苔 Spruceanthus kiushianus (Horik.) X.Q.Shi

凭证标本：黄猄洞天坑，唐启明 20180612-311D；悬崖天坑，唐启明等 20210721-42B。

Specimens examined: Huangjingdong Tiankeng, *Tang 20180612-311D*; Xuanya Tiankeng, *Tang et al. 20210721-42B*.

124. 疣叶多褶苔 Spruceanthus mamillilobulus (Herzog) Verd.

凭证标本：二沟，*韦玉梅等 201109-33*。

Representative specimens examined: Ergou, *Wei et al. 201109-33*.

125. 多褶苔 Spruceanthus semirepandus (Nees) Verd.

凭证标本：大棚屯，*唐启明 20190520-385*；盘古王，*唐启明 20190519-330A*。

Representative specimens examined: Dapeng Tun, *Tang 20190520-385*; Panguwang, *Tang 20190519-330A*.

126. 南亚毛鳞苔 Thysananthus repletus (Taylor) Sukkharak et Gradst

凭证标本：白岩坨屯，*韦玉梅等 191020-430*；黄猄洞天坑，*韦玉梅等 201104-43*。

Representative specimens examined: Baiyantuo Tun, *Wei et al. 191020-430*; Huangjingdong Tiankeng, *Wei et al. 201104-43*.

127. 异鳞苔 Tuzibeanthus chinensis (Steph.) Mizut.

凭证标本：白岩坨屯，*韦玉梅等 191020-426*；达陇坪屯，*韦玉梅等 191016-224*；大宴坪天坑漏斗，*唐启明 & 韦玉梅 20191017-255*；吊井天坑，*唐启明等 20210722-74*；黄猄洞天坑，*韦玉梅等 191012-45*；蓝家湾天坑，*韦玉梅等 191018-321*；老屋基天坑，*唐启明等 20210721-101*；里郎天坑，*唐启明等 20210723-21*；盘古王，*唐启明 & 张仕艳 20181004-455B*；深洞，*韦玉梅等 191017-270*；下岩洞屯，*唐启明 & 张仕艳 20181001-209*；悬崖天坑，*唐启明等 20210721-16*；中井天坑，*唐启明 & 张仕艳 20181007-546B*；中井屯霄罗湾洞穴，*唐启明 20190515-17*。

Representative specimens examined: Baiyantuo Tun, *Wei et al. 191020-426*; Dalongping Tun, *Wei et al. 191016-224*; Dayanping Tiankeng, *Tang & Wei 20191017-255*; Diaojing Tiankeng, *Tang et al. 20210722-74*; Huangjingdong Tiankeng, *Wei et al. 191012-45*; Lanjiawan Tiankeng, *Wei et al. 191018-321*; Laowuji Tiankeng, *Tang et al. 20210721-101*; Lilang Tiankeng, *Tang et al. 20210723-21*; Panguwang, *Tang & Zhang 20181004-455B*; Shendong, *Wei et al. 191017-270*; Xiayandong Tun, *Tang & Zhang 20181001-209*; Xuanya Tiankeng, *Tang et al. 20210721-16*; Zhongjing Tiankeng, *Tang & Zhang 20181007-546B*; Xiaoluowan Cave, *Tang 20190515-17*.

128. 圆叶唇鳞苔 Cheilolejeunea intertexta (Lindenb.) Steph.

凭证标本：黄猄洞天坑，*韦玉梅等 191012-41*；蓝家湾天坑，*韦玉梅等 201114-6*。

Representative specimens examined: Huangjingdong Tiankeng, *Wei et al. 191012-41*; Lanjiawan Tiankeng, *Wei et al. 201114-6*.

129. 粗茎唇鳞苔 Cheilolejeunea trapezia (Nees) Kachroo et R.M.Schust.

凭证标本：草王山，*韦玉梅等 191013-109C*；蓝家湾天坑，*韦玉梅等 191019-375*。

Representative specimens examined: Caowangshan, *Wei et al. 191013-109C*; Lanjiawan Tiankeng, *Wei et al. 191019-375*.

130. 卷边唇鳞苔 **Cheilolejeunea xanthocarpa** (Lehm. et Lindenb.) Malombe

凭证标本：草王山，*韦玉梅等 201108-31A*。

Specimens examined: Caowangshan, *Wei et al. 201108-31A*.

131. 单胞疣鳞苔 **Cololejeunea kodamae** Kamim.

凭证标本：蓝家湾天坑，*韦玉梅等 191019-398A*。

Specimens examined: Lanjiawan Tiankeng, *Wei et al. 191019-398A*.

132. 狭瓣疣鳞苔 **Cololejeunea lanciloba** Steph.

凭证标本：二沟，*韦玉梅等 201109-24*。

Specimens examined: Ergou, *Wei et al. 201109-24*.

133. 阔瓣疣鳞苔 **Cololejeunea latilobula** (Herzog) Tixier

凭证标本：黄猄洞天坑，*韦玉梅等 201104-52*；拉雅沟，*唐启明 & 张仕艳 20180930-50B*；蓝家湾天坑，*韦玉梅等 201115-1B*；瞭望台，*唐启明 & 张仕艳 20181006-524*；旁墙屯，*韦玉梅等 191019-363*。

Representative specimens examined: Huangjingdong Tiankeng, *Wei et al. 201104-52*; Layagou, *Tang & Zhang 20180930-50B*; Lanjiawan Tiankeng, *Wei et al. 201115-1B*; Liaowangtai, *Tang & Zhang 20181006-524*; Pangqiang Tun, *Wei et al. 191019-363*.

134. 阔体疣鳞苔 **Cololejeunea latistyla** R.L.Zhu

凭证标本：黄猄洞天坑，*韦玉梅等 191012-51*；盘古王，*韦玉梅等 201110-4*；下棚屯，*韦玉梅等 191015-182*。

Representative specimens examined: Huangjingdong Tiankeng, *Wei et al. 191012-51*; Panguwang, *Wei et al. 201110-4*; Xiapeng Tun, *Wei et al. 191015-182*.

135. 鳞叶疣鳞苔 **Cololejeunea longifolia** (Mitt.) Benedix ex Mizut.

凭证标本：大宴坪天坑漏斗，*韦玉梅等 201105-29*；吊井天坑，*唐启明等 20210722-47*；黄猄洞天坑，*韦玉梅等 201104-59*；蓝家湾天坑，*韦玉梅等 201115-1A*；李家坨屯，*唐启明 & 韦玉梅 20191018-280*；盘古王，*韦玉梅等 201110-13*；旁墙屯，*韦玉梅等 191019-367*；山干屯，*唐启明 20190522-452*；深洞，*韦玉梅等 191017-241A*；下棚屯，*韦玉梅等 191015-174*；中井屯竖井，*唐启明 20190517-212*；中井屯，*韦玉梅等 191016-217C*。

Representative specimens examined: Dayanping Tiankeng, *Wei et al. 201105-29*; Diaojing Tiankeng, *Tang et al. 20210722-47*; Huangjingdong Tiankeng, *Wei et al. 201104-59*; Lanjiawan Tiankeng, *Wei et al. 201115-1A*; Lijiatuo Tun, *Tang & Wei 20191018-280*; Panguwang, *Wei et al. 201110-13*; Pangqiang Tun, *Wei et al. 191019-367*; Shangan Tun, *Tang 20190522-452*; Shendong, *Wei et al. 191017-241A*; Xiapeng Tun, *Wei et al. 191015-174*; Zhongjing Tun, *Tang 20190517-212*; Zhongtuo Tun, *Wei et al. 191016-217C*.

136. 大瓣疣鳞苔 Cololejeunea magnilobula (Horik.) S.Hatt.

凭证标本： 草王山，*韦玉梅等 201108-31C*。

Representative specimens examined: Caowangshan, *Wei et al. 201108-31C*.

137. 粗柱疣鳞苔 Cololejeunea ornata A.Evans

凭证标本： 大宴坪天坑漏斗，*韦玉梅等 201105-25*；黄猄洞天坑，*唐启明 20180612-325A*；蓝家湾天坑，*韦玉梅等 201114-4*。

Representative specimens examined: Dayanping Tiankeng, *Wei et al. 201105-25*; Huangjingdong Tiankeng, *Tang 20180612-325A*; Lanjiawan Tiankeng, *Wei et al. 201114-4*.

138. 粗齿疣鳞苔 Cololejeunea planissima (Mitt.) Abeyw.

凭证标本： 黄猄洞天坑，*韦玉梅等 201104-55*；蓝家湾天坑，*韦玉梅等 191018-300*；旁墙屯，*韦玉梅等 191019-365*；深洞，*韦玉梅等 191017-239B*；二沟，*韦玉梅等 201109-36A*；一沟，*韦玉梅等 191014-146A*。

Representative specimens examined: Huangjingdong Tiankeng, *Wei et al. 201104-55*; Lanjiawan Tiankeng, *Wei et al. 191018-300*; Pangqiang Tun, *Wei et al. 191019-365*; Shendong, *Wei et al. 191017-239B*; Ergou, *Wei et al. 201109-36A*; Yigou, *Wei et al. 191014-146A*.

139. 尖叶疣鳞苔 Cololejeunea pseudocristallina P.C.Chen et P.C.Wu

凭证标本： 大宴坪天坑漏斗，*韦玉梅等 201105-41*；蓝家湾天坑，*韦玉梅等 201115-1C*。

Representative specimens examined: Dayanping Tiankeng, *Wei et al. 201105-41*; Lanjiawan Tiankeng, *Wei et al. 201115-1C*.

140. 拟疣鳞苔 Cololejeunea raduliloba Steph.

凭证标本： 二沟，*唐启明等 20210715-10*；一沟，*韦玉梅等 201109-30*。

Representative specimens examined: Ergou, *Tang et al. 20210715-10*; Yigou, *Wei et al. 201109-30*.

141. 全缘疣鳞苔 Cololejeunea schwabei Herzog

凭证标本： 拉雅沟，*唐启明 & 张仕艳 20180930-50A*；二沟，*韦玉梅等 201109-34*。

Representative specimens examined: Layagou, *Tang & Zhang 20180930-50A*; Ergou, *Wei et al. 201109-34*.

142. 卵叶疣鳞苔 Cololejeunea shibiensis Mizut.

凭证标本： 逻家田屯，*唐启明 20190521-433B*；香当天坑，*唐启明 20190516-162B*。

Representative specimens examined: Luojiatian Tun, *Tang 20190521-433B*; Xiangdang Tiankeng, *Tang 20190516-162B*.

143. 刺疣鳞苔 Cololejeunea spinosa (Horik.) Pandé et R.N.Misra

凭证标本：蓝家湾天坑，*唐启明 & 韦玉梅 20191019-370B*，*韦玉梅等 201114-5*。

Representative specimens examined: Lanjiawan Tiankeng, *Tang & Wei 20191019-370B*, *Wei et al. 201114-5*.

144. 疣瓣疣鳞苔 Cololejeunea subkodamae Mizut.

凭证标本：蓝家湾天坑，*韦玉梅等 191018-284A*，*韦玉梅等 201114-3B*。

Representative specimens examined: Lanjiawan Tiankeng, *Wei et al191018-284A*, *Wei et al. 201114-3B*.

145. 管叶苔 Colura calyptrifolia (Hook.) Dumort.

凭证标本：草王山，*韦玉梅等 191013-123*。

Representative specimens examined: Caowangshan, *Wei et al. 191013-123*.

146. 细角管叶苔 Colura tenuicornis (A. Evans) Steph

凭证标本：草王山，*韦玉梅等 221114-7B*。

Representative specimens examined: Caowangshan, *Wei et al. 221114-7B*.

147. 狭叶角鳞苔 Drepanolejeunea angustifolia (Mitt.) Grolle

凭证标本：草王山，*韦玉梅等 221112-37*，*韦玉梅等 221112-83*，*韦玉梅等 221114-11*，*韦玉梅等 221114-22*。

Representative specimens examined: Caowangshan, *Wei et al. 221112-37*, *Wei et al. 221112-83*, *Wei et al. 221114-11*, *Wei et al. 221114-22*.

148. 日本角鳞苔 Drepanolejeunea erecta (Steph.) Mizut.

凭证标本：草王山，*韦玉梅等 201108-19*；黄猄洞天坑，*韦玉梅等 201104-29*；蓝家湾天坑，*韦玉梅等 201114-2*；盘古王，*唐启明 20190519-330B*；全达村，*韦玉梅等 201108-47*；二沟，*韦玉梅等 201109-7A*。

Representative specimens examined: Caowangshan, *Wei et al. 201108-19*; Huangjingdong Tiankeng, *Wei et al. 201104-29*; Lanjiawan Tiankeng, *Wei et al. 201114-2*; Panguwang, *Tang 20190519-330B*; Quanda Village, *Wei et al. 201108-47*; Ergou, *Wei et al. 201109-7A*.

149. 单齿角鳞苔 Drepanolejeunea ternatensis (Gottsche) Schiffn.

凭证标本：草王山，*韦玉梅等 221112-83B*，*韦玉梅等 221114-7A*。

Representative specimens examined: Caowangshan, *Wei et al. 221112-83B*, *Wei et al. 221114-7A*.

150. 短叶角鳞苔 Drepanolejeunea vesiculosa (Mitt.) Steph.

凭证标本：草王山，*韦玉梅等 221112-45*，*韦玉梅等 221112-71*，*韦玉梅等 221112-84A*。

Representative specimens examined: Caowangshan, *Wei et al. 221112-45*, *Wei et al. 221112-71*, *Wei et al. 221112-84A*.

151. 狭瓣细鳞苔 Lejeunea anisophylla Mont.

凭证标本：大宴坪天坑漏斗，*韦玉梅等 201105-14A*；黄猄洞天坑，*唐启明等 20210717-30，黄萍等 210717-13*；深洞，*韦玉梅等 191017-239A*；一沟，*韦玉梅等 191014-141*；中井屯，*唐启明 20190517-182，黄萍等 210715-4*。

Representative specimens examined: Dayanping Tiankeng, *Wei et al. 201105-14A*; Huangjingdong Tiankeng, *Tang et al. 20210717-30, Huang et al. 210717-13*; Shendong, *Wei et al. 191017-239A*; Yigou, *Wei et al. 191014-141*; Zhongjing Tun, *Tang 20190517-182, Huang et al. 210715-4*.

152. 拟细鳞苔 Lejeunea apiculata Sande Lac.

凭证标本：草王山，*唐启明 20190518-292C*。

Specimens examined: Caowangshan, *Tang 20190518-292C*.

153. 双齿细鳞苔 Lejeunea bidentula Herzog

凭证标本：草王山，*韦玉梅等 191013-127，唐启明等 20190518-288B*。

Specimens examined: Caowangshan, *Wei et al. 191013-127, Tang et al. 20190518-288B*.

154. 瓣叶细鳞苔 Lejeunea cocoes Mitt.

凭证标本：草王山，*唐启明 20190518-292A*；达陇坪屯，*韦玉梅等 191016-228*；黄猄洞天坑，*韦玉梅等 201104-17A*；拉雅沟，*唐启明 & 张仕艳 20180930-64B*；蓝家湾天坑，*韦玉梅等 191018-299*；李家坨屯，*唐启明 & 韦玉梅 20191018-276*；龙坪村，*韦玉梅等 201116-10A*；全达村，*韦玉梅等 201108-39*；深洞，*韦玉梅等 191017-255*；二沟，*韦玉梅等 201109-35*；一沟，*韦玉梅等 191014-152*。

Representative specimens examined: Caowangshan, *Tang 20190518-292A*; Dalongping Tun, *Wei et al. 191016-228*; Huangjingdong Tiankeng, *Wei et al. 201104-17A*; Layagou, *Tang & Zhang 20180930-64B*; Lanjiawan Tiankeng, *Wei et al. 191018-299*; Lijiatuo Tun, *Tang & Wei 20191018-276*; Longping Village, *Wei et al. 201116-10A*; Quanda Village, *Wei et al. 201108-39*; Shendong, *Wei et al. 191017-255*; Ergou, *Wei et al. 201109-35*; Yigou, *Wei et al. 191014-152*.

155. 弯叶细鳞苔 Lejeunea curviloba Steph.

凭证标本：草王山，*唐启明 & 张仕艳 20181002-263，韦玉梅等 191013-103，唐启明等 20190518-292B*；白岩垱漏斗，*黄萍等 210718-36*。

Representative specimens examined: Caowangshan, *Tang & Zhang 20181002-263, Wei et al. 191013-103, Tang et al. 20190518-292B*; Baiyandang, *Huang 210718-36*.

156. 神山细鳞苔 Lejeunea eifrigii Mizut.

凭证标本：深洞，*韦玉梅等 191017-260*。
Specimens examined: Shendong, *Wei et al. 191017-260*.

157. 黄色细鳞苔 Lejeunea flava (Sw.) Nees

凭证标本：草王山，*韦玉梅等 201108-32*；黄猄洞天坑，*韦玉梅等 201104-57*；盘古王，*唐启明 & 张仕艳 20181004-412A*；悬崖天坑，*韦玉梅等 201107-42*；中井屯，*韦玉梅等 191016-217B*。
Representative specimens examined: Caowangshan, *Wei et al. 201108-32*; Huangjingdong Tiankeng, *Wei et al. 201104-57*; Panguwang, *Tang & Zhang 20181004-412A*; Xuanya Tiankeng, *Wei et al. 201107-42*; Zhongtuo Tun, *Wei et al. 191016-217B*.

158. 巨齿细鳞苔 Lejeunea kodamae Ikegami et Inoue

凭证标本：吊井天坑，*唐启明等 20210722-47*；达陇坪屯，*韦玉梅等 191016-226*；大棚屯，*唐启明 20190520-388*；大宴坪天坑漏斗，*韦玉梅等 201105-19*；黄猄洞天坑，*韦玉梅等 191012-49*；老屋基天坑，*唐启明等 20210721-75*；李家坨屯，*韦玉梅等 201116-3*；里郎天坑，*唐启明等 20210723-33*；逻家田屯，*韦玉梅等 191020-408*；盘古王，*唐启明 20190519-339*；旁墙屯，*韦玉梅等 191019-352*；深洞，*韦玉梅等 191017-271*；塘英村，*韦玉梅等 191015-161*；下棚屯，*韦玉梅等 191015-183*；悬崖天坑，*唐启明等 20210721-4*；中井屯，*韦玉梅等 191016-209*。
Representative specimens examined: Diaojing Tiankeng, *Tang et al. 20210722-47*; Dalongping Tun, *Wei et al. 191016-226*; Dapeng Tun, *Tang 20190520-388*; Dayanping Tiankeng, *Wei et al. 201105-19*; Huangjingdong Tiankeng, *Wei et al. 191012-49*; Laowuji Tiankeng, *Tang et al. 20210721-75*; Lijiatuo Tun, *Wei et al. 201116-3*; Lilang Tiankeng, *Tang et al. 20210723-33*; Luojiatian Tun, *Wei et al. 191020-408*; Panguwang, *Tang 20190519-339*; Pangqiang Tun, *Wei et al. 191019-352*; Shendong, *Wei et al. 191017-271*; Tangying Village, *Wei et al. 191015-161*; Xiapeng Tun, *Wei et al. 191015-183*; Xuanya Tiankeng, *Tang et al. 20210721-4*; Zhongtuo Tun, *Wei et al. 191016-209*.

159. 科诺细鳞苔 Lejeunea konosensis Mizut.

凭证标本：大宴坪天坑漏斗，*唐启明等 20210718-116*；黄猄洞天坑，*韦玉梅等 201104-41*；蓝家湾天坑，*韦玉梅等 191018-279*；下岩洞屯，*唐启明 & 张仕艳 20181001-215B*；悬崖天坑，*韦玉梅等 201107-26*。
Representative specimens examined: Dayanping Tiankeng, *Tang et al. 20210718-116*; Huangjingdong Tiankeng, *Wei et al. 201104-41*; Lanjiawan Tiankeng, *Wei et al. 191018-279*; Xiayandong Tun, *Tang & Zhang 20181001-215B*; Xuanya Tiankeng, *Wei et al. 201107-26*.

160. 麦氏细鳞苔 Lejeunea micholitzii Mizut.

凭证标本：蓝家湾天坑，*韦玉梅等 201114-7*。
Representative specimens examined: Lanjiawan Tiankeng, *Wei et al. 201114-7*.

161. 暗绿细鳞苔 Lejeunea obscura Mitt.

凭证标本：草王山，*韦玉梅等 191013-125*；大宴坪天坑漏斗，*韦玉梅等 201105-14A*；吊井天坑，*唐启明等 20210722-102*；黄猄洞天坑，*韦玉梅等 201104-47B*；九十九堡，*韦玉梅等 201111-16*；蓝家湾天坑，*韦玉梅等 201115-1D*；逻家田屯，*韦玉梅等 191020-415*；盘古王，*韦玉梅等 201110-6*；旁墙屯，*韦玉梅等 191019-361*；深洞，*韦玉梅等 191017-246*；下岩洞屯，*唐启明 & 张仕艳 20181001-212*；香当天坑，*韦玉梅等 201107-2*；二沟，*韦玉梅等 201109-31*；一沟，*韦玉梅等 191014-149A*；中井天坑，*唐启明 & 张仕艳 20181007-569A*；中井屯竖井，*唐启明 20190517-222*；中井屯，*韦玉梅等 191016-216*。

Representative specimens examined: Caowangshan, *Wei et al. 191013-125*; Dayanping Tiankeng, *Wei et al. 201105-14A*; Diaojing Tiankeng, *Tang et al. 20210722-102*; Huangjingdong Tiankeng, *Wei et al. 201104-47B*; Jiushijiubao, *Wei et al. 201111-16*; Lanjiawan Tiankeng, *Wei et al. 201115-1D*; Luojiatian Tun, *Wei et al. 191020-415*; Panguwang, *Wei et al. 201110-6*; Pangqiang Tun, *Wei et al. 191019-361*; Shendong, *Wei et al. 191017-246*; Xiayandong Tun, *Tang & Zhang 20181001-212*; Xiangdang Tiankeng, *Wei et al. 201107-2*; Ergou, *Wei et al. 201109-31*; Yigou, *Wei et al. 191014-149A*; Zhongjing Tiankeng, *Tang & Zhang 20181007-569A*; Zhongjing Tun, *Tang 20190517-222*; Zhongtuo Tun, *Wei et al. 191016-216*.

162. 角齿细鳞苔 Lejeunea otiana S.Hatt.

凭证标本：拉雅沟，*唐启明 & 张仕艳 20180930-87*；深洞，*韦玉梅等 191017-249*；二沟，*唐启明等 20210715-4*；一沟，*韦玉梅等 191014-137*。

Representative specimens examined: Layagou, *Tang & Zhang 20180930-87*; Shendong, *Wei et al. 191017-249*; Ergou, *Tang et al. 20210715-4*; Yigou, *Wei et al. 191014-137*.

163. 疣萼细鳞苔 Lejeunea tuberculosa Steph.

凭证标本：草王山，*韦玉梅等 191013-102*；大棚屯，*唐启明 20190520-394B*；盘古王，*韦玉梅等 201110-25*；全达村，*韦玉梅等 201108-38*；下棚屯，*唐启明 & 韦玉梅 20191015-192B*。

Representative specimens examined: Caowangshan, *Wei et al. 191013-102*; Dapeng Tun, *Tang 20190520-394B*; Panguwang, *Wei et al. 201110-25*; Quanda Village, *Wei et al. 201108-38*; Xiapeng Tun, *Tang & Wei 20191015-192B*.

164. 巴氏薄鳞苔 Leptolejeunea balansae Steph.

凭证标本：黄猄洞天坑，*韦玉梅等 201104-23*，*韦玉梅等 191012-38*，*韦玉梅等 191012-72*，*韦玉梅等 191012-79*。

Representative specimens examined: Huangjingdong Tiankeng, *Wei et al. 201104-23*, *Wei et al. 191012-38*, *Wei et al. 191012-72*, *Wei et al. 191012-79*.

165. 斑叶纤鳞苔 Microlejeunea punctiformis (Taylor) Steph.

凭证标本：草王山，*唐启明 & 张仕艳 20181002-343B*；黄猄洞天坑，*唐启明 20180612-319A*；逻家田屯，*唐启明 20190521-433C*；盘古王，*唐启明 & 张仕艳 20181004-458*；旁墙屯，*韦玉梅等 191019-372B*；下棚屯，*韦玉梅等 191015-175*；中井屯竖井，*唐启明 20190517-228C*；中井屯，*韦玉梅等 191016-218*，*韦玉梅等 201105-33*。

Representative specimens examined: Caowangshan, *Tang & Zhang 20181002-343B*; Huangjingdong Tiankeng, *Tang 20180612-319A*; Luojiatian Tun, *Tang 20190521-433C*; Panguwang, *Tang & Zhang 20181004-458*; Pangqiang Tun, *Wei et al. 191019-372B*; Xiapeng Tun, *Wei et al. 191015-175*; Zhongjing Tun, *Tang 20190517-228C*; Zhongtuo Tun, *Wei et al. 191016-218*, *Wei et al. 201105-33*.

166. 疏叶纤鳞苔 **Microlejeunea ulicina** (Taylor) Steph.

凭证标本：香当天坑，*唐启明 20190516-168B*；旁墙屯，*韦玉梅等 191019-350*。

Representative specimens examined: Xiangdang Tiankeng, *Tang 20190516-168B*; Pangqiang Tun, *Wei et al. 191019-350*.

167. 圆叶拟多果苔 **Myriocoleopsis minutissima** (Sm.) R.L.Zhu

凭证标本：大宴坪天坑漏斗，*韦玉梅等 201105-23*；吊井天坑，*唐启明等 20210720-19*；黄猄洞天坑，*韦玉梅等 191012-71*；拉洞天坑，*唐启明 20190517-185*；李家坨屯，*韦玉梅等 201116-9*；旁墙屯，*韦玉梅等 191019-372C*；下棚屯，*韦玉梅等 191015-193*；香当天坑，*唐启明 20190516-162A*；中井屯，*韦玉梅等 191016-207*。

Representative specimens examined: Dayanping Tiankeng, *Wei et al. 201105-23*; Diaojing Tiankeng, *Tang et al. 20210720-19*; Huangjingdong Tiankeng, *Wei et al. 191012-71*; Ladong Tiankeng, *Tang 20190517-185*; Lijiatuo Tun, *Wei et al. 201116-9*; Pangqiang Tun, *Wei et al. 191019-372C*; Xiapeng Tun, *Wei et al. 191015-193*; Xiangdang Tiankeng, *Tang 20190516-162A*; Zhongtuo Tun, *Wei et al. 191016-207*.

168. 大绿片苔 **Aneura maxima** (Schiffn.) Steph.

凭证标本：草王山，*唐启明等 20201108-219*；下岩洞屯，*唐启明 & 张仕艳 20181001-119C1*。

Representative specimens examined: Caowangshan, *Tang et al. 20201108-219*; Xiayandong Tun, *Tang & Zhang 20181001-119C1*.

169. 波叶片叶苔 **Riccardia chamedryfolia** (With.) Grolle

凭证标本：黄猄洞天坑，*唐启明等 20210717-26*。

Specimens examined: Huangjingdong Tiankeng, *Tang 20210717-26*.

170. 黄片叶苔 **Riccardia flavovirens** Furuki

凭证标本：盘古王，*韦玉梅等 201110-49B*。

Specimens examined: Panguwang, *Wei et al. 201110-49B*.

171. 长崎片叶苔 Riccardia nagasakiensis (Steph.) S.Hatt.

凭证标本：盘古王，*韦玉梅等 201110-49A*。

Specimens examined: Panguwang, *Wei et al. 201110-49A*.

172. 掌状片叶苔 Riccardia palmata (Hedw.) Carruth.

凭证标本：草王山，*韦玉梅等 201108-25*，*韦玉梅等 191013-122*。

Representative specimens examined: Caowangshan, *Wei et al. 201108-25*, *Wei et al. 191013-122*.

173. 纤细片叶苔 Riccardia pusilla Grolle

凭证标本：草王山，*韦玉梅等 191013-111*；深洞，*韦玉梅等 191017-245*。

Representative specimens examined: Caowangshan, *Wei et al. 191013-111*; Shendong, *Wei et al. 191017-245*.

174. 狭尖叉苔 Metzgeria consanguinea Schiffn.

凭证标本：草王山，*韦玉梅等 201108-17*，*唐启明等 20190518-288A*；大棚屯，*唐启明 20190520-384*。

Representative specimens examined: Caowangshan, *Wei et al. 201108-17*, *Tang et al. 20190518-288A*; Dapeng Tun, *Tang 20190520-384*.

175. 林氏叉苔 Metzgeria lindbergii Schiffn.

凭证标本：草王山，*唐启明 & 张仕艳 20181002-284*；大宴坪天坑漏斗，*韦玉梅等 201105-18*；逻家田屯，*韦玉梅等 191020-409*；盘古王，*唐启明 & 张仕艳 20181005-465B*；下棚屯，*唐启明 & 韦玉梅 20191015-171A*。

Representative specimens examined: Caowangshan, *Tang & Zhang 20181002-284*; Dayanping Tiankeng, *Wei et al. 201105-18*; Luojiatian Tun, *Wei et al. 191020-409*; Panguwang, *Tang & Zhang 20181005-465B*; Xiapeng Tun, *Tang & Wei 20191015-171A*.

176. 塔拉加角苔 Anthoceros telaganus Steph.

凭证标本：李家坨屯，*韦玉梅等 201116-1*。

Representative specimens examined: Lijiatuo Tun, *Wei et al. 201116-1*.

177. 东亚短角苔 Notothylas javanica (Sande Lac.) Gottsche

凭证标本：逻家田屯，*韦玉梅等 191020-401*；悬崖天坑，*韦玉梅等 201107-16*。

Representative specimens examined: Luojiatian Tun, *Wei et al. 191020-401*; Xuanya Tiankeng, *Wei et al. 201107-16*.

178. 短角苔 Notothylas orbicularis (Schwein.) Sull.

凭证标本：中井屯，*唐启明等 20201105-79*。

Specimens examined: Zhongjing Tun, *Tang et al. 20201105-79*.

179. 高领黄角苔 Phaeoceros carolinianus (Michx.) Prosk.

凭证标本：草王山，*唐启明 & 张仕艳 20181002-324*；大宴坪天坑漏斗，*韦玉梅等 201105-11*；九十九堡，*韦玉梅等 201111-14*；拉雅沟，*唐启明 & 张仕艳 20180930-104*；香当天坑，*唐启明等 20201107-173*；李家坨屯，*唐启明 & 韦玉梅 20191018-271*；逻家田屯，*韦玉梅等 191020-400*；落花生屯，*韦玉梅等 191017-235*；旁墙屯，*韦玉梅等 191018-275*；全达村，*韦玉梅等 201108-40*；山干屯，*唐启明 20190522-453*；塘英村，*韦玉梅等 191013-134*；下棚屯，*韦玉梅等 191015-178*；下岩洞屯，*唐启明 & 张仕艳 20181001-227A*；二沟，*韦玉梅等 201109-19*。

Representative specimens examined: Caowangshan, *Tang & Zhang 20181002-324*; Dayanping Tiankeng, *Wei et al. 201105-11*; Jiushijiubao, *Wei et al. 201111-14*; Layagou, *Tang & Zhang 20180930-104*; Xiangdang Tiankeng, *Tang et al. 20201107-173*; Lijiatuo Tun, *Tang & Wei 20191018-271*; Luojiatian Tun, *Wei et al. 191020-400*; Luohuasheng Tun, *Wei et al. 191017-235*; Pangqiang Tun, *Wei et al. 191018-275*; Quanda Village, *Wei et al. 201108-40*; Shangan Tun, *Tang 20190522-453*; Tangying Village, *Wei et al. 191013-134*; Xiapeng Tun, *Wei et al. 191015-178*; Xiayandong Tun, *Tang & Zhang 20181001-227A*; Ergou, *Wei et al. 201109-19*.

180. 小黄角苔 Phaeoceros exiguus (Steph.) J.Haseg.

凭证标本：悬崖屯，*唐启明等 20210721-65*；盘古王，*韦玉梅等 201110-7*；悬崖天坑，*韦玉梅等 201107-17A*。

Representative specimens examined: Xuanya Tun, *Tang et al. 20210721-65*; Panguwang, *Wei et al. 201110-7*; Xuanya Tiankeng, *Wei et al. 201107-17A*.

181. 东亚大角苔 Megaceros flagellaris (Mitt.) Steph.

凭证标本：盘古王，*韦玉梅等 201110-40*。

Specimens examined: Panguwang, *Wei et al. 201110-40*.

中文名索引
INDEX OF CHINESE NAMES

A

阿萨羽苔 106
埃氏羽苔 104
暗绿细鳞苔 332
暗色蛇苔 16

B

巴氏薄鳞苔 338
白鳞苔 270
斑叶细鳞苔 340
斑叶纤鳞苔 340
瓣叶细鳞苔 318
薄壁大萼苔 76
扁萼苔 196
波叶片叶苔 348

C

齿边光萼苔 188
刺边扁萼苔 202
刺叶护蒴苔 66
刺叶羽苔 130
刺疣鳞苔 296
丛生光萼苔细柄变种 162
丛生光萼苔心叶变种 162
丛生光萼苔原变种 160
粗齿拟大萼苔 82
粗齿疣鳞苔 286
粗齿原鳞苔 256
粗茎唇鳞苔 268
粗裂地钱风兜亚种 20
粗柱疣鳞苔 284
长角剪叶苔 100
长崎片叶苔 352
长叶唇鳞苔 268

D

达乌里耳叶苔 212
大瓣扁萼苔 194
大瓣疣鳞苔 282
大耳羽苔 132
大绿片苔 346
大蒴耳叶苔 234
大叶冠鳞苔 250
大叶光萼苔 168
大隅耳叶苔 228
带叶苔 46
单胞疣鳞苔 272
单齿角鳞苔 308
单纹花萼苔 6
单月苔 26
德氏羽苔 112
地钱土生亚种 22
东亚大角苔 374
东亚地钱 18
东亚短角苔 364
东亚对耳苔（雌株）72
东亚对耳苔 74
东亚多褶苔 256
短齿羽苔 134
短萼狭叶苔 50
短角苔 368
短叶角鳞苔 310
钝鳞紫背苔 8
钝叶光萼苔鳞叶变种 178
多瓣光萼苔 190
多瓣苔 190
多褶苔 260

F

风兜地钱 20

G

高领黄角苔 370
管叶苔 300
光苔 28

H

合叶裂齿苔 80
黑耳叶苔 208
黑冠鳞苔 252
厚壁钱苔 38
厚角杯囊苔 60
花边钱苔 40
花叶溪苔 48
黄片叶苔 350
黄色杯囊苔 60
黄色细鳞苔 324
喙瓣耳叶苔 232
喙尖耳叶苔 222

J

基齿光萼苔 174
加萨花萼苔 4
加萨羽苔 120
甲克苔 88
假苞苔 60
尖瓣光萼苔东亚亚种 158
尖瓣光萼苔原亚种 156
尖舌扁萼苔 192
尖叶扁萼苔 200
尖叶齿萼苔 148
尖叶裂萼苔 148

尖叶疣鳞苔 288
角齿细鳞苔 334
截叶管口苔 64
截叶叶苔 64
巨齿细鳞苔 326
卷边唇鳞苔 270
卷边花萼苔 6

K
科诺细鳞苔 328
柯氏合叶苔 90
盔瓣耳叶苔 224
昆明羽苔 122
阔瓣疣鳞苔 276
阔体疣鳞苔 278

L
亮叶光萼苔 176
列胞耳叶苔 220
裂萼苔 138
裂叶羽苔 114
林氏叉苔 360
鳞叶拟大萼苔 86
鳞叶筒萼苔 86
鳞叶疣鳞苔 280
鹿角苔 48
卵叶鞭苔 94
卵叶疣鳞苔 294
卵叶羽苔 126
裸茎羽苔 116

M
麦氏细鳞苔 330
毛边光萼苔齿叶变种 184
毛边光萼苔狭叶变种 182
毛边光萼苔原变种 180
毛地钱 24
密叶光萼苔长叶亚种 166
密叶光萼苔原亚种 164

N
南亚被蒴苔 58

南亚鞭鳞苔 262
南亚顶鳞苔 246
南亚毛鳞苔 262
南亚瓦鳞苔 246
南亚异萼苔 146
尼泊尔耳叶苔 226
尼泊尔羽苔 124
拟细鳞苔 314
拟异叶齿萼苔 150
拟疣鳞苔 290

O
欧耳叶苔长叶变种 240

P
偏叶管口苔 62
偏叶叶苔 62
平叶异萼苔 144

Q
拳叶苔 78
全缘护蒴苔 68
全缘疣鳞苔 292

R
日本光萼苔 172
日本角鳞苔 306
日本小叶苔 42
容氏羽苔 118
柔叶花萼苔 2

S
三裂鞭苔 96
陕西耳叶苔 238
蛇苔 16
神山细鳞苔 322
十字花萼苔 2
石地钱 12
石生耳叶苔 218
疏叶齿萼苔 152
疏叶裂萼苔 152
疏叶细鳞苔 342

疏叶纤鳞苔 342
树生羽苔 110
树羽苔 102
双齿护蒴苔 70
双齿细鳞苔 316
双齿异萼苔 142
四齿异萼苔 140

T
塔拉加角苔 362
汤泽耳叶苔 244

W
瓦氏花萼苔 6
瓦叶唇鳞苔 268
弯叶大萼苔 76
弯叶细鳞苔 320
微齿耳叶苔 236
韦氏羽苔 136
尾尖光萼苔 170
无纹紫背苔 10
无翼钱苔 32

X
稀枝钱苔 34
细瓣耳叶苔 216
细角管叶苔 302
细茎耳叶苔 210
狭瓣细鳞苔 312
狭瓣疣鳞苔 274
狭尖叉苔 358
狭鳞苔 314
狭叶花萼苔 6
狭叶角鳞苔 304
狭叶叶苔 50
纤细片叶苔 356
小瓣光萼苔 186
小杯囊苔 60
小黄角苔 372
小睫毛苔 92
小蛇苔 14
小叶拟大萼苔 84

415

小叶苔 44
楔瓣地钱东亚亚种 18
星苞扁萼苔 206
喜马拉雅叉苔 360

Y

芽胞扁萼苔 204
芽胞齿萼苔 154
芽胞光苔 30
芽胞裂萼苔 154
羊角耳叶苔喙尖变种 222
异瓣裂叶苔 56
异鳞苔 264

印尼钱苔 36
疣瓣疣鳞苔 298
疣萼细鳞苔 336
疣叶多褶苔 258
羽状羽苔 102
玉山裂叶苔 56
玉山无褶苔 56
圆头羽苔 128
圆叶唇鳞苔 266
圆叶拟多果苔 344
圆叶疣鳞苔 344
云南耳叶苔密叶变种 242

Z

掌状片叶苔 354
爪哇扁萼苔 198
指叶苔 98
秩父无褶苔 52
中华顶鳞苔 248
中华无褶苔 54
中华斜裂苔 54
中华羽苔 108
钟瓣耳叶苔 230
皱萼苔 254
皱叶耳叶苔 214

拉丁名索引
INDEX OF SCIENTIFIC NAMES

A

Acrolejeunea sandvicensis **246**

Acrolejeunea sinensis **248**

Aneura maxima **346**

Anthoceros telaganus **362**

Apopellia endiviifolia **48**

Archilejeunea kiushiana **256**

Asterella angusta **6**

Asterella cruciata **2**

Asterella khasyana **4**

Asterella mitsuminensis **2**

Asterella monospiris **6**

Asterella reflexa **6**

Asterella wallichiana **6**

B

Bazzania angustistipula **94**

Bazzania tridens **96**

Blepharostoma minus **92**

C

Calypogeia arguta **66**

Calypogeia japonica **68**

Calypogeia tosana **70**

Cephalozia hamatiloba **76**

Cephalozia otaruensis **76**

Cephaloziella dentata **82**

Cephaloziella kiaeri **86**

Cephaloziella microphylla **84**

Cheilolejeunea imbricata **268**

Cheilolejeunea intertexta **266**

Cheilolejeunea longiloba **268**

Cheilolejeunea trapezia **268**

Cheilolejeunea xanthocarpa **270**

Chiastocaulon dendroides **102**

Chiloscyphus cuspidatus **148**

Chiloscyphus itoanus **152**

Chiloscyphus minor **154**

Chiloscyphus polyanthos **138**

Cololejeunea kodamae **272**

Cololejeunea lanciloba **274**

Cololejeunea latilobula **276**

Cololejeunea latistyla **278**

Cololejeunea longifolia **280**

Cololejeunea magnilobula **282**

Cololejeunea minutissima **344**

Cololejeunea ornata **284**

Cololejeunea planissima **286**

Cololejeunea pseudocristallina **288**

Cololejeunea raduliloba **290**

Cololejeunea schwabei **292**

Cololejeunea shibiensis **294**

Cololejeunea spinosa **296**

Cololejeunea subkodamae **298**

Colura calyptrifolia **300**

Colura tenuicornis **302**

Conocephalum conicum **16**

Conocephalum japonicum **14**

Conocephalum salebrosum **16**

Cyathodium cavernarum **28**

Cyathodium tuberosum **30**

Cylindrocolea kiaeri **86**

D

Drepanolejeunea angustifolia **304**

Drepanolejeunea erecta **306**

Drepanolejeunea ternatensis **308**

Drepanolejeunea vesiculosa **310**

Dumortiera hirsuta **24**

F

Fossombronia japonica **42**
Fossombronia pusilla **44**
Frullania acutiloba **222**
Frullania amplicrania **208**
Frullania bolanderi **210**
Frullania davurica **212**
Frullania ericoides **214**
Frullania hypoleuca **216**
Frullania inflata **218**
Frullania moniliata **220**
Frullania monocera var. *acutiloba* **222**
Frullania muscicola **224**
Frullania nepalensis **226**
Frullania osumiensis **228**
Frullania parvistipula **230**
Frullania pedicellata **232**
Frullania physantha **234**
Frullania rhytidantha **236**
Frullania schensiana **238**
Frullania tamarisci var. *elongatistipula* **240**
Frullania yunnanensis var. *siamensis* **242**
Frullania yuzawana **244**

H

Herbertus dicranus **100**
Heteroscyphus argutus **140**
Heteroscyphus coalitus **142**
Heteroscyphus planus **144**
Heteroscyphus zollingeri **146**

J

Jackiella javanica **88**
Jamesoniella nipponica **72, 74**
Jungermannia comata **62**
Jungermannia subulata **50**
Jungermannia truncata **64**

L

Lejeunea anisophylla **312**
Lejeunea apiculata **314**
Lejeunea bidentula **316**

Lejeunea cocoes **318**
Lejeunea curviloba **320**
Lejeunea eifrigii **322**
Lejeunea flava **324**
Lejeunea kodamae **326**
Lejeunea konosensis **328**
Lejeunea micholitzii **330**
Lejeunea obscura **332**
Lejeunea otiana **334**
Lejeunea punctiformis **340**
Lejeunea tuberculosa **336**
Lejeunea ulicina **342**
Lepidozia reptans **98**
Leptolejeunea balansae **338**
Leucolejeunea xanthocarpa **270**
Liochlaena subulata **50**
Lophocolea bidentata **148**
Lophocolea concreta **150**
Lophocolea itoana **152**
Lophocolea minor **154**
Lopholejeunea eulopha **250**
Lopholejeunea nigricans **252**
Lophozia diversiloba **56**
Lophozia morrisoncola **56**

M

Macvicaria ulophylla **190**
Marchantia diptera **20**
Marchantia emarginata subsp. *tosata* **18**
Marchantia paleacea subsp. *diptera* **20**
Marchantia polymorpha subsp. *ruderalis* **22**
Marchantia tosana **18**
Mastigolejeunea repleta **262**
Megaceros flagellaris **372**
Mesoptychia chichibuensis **52**
Mesoptychia chinensis **54**
Mesoptychia morrisoncola **56**
Metzgeria consanguinea **358**
Metzgeria himalayensis **360**
Metzgeria lindbergii **360**
Microlejeunea punctiformis **340**
Microlejeunea ulicina **342**

Monosolenium tenerum **26**
Myriocoleopsis minutissima **344**

N

Nardia assamica **58**
Notoscyphus collenchymatosus **60**
Notoscyphus lutescens **60**
Notoscyphus parvus **60**
Notothylas japonica **366**
Notothylas orbicularis **368**
Nowellia curvifolia **78**

O

Odontoschisma denudatum **80**

P

Pallavicinia lyellii **46**
Pellia endiviifolia **48**
Phaeoceros carolinianus **368**
Phaeoceros exiguus **372**
Plagiochasma appendiculatum **8**
Plagiochasma intermedium **10**
Plagiochila akiyamae **104**
Plagiochila assamica **106**
Plagiochila chinensis **108**
Plagiochila corticola **110**
Plagiochila delavayi **112**
Plagiochila dendroides **102**
Plagiochila furcifolia **114**
Plagiochila gymnoclada **116**
Plagiochila junghuhniana **118**
Plagiochila khasiana **120**
Plagiochila kunmingensis **122**
Plagiochila nepalensis **124**
Plagiochila ovalifolia **126**
Plagiochila parvifolia **128**
Plagiochila sciophila **130**
Plagiochila subtropica **132**
Plagiochila vexans **134**
Plagiochila wightii **136**
Porella acutifolia **156**
Porella acutifolia subsp. *tosana* **158**

Porella caespitans **160**
Porella caespitans var. *cordifolia* **162**
Porella caespitans var. *setigera* **162**
Porella densifolia **164**
Porella densifolia subsp. *appendiculata* **166**
Porella grandifolia **168**
Porella handelii **170**
Porella japonica **172**
Porella madagascariensis **174**
Porella nitens **176**
Porella obtusata var. *macroloba* **178**
Porella perrottetiana **180**
Porella perrottetiana var. *angustifolia* **182**
Porella perrottetiana var. *ciliatodentata* **184**
Porella plumosa **186**
Porella stephaniana **188**
Porella ulophylla **190**
Ptychanthus striatus **254**

R

Radula acuminata **192**
Radula cavifolia **194**
Radula complanata **196**
Radula javanica **198**
Radula kojana **200**
Radula lacerata **202**
Radula lindenbergiana **204**
Radula stellatogemmipara **206**
Reboulia hemisphaerica **12**
Riccardia chamaedryfolia **348**
Riccardia flavovirens **350**
Riccardia nagasakiensis **352**
Riccardia palmata **354**
Riccardia pusilla **356**
Riccia billardieri **32**
Riccia hueberiana **34**
Riccia junghuhniana **36**
Riccia oryzicola **38**
Riccia rhenana **40**

S

Scapania koponenii **90**

Solenostoma comatum **62**

Solenostoma truncatum **64**

Spruceanthus kiushianus **256**

Spruceanthus mamillilobulus **258**

Spruceanthus semirepandus **260**

Stenolejeunea apiculata **314**

Syzygiella nipponica **72**, **74**

T

Thysananthus repletus **262**

Trocholejeunea sandvicensis **246**

Tuzibeanthus chinensis **264**